AC POWER SYSTEMS

AC POWER SYSTEMS

JERRY C. WHITAKER

CRC Press

Boca Raton   Ann Arbor   Boston

Library of Congress Catalog Card Number 91-73336

Direct all inquiries to CRC Press, Inc., 2000 Corporate Blvd., N.W., Boca Raton, Florida 33431.

International Standard Book Number 0-8493-7412-X
Printed in the United States of America

This book is dedicated to
Laura

PREFACE

Disturbances on the ac power line are what headaches are made of. Outages, surges, sags, transients: They all combine to create an environment that can damage or destroy sensitive load equipment. They can take your system down and leave you with a complicated and expensive repair job.

Ensuring that the equipment at your facility receives clean ac power has always been important. But now, with microcomputers being integrated into a wide variety of electronic products, the question of ac power quality is more critical than ever. The high-speed logic systems prevalent today can garble or lose data because of power-supply disturbances or interruptions. And if the operational problems are not enough, there is the usually difficult task of equipment troubleshooting and repair that follows a utility system fault.

This book examines the key elements of ac power use for commercial and industrial customers. The roots of ac power-system problems are identified, and effective solutions are detailed. The book follows a logical progression from generating ac energy to the protection of life and property. General topics include:

Power-System Operation: Every electronic installation requires a steady supply of clean power to function properly. The ac power line into a facility is, in fact, the lifeblood of any operation. It is also, however, a frequent source of equipment malfunctions and component failures.

This book details the process of generating ac energy and distributing it to end-users. The causes of power-system disturbances are detailed, and the characteristics of common fault conditions are outlined.

Protecting Equipment Loads: Power quality is a moving target. Utility companies work hard to maintain acceptable levels of performance. However, the wide variety of loads and unpredictable situations make this job difficult. Users cannot expect power suppliers to solve all of their problems. Responsibility for protecting sensitive loads clearly rests with the end-user.

Several chapters are devoted to this important topic. Power-system protection options are outlined, and their relative benefits discussed. Evaluating the many tradeoffs involved in protection system design requires a thorough knowledge of the operating principles.

How Much Protection?:   The degree of protection afforded a facility is generally a compromise between the line abnormalities that will account for most of the expected problems and the amount of money available to spend on that protection. Each installation is unique and requires an assessment of the importance of keeping the system up and running at all times as well as the threat of disturbances posed by the ac feed to the plant.

The author firmly believes that the degree of protection provided a power-distribution system should match the threat of system failure. In this publication, all alternatives are examined with an eye toward deciding how much protection really is needed, and how much money can be justified for ac protection hardware.

Grounding: The attention given to the design and installation of a facility ground system is a key element in the day-to-day reliability of any plant. A well-planned ground network is invisible to the engineering staff. A marginal ground system, however, will cause problems on a regular basis. Although most engineers view grounding mainly as a method to protect equipment from damage or malfunction, the most important element is operator safety. The 120 V or 208 V ac line current that powers most equipment can be dangerous — even deadly — if improperly handled. Grounding of equipment and structures provides protection against wiring errors or faults that could endanger human life.

Grounding concepts and practices are examined in detail. Clear, step-by-step guidelines are given for designing and installing an effective ground system to achieve good equipment performance, and to provide for operator safety.

Standby Power: Blackouts are, without a doubt, the most troublesome utility company problem that a facility will have to deal with. Statistics show that power failures are, generally speaking, a rare occurrence in most areas of the country. They also are short in duration. Typical failure rates are not usually cause for concern to commercial users, except where computer-based operations, transportation control systems, medical facilities, and communications sites are concerned. When the continuity of operation is critical, redundancy must be carried throughout the system.

All of the practical standby power systems are examined in this book. The advantages and disadvantages of each approach are given, and examples are provided of actual installations.

Safety: Safety is critically important to engineering personnel who work around powered hardware, and who may work under time pressures. Safety is not something to be taken lightly. The voltages contained in the ac power system are high enough to kill through electrocution.

The author takes safety seriously. A full chapter is devoted to the topic. Safety requires not only the right equipment, but operator training as well. Safety is, in the final analysis, a state of mind.

The utility company ac feed contains not only the 60 Hz power needed to run the facility, but also a variety of voltage disturbances, which can cause problems ranging from process control interruptions to life-threatening situations. Protection against ac line disturbances is a science that demands attention to detail. This work is not inexpensive. It is not something that can be accomplished overnight. Facilities will, however, wind up paying for protection one way or another, either *before* problems occur or *after* problems occur. Power-system protection is a *systems problem* that extends from the utility company ac input to the circuit boards in each piece of hardware. There is nothing magical about effective systems protection. Disturbances on the ac line can be suppressed if the protection method used has been designed carefully and installed properly. That is the goal of this book.

Jerry C. Whitaker

CONTENTS

PREFACE   vii

1. POWER DISTRIBUTION AND CONTROL   1

  1.1 INTRODUCTION   1
      1.1.1 Defining Terms   1
      1.1.2 Vector Representations   4
      1.1.3 Power Relationship in AC Circuits   4

  1.2 ELEMENTS OF THE AC POWER SYSTEM   5
      1.2.1 Power Transformers   5
      1.2.2 Power Generators   14
      1.2.3 Capacitors   17
      1.2.4 Transmission Circuits   21
      1.2.5 Control and Switching Systems   21

  1.3 UTILITY AC POWER SYSTEM   22
      1.3.1 Power Distribution   23
      1.3.2 Power Factor   25
      1.3.3 Utility Company Interfacing   32
      1.3.4 Phase-to-Phase Balance   33
      1.3.5 Reliability Considerations   34

  1.4 BIBLIOGRAPHY   34

xi

2. THE ORIGINS OF AC LINE DISTURBANCES   37

2.1   INTRODUCTION   37

2.2   NATURALLY OCCURRING DISTURBANCES   37

    2.2.1   Sources of Atmospheric Energy   38
    2.2.2   Characteristics of Lightning   41
    2.2.3   Lightning Protection   44
    2.2.4   Electrostatic Discharge   47
    2.2.5   EMP Radiation   49
    2.2.6   Coupling Transient Energy   50

2.3   EQUIPMENT-CAUSED TRANSIENT DISTURBANCES   53

    2.3.1   Utility System Faults   54
    2.3.2   Switch Contact Arcing   54
    2.3.3   Telephone System Transients   56
    2.3.4   Carrier Storage   58

2.4   TRANSIENT-GENERATED NOISE   59
    2.4.1   Noise Sources   59

2.5   POWER-DISTURBANCE CLASSIFICATIONS   61
    2.5.1   Standards of Measurement   62

2.6   ASSESSING THE THREAT   66

    2.6.1   Digital Measurement Instruments   66
    2.6.2   A/D Conversion   68
    2.6.3   Digital Monitor Features   69
    2.6.4   Capturing Transient Waveforms   70
    2.6.5   Case in Point   70

2.7   BIBLIOGRAPHY   72

3. THE EFFECTS OF TRANSIENT DISTURBANCES   75

3.1   INTRODUCTION   75

3.2   SEMICONDUCTOR FAILURE MODES   76

    3.2.1   Device Ruggedness   76
    3.2.2   Forward Bias Safe Operating Area   77
    3.2.3   Reverse Bias Safe Operating Area   78
    3.2.4   Power-Handling Capability   78
    3.2.5   Semiconductor Derating   79

| | 3.2.6 | Failure Mechanisms | 79 |
| | 3.2.7 | Avalanche-Related Failure | 81 |

3.3	MOSFET DEVICES		83
	3.3.1	Safe Operating Area	83
	3.3.2	MOSFET Failure Modes	86

3.4	THYRISTOR COMPONENTS		88
	3.4.1	Failure Modes	89
	3.4.2	Application Considerations	90

| 3.5 | ESD FAILURE MODES | | 91 |
| | 3.5.1 | Failure Mechanisms | 92 |

| 3.6 | SEMICONDUCTOR DEVELOPMENT | | 96 |
| | 3.6.1 | Chip Protection | 99 |

| 3.7 | SWITCH-ARCING EFFECTS | | 101 |
| | 3.7.1 | Insulation Breakdown | 101 |

| 3.8 | BIBLIOGRAPHY | | 102 |

| **4.** | **POWER-SYSTEM COMPONENTS** | | **105** |

| 4.1 | INTRODUCTION | | 105 |
| | 4.1.1 | Power-Supply Components | 105 |

4.2	RELIABILITY OF POWER RECTIFIERS		107
	4.2.1	Operating Rectifiers in Series	107
	4.2.2	Operating Rectifiers in Parallel	109
	4.2.3	Silicon Avalanche Rectifiers	109

4.3	THYRISTOR SERVO SYSTEMS		111
	4.3.1	Inductive Loads	113
	4.3.2	Applications	114
	4.3.3	Triggering Circuits	115
	4.3.4	Fusing	117
	4.3.5	Control Flexibility	118

4.4	TRANSFORMER FAILURE MODES		118
	4.4.1	Thermal Considerations	119
	4.4.2	Voltage Considerations	119
	4.4.3	Mechanical Considerations	120

4.5   CAPACITOR FAILURE MODES   121

    4.5.1   Electrolytic Capacitors   121
    4.5.2   Mechanical Failure   121
    4.5.3   Electrolyte Failures   123
    4.5.4   Capacitor Life Span   124
    4.5.5   Tantalum Capacitors   125

4.6   FAULT PROTECTORS   126

    4.6.1   Fuses   126
    4.6.2   Circuit Breakers   127
    4.6.3   Semiconductor Fuses   128
    4.6.4   Application Considerations   128

4.7   BIBLIOGRAPHY   131

5.   POWER-SYSTEM PROTECTION ALTERNATIVES   133

5.1   INTRODUCTION   133

    5.1.1   The Key Tolerance Envelope   134
    5.1.2   Assessing the Lightning Hazard   135
    5.1.3   Transient Protection Alternatives   136

5.2   MOTOR-GENERATOR SET   141

    5.2.1   System Configuration   144
    5.2.2   Motor-Design Considerations   146
    5.2.3   Maintenance Considerations   149
    5.2.4   Motor-Generator UPS   149

5.3   UNINTERRUPTIBLE POWER SYSTEMS   153

    5.3.1   UPS Configuration   154
    5.3.2   Power-Conversion Methods   156
    5.3.3   Redundant Operation   162
    5.3.4   Output Transfer Switch   164
    5.3.5   Battery Supply   164

5.4   DEDICATED PROTECTION SYSTEMS   166

    5.4.1   Ferroresonant Transformer   166
    5.4.2   Isolation Transformer   169
    5.4.3   Tap-Changing Regulator   170
    5.4.4   Line Conditioner   175

5.5   BIBLIOGRAPHY   178

6. FACILITY-PROTECTION METHODS — 179

6.1 INTRODUCTION — 179
  6.1.1 Filter Devices — 179
  6.1.2 Crowbar Devices — 182
  6.1.3 Voltage-Clamping Devices — 183
  6.1.4 Selecting Protection Components — 186
  6.1.5 Performance Testing — 187

6.2 FACILITY PROTECTION — 190
  6.2.1 Facility Wiring — 191
  6.2.2 Utility Service Entrance — 193

6.3 POWER-SYSTEM PROTECTION — 195
  6.3.1 Staging — 197
  6.3.2 Design Cautions — 198
  6.3.3 Single-Phasing — 199

6.4 CIRCUIT-LEVEL APPLICATIONS — 201
  6.4.1 Protecting Low-Voltage Supplies — 202
  6.4.2 Protecting High-Voltage Supplies — 203
  6.4.3 Protecting Logic Circuits — 205
  6.4.4 Protecting Telco Lines — 205
  6.4.5 Inductive Load Switching — 207
  6.4.6 Device Application Cautions — 207

6.5 BIBLIOGRAPHY — 209

7. FACILITY GROUNDING — 211

7.1 INTRODUCTION — 211

7.2 ESTABLISHING AN EARTH GROUND — 212
  7.2.1 Grounding Interface — 212
  7.2.2 Chemical Ground Rods — 215
  7.2.3 Ufer Ground System — 219

7.3 BONDING GROUND-SYSTEM ELEMENTS — 222
  7.3.1 Cadwelding — 222
  7.3.2 Ground-System Inductance — 225
  7.3.3 Grounding Tower Elements — 225
  7.3.4 Ground-Wire Dressing — 226
  7.3.5 Facility Ground Interconnection — 227

7.4   GROUNDING ON BARE ROCK   230

    7.4.1   Rock-Based Radial Elements   230

7.5   TRANSMISSION-SYSTEM GROUNDING   231

    7.5.1   Transmission Line   231
    7.5.2   Cable Considerations   233
    7.5.3   Satellite Antenna Grounding   233

7.6   DESIGNING A BUILDING GROUND SYSTEM   235

    7.6.1   Bulkhead Panel   237
    7.6.2   Bulkhead Grounding   241
    7.6.3   Lightning Protectors   242
    7.6.4   Typical Installation   243
    7.6.5   Checklist for Proper Grounding   244

7.7   BIBLIOGRAPHY   247

8.   AC SYSTEM GROUNDING PRACTICES   249

8.1   INTRODUCTION   249

    8.1.1   Building Codes   249

8.2   SINGLE-POINT GROUND   250

    8.2.1   Facility Ground System   250
    8.2.2   Power-Center Grounding   255
    8.2.3   Isolation Transformers   256
    8.2.4   Grounding Equipment Racks   257

8.3   GROUNDING SIGNAL-CARRYING CABLES   260

    8.3.1   Analyzing Noise Currents   261
    8.3.2   Types of Noise   263
    8.3.3   Skin Effect   265
    8.3.4   Patch-Bay Grounding   265
    8.3.5   Input/Output Circuits   266
    8.3.6   Cable Routing   270
    8.3.7   Overcoming Ground-System Problems   270

8.4   BIBLIOGRAPHY   271

9. STANDBY POWER SYSTEMS 273

9.1 INTRODUCTION 273
   9.1.1 Blackout Effects 273

9.2 STANDBY POWER OPTIONS 274
   9.2.1 Dual Feeder System 274
   9.2.2 Peak Power Shaving 276
   9.2.3 Advanced System Protection 276
   9.2.4 Choosing a Generator 278
   9.2.5 UPS Systems 281
   9.2.6 Standby Power-System Noise 281
   9.2.7 Critical System Bus 282

9.3 THE EFFICIENT USE OF ENERGY 284
   9.3.1 Energy Usage 284
   9.3.2 Peak Demand 285
   9.3.3 Load Factor 287
   9.3.4 Power Factor 287

9.4 PLANT MAINTENANCE 288
   9.4.1 Switchgear Maintenance 288
   9.4.2 Ground-System Maintenance 289

9.5 BIBLIOGRAPHY 292

10. SAFETY AND PROTECTION SYSTEMS 293

10.1 INTRODUCTION 293
   10.1.1 Facility Safety Equipment 293

10.2 ELECTRIC SHOCK 296
   10.2.1 Effects on the Human Body 296
   10.2.2 Circuit-Protection Hardware 298
   10.2.3 Working with High Voltage 302
   10.2.4 First Aid Procedures 302

10.3 POLYCHLORINATED BIPHENYLS 303
   10.3.1 Health Risk 304
   10.3.2 Governmental Action 305

10.3.3   PCB Components   305
10.3.4   Identifying PCB Components   308
10.3.5   Labeling PCB Components   308
10.3.6   Record Keeping   309
10.3.7   Disposal   310
10.3.8   Proper Management   311

10.4 OSHA SAFETY REQUIREMENTS   311

10.4.1   Protective Covers   311
10.4.2   Identification and Marking   312
10.4.3   Extension Cords   313
10.4.4   Grounding   313
10.4.5   Management Responsibility   315

10.5 BIBLIOGRAPHY   317

INDEX   319

1

POWER DISTRIBUTION AND CONTROL

1.1 INTRODUCTION

Every electronic installation requires a steady supply of clean power to function properly. Recent advances in technology have made the question of alternating current (ac) power quality even more important, as microcomputers are integrated into a wide variety of electronic products. The high-speed logic systems prevalent today can garble or lose data because of power-supply disturbances or interruptions.

The ac power line into a facility is the lifeblood of any operation. It is also, however, a frequent source of equipment malfunctions and component failures. The utility company ac feed contains not only the 60 Hz power needed to run the facility, but also a variety of voltage sags, surges, and transients. These abnormalities cause different problems for different types of equipment.

1.1.1 Defining Terms

To explain the ac power-distribution system, and how to protect sensitive loads from damage resulting from disturbances, it is necessary to define key terms.

- *Alternator*. An ac generator.
- *Circular mil*. The unit of measurement for current-carrying conductors. One mil is equal to 0.001 inches (0.025 millimeters). One circular mil is equal to a circle whose diameter is 0.001 inches. The area of a circle with a 1-inch diameter is 1,000,000 circular mils.
- *Common-mode noise*. Unwanted signals in the form of voltages appearing between the local ground reference and each of the power conductors, including neutral and the equipment ground.

1

- *Cone of protection* (lightning). The space enclosed by a cone formed with its apex at the highest point of a lightning rod or protecting tower, the diameter of the base of the cone having a definite relationship to the height of the rod or tower. When overhead ground wires are used, the space protected is referred to as a *protected zone*.
- *Cosmic rays*. Charged particles (ions) emitted by all radiating bodies in space.
- *Coulomb*. A unit of electric charge. The coulomb is the quantity of electric charge that passes the cross section of a conductor when the current is maintained constant at one ampere.
- *Counter-electromotive force*. The effective electromotive force within a system that opposes the passage of current in a specified direction.
- *Counterpoise*. A conductor or system of conductors arranged (typically) below the surface of the earth and connected to the footings of a tower or pole to provide grounding for the structure.
- *Demand meter*. A measuring device used to monitor the power demand of a system; it compares the peak power of the system with the average power.
- *Dielectric* (ideal). An insulating material in which all of the energy required to establish an electric field in the dielectric is recoverable when the field or impressed voltage is removed. A perfect dielectric has zero conductivity, and all absorption phenomena are absent. A complete vacuum is the only known perfect dielectric.
- *Eddy currents*. The currents that are induced in the body of a conducting mass by the time variations of magnetic flux.
- *Efficiency* (electric equipment). Output power divided by input power, expressed as a percentage.
- *Generator*. A machine that converts mechanical power into electrical power. (In this publication, the terms "alternator" and "generator" will be used interchangeably.)
- *Ground loop*. Sections of conductors shared by two different electronic and/or electric circuits, usually referring to circuit return paths.
- *Horsepower*. The basic unit of mechanical power. One horsepower (hp) equals 550 foot-pounds per second or 746 watts.
- *HVAC*. Abbreviation for "heating, ventilation, and air conditioning" system.
- *Hysteresis loss* (magnetic, power, and distribution transformer). The energy loss in magnetic material that results from an alternating magnetic field as the elementary magnets within the material seek to align themselves with the reversing field.
- *Impedance*. A linear operator expressing the relationship between voltage and current. The inverse of impedance is *admittance*.
- *Induced voltage*. A voltage produced around a closed path or circuit by a time rate of change in a magnetic flux linking that path when there is no relative motion between the path or circuit and the magnetic flux.
- *Joule*. A unit of energy equal to one watt-second.

- *Life safety system.* Systems designed to protect life and property, such as emergency lighting, fire alarms, smoke exhaust and ventilating fans, and site security.
- *Lightning flash.* An electrostatic atmospheric discharge. The typical duration of a lightning flash is approximately 0.5 seconds. A single flash is made up of various discharge components, usually including three or four high-current pulses called *strokes.*
- *Metal-oxide varistor.* A solid-state voltage clamping device used for transient-suppression applications.
- *Normal-mode noise.* Unwanted signals in the form of voltages appearing in line-to-line and line-to-neutral signals.
- *Permeability.* A general term used to express relationships between magnetic induction and magnetizing force. These relationships are either: (1) *absolute permeability*, which is the quotient of a change in magnetic induction divided by the corresponding change in magnetizing force; or (2) *specific* (relative) *permeability*, which is the ratio of absolute permeability to the magnetic constant.
- *Power factor.* The ratio of total watts to the total rms (root-mean-square) volt-amperes in a given circuit. Power factor = W/VA.
- *Radio frequency interference.* Noise resulting from the interception of transmitted radio frequency energy.
- *Reactance.* The imaginary part of impedance.
- *Reactive power.* The quantity of "unused" power that is developed by reactive components (inductive or capacitive) in an ac circuit or system.
- *Safe operating area.* Semiconductor device parameters, usually provided in chart form, that outline the maximum permissible limits of operation.
- *Saturation* (in a transformer). The maximum intrinsic value of induction possible in a material.
- *Self inductance.* The property of an electric circuit whereby a change of current induces an electromotive force in that circuit.
- *Single-phasing.* A fault condition in which one of the three legs in a three-phase power system becomes disconnected, usually because of an open fuse or fault condition.
- *Solar wind.* Charged particles from the sun that continuously bombard the surface of the earth.
- *Transient disturbance.* A voltage pulse of high energy and short duration impressed upon the ac waveform. The overvoltage pulse may be one to 100 times the normal ac potential and may last up to 15 ms. Rise times measure in the nanosecond range.
- *Uninterruptible power system.* An ac power-supply system that is used for computers and other sensitive loads to: (1) protect the load from power interruptions, and (2) protect the load from transient disturbances.
- *Voltage regulation.* The deviation from a nominal voltage, expressed as a percentage of the nominal voltage.

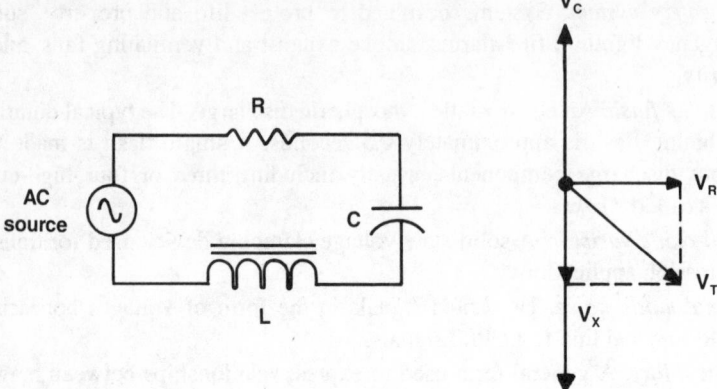

Figure 1.1   Voltage vectors in a series *RLC* circuit.

1.1.2   Vector Representations

Vectors are used commonly in ac circuit analysis to represent voltage or current values. Rather than using waveforms to show phase relationships, it is accepted practice to use vector representations (sometimes called *phasor diagrams*). To begin a vector diagram, a horizontal line is drawn, its left end being the reference point. Rotation in a counterclockwise direction from the reference point is considered to be positive. Vectors may be used to compare voltage drops across the components of a circuit containing resistance, inductance, and/or capacitance. Figure 1.1 shows the vector relationship in a series *RLC* circuit, and Figure 1.2 shows a parallel *RLC* circuit.

1.1.3   Power Relationship in AC Circuits

In a dc circuit, power is equal to the product of voltage and current. This formula also is true for purely resistive ac circuits. However, when a reactance — either inductive

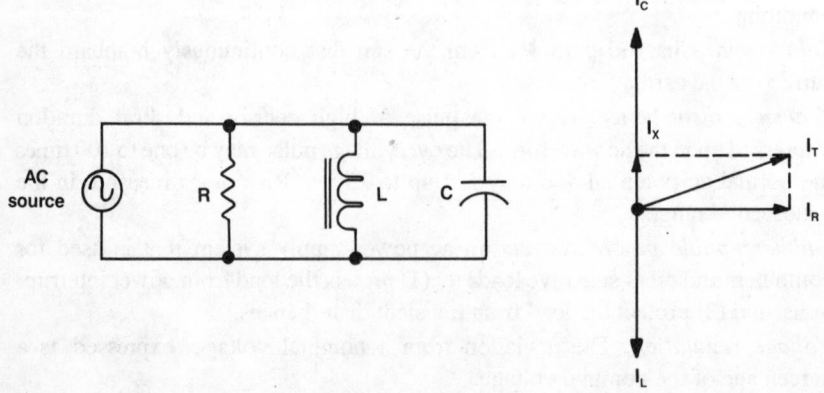

Figure 1.2   Current vectors in a parallel *RLC* circuit.

or capacitive — is present in an ac circuit, the dc power formula does not apply. The product of voltage and current is, instead, expressed in volt-amperes (VA) or kilovolt-amperes (kVA). This product is known as the *apparent power*. When meters are used to measure power in an ac circuit, the apparent power is the voltage reading multiplied by the current reading. The actual power that is converted to another form of energy by the circuit is measured with a wattmeter, and is referred to as the *true power*. In ac power-system design and operation, it is desirable to know the ratio of true power converted in a given circuit to the apparent power of the circuit. This ratio is referred to as the *power factor*.

1.2 ELEMENTS OF THE AC POWER SYSTEM

The process of generating, distributing, and controlling the large amounts of power required for a municipality or geographic area is highly complex. However, each system, regardless of its complexity, is composed of the same basic elements with the same basic goal: Deliver ac power where it is needed by customers. The primary elements of an ac power system can be divided into the following general areas of technology:

- Power transformers
- Power generators
- Capacitors
- Transmission circuits
- Control and switching systems, including voltage regulators, protection devices, and fault isolation devices

The path that electrical power takes to end-users begins at a power plant, where electricity is generated by one of several means and is then stepped-up to a high voltage (500 kV is common) for transmission on high-tension lines. Step-down transformers reduce the voltage to levels appropriate for local distribution and eventual use by customers. Figure 1.3 shows how these elements interconnect to provide ac power to consumers.

1.2.1 Power Transformers

The transformer is the basis of all ac power-distribution systems. In 1831, English physicist Michael Faraday demonstrated the phenomenon of *electromagnetic induction*. The concept is best understood in terms of lines of force, a convention Faraday introduced to describe the direction and strength of a magnetic field. The lines of force for the field generated by a current in a loop of wire are shown in Figure 1.4. When a second, independent loop of wire is immersed in a changing magnetic field, a voltage will be induced in the loop. The voltage will be proportional to the time rate of change of the number of force lines enclosed by the loop. If the loop has two turns, such induction occurs in each turn, and twice the voltage results. If the loop has three turns,

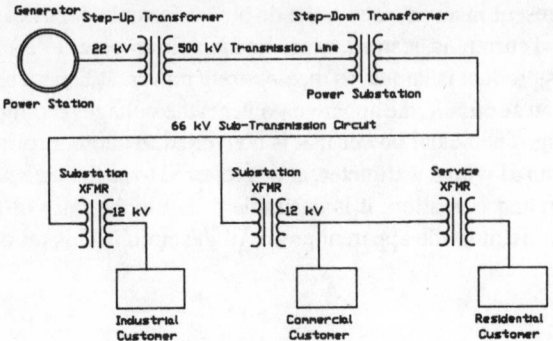

Figure 1.3 Typical electrical power-generation and -distribution system. Although this schematic diagram is linear, in practice power lines branch at each voltage reduction to establish the distribution network.

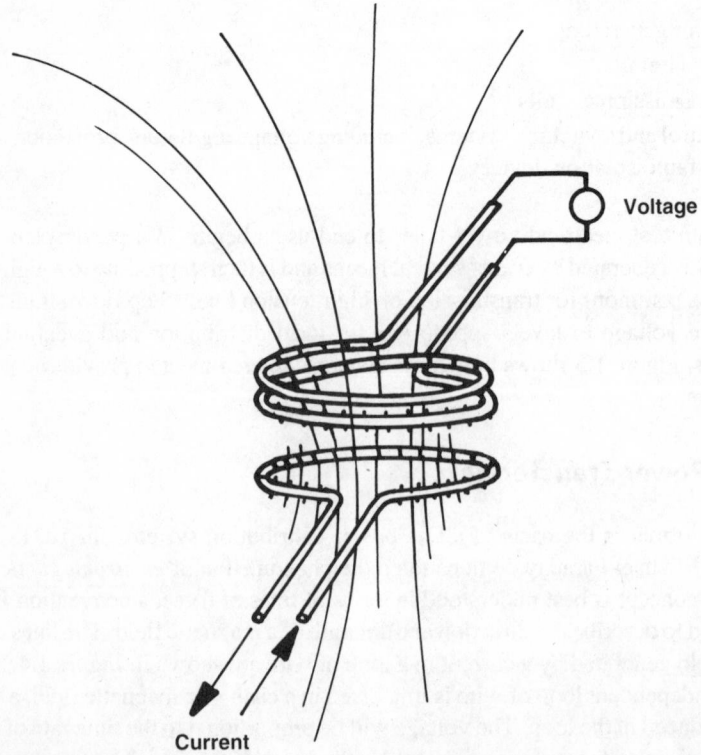

Figure 1.4 The basic principles of electromagnetic induction.

3 times the voltage results, and so on. The concurrent phenomena of mutual induction between the coils and self-induction in each coil form the basis of transformer action.

For a power transformer to do its job effectively, the coils must be coupled tightly and must have high self-induction. That is, almost all the lines of force enclosed by the primary also must be enclosed by the secondary, and the number of force lines produced by a given rate of change of current must be high. Both conditions can be met by wrapping the primary and secondary coils around an iron core, as Faraday did in his early experiments. Iron increases the number of lines of force generated in the transformer by a factor of about 10,000. This property of iron is referred to as *permeability*. The iron core also contains the lines so that the primary and secondary coils can be separated spatially and still closely coupled magnetically.

With the principles of the transformer firmly established, American industrialist George Westinghouse and his associates made several key refinements that made practical transformers possible. The iron core was constructed of thin sheets of iron cut in the shape of the letter *E*. Coils of insulated copper wire were wound and placed over the center element of the core. Straight pieces of iron were laid across the ends of the arms to complete the magnetic circuit. This construction still is common today. Figure 1.5 shows a common *E*-type transformer. Note how the low-voltage and high-voltage windings are stacked on top of each other. An alternative configuration, in which the low-voltage and high-voltage windings are located on separate arms of a core box, is shown in Figure 1.6.

In an ideal transformer, all lines of force pass through all the turns in both coils. Because a changing magnetic field produces the same voltage in each turn of the coil, the total voltage induced in a coil is proportional to the total number of turns. If no energy is lost in the transformer, the power available in the secondary is equal to the power fed into the primary. In other words, the product of current and voltage in the primary is equal to the product of current and voltage in the secondary. Thus, the two currents are inversely proportional to the two voltages, and therefore, inversely proportional to the turns ratio between the coils. This expression of power and current in a transformer is true only for an ideal transformer. Practical limitations prevent the perfect transformer from being constructed.

The key properties of importance in transformer core design include:

- Permeability
- Saturation
- Resistivity
- Hysteresis loss

Permeability, as discussed previously, refers to the number of lines of force a material produces in response to a given magnetizing influence. Saturation identifies the point at which the ability of the core to carry a magnetic force reaches a plateau. These two properties define the power-handling capability of the core element. Electrical resistivity is desirable in the core because it minimizes energy losses resulting from *eddy currents*. In contrast, hysteresis undermines the efficiency of a transformer. Because of the interactions among groups of magnetized atoms, losses are incurred as the frequency of the changing magnetic field is increased. Throughout

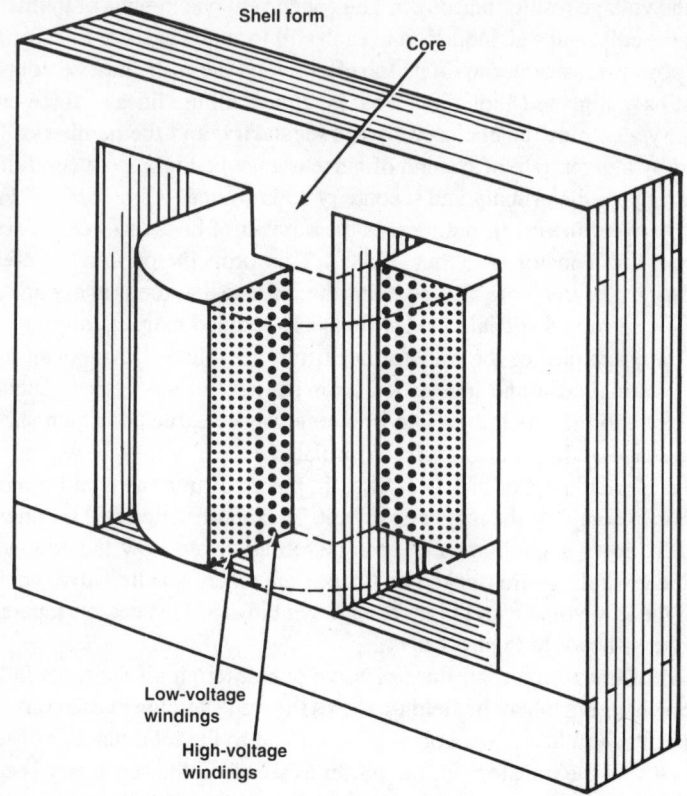

Shell form

Core

Low-voltage
windings

High-voltage
windings

Figure 1.5   Physical construction of an E-shaped core transformer. The low- and high-voltage windings are stacked as shown.

the history of transformer development, the goal of the design engineer has been to increase permeability, saturation, and resistivity, while decreasing hysteresis losses. A variety of core materials, including silicon iron in various forms, have been used over the years. Transformer efficiency is defined as follows:

$$\text{Efficiency (\%)} = \frac{P_{out}}{P_{in}} \bullet 100$$

Where:

P_{out} = transformer power output in watts
P_{in} = transformer power input in watts

Also bearing on transformer performance are electrical insulation and the cooling system used. These two elements are intimately related because the amount of heat that the core and conductors generate determines the longevity of the insulation; the

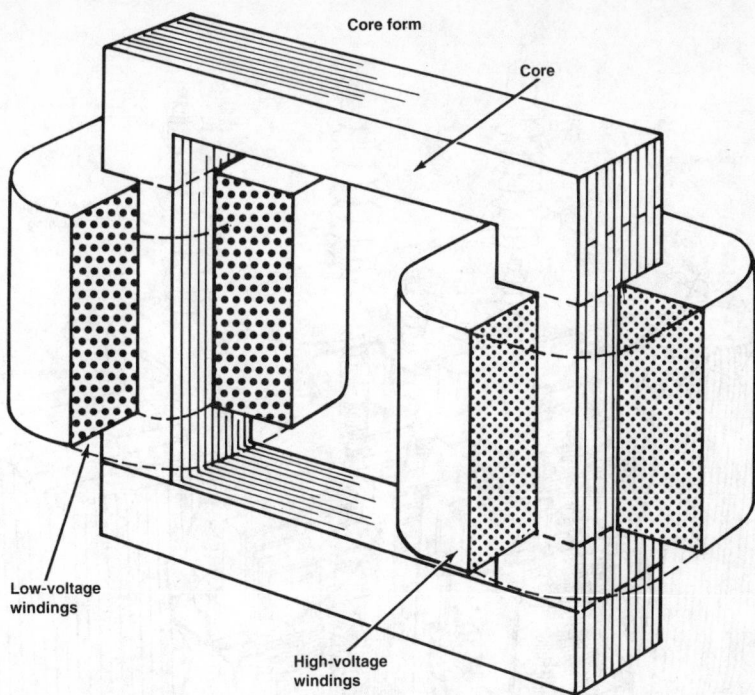

Figure 1.6 Transformer construction using a box core with physical separation between the low- and high-voltage windings.

insulation itself — whether solid, liquid, or gas — serves to carry off some portion of the heat produced. Temperatures inside a commercial transformer may reach 100°C, the boiling point of water. Under such conditions, deterioration of insulating materials can limit the useful lifetime of the device. Although oils are inexpensive and effective as insulators and coolants, some oils are flammable, making them unacceptable for units placed inside buildings. Chlorinated hydrocarbon liquids (PCBs) were used extensively from the 1930s to the late 1970s, but evidence of long-term toxic effects prompted a ban on their use. Some transformers rely on air- or nitrogen-gas-based insulators. Such devices can be installed indoors. The breakdown strength of gas sometimes is enhanced through the addition of small quantities of fluorocarbons. Other dry transformers depend on cast-resin insulation made of polymerizing liquids that harden into high-integrity solids. Progress in heat removal is largely responsible for reducing the overall size of the transformer assembly.

Modern high-power commercial transformers may operate at voltages of 750 kV or more and can handle more than 1000 kVA. The expected lifetime of a commercial power transformer ranges from 24 to 40 years. A typical three-phase oil-cooled transformer is shown in Figure 1.7.

Counter-Electromotive Force (counter-emf). All transformers, generators, and motors exhibit the property of inductance. This property is the result of a counter-emf

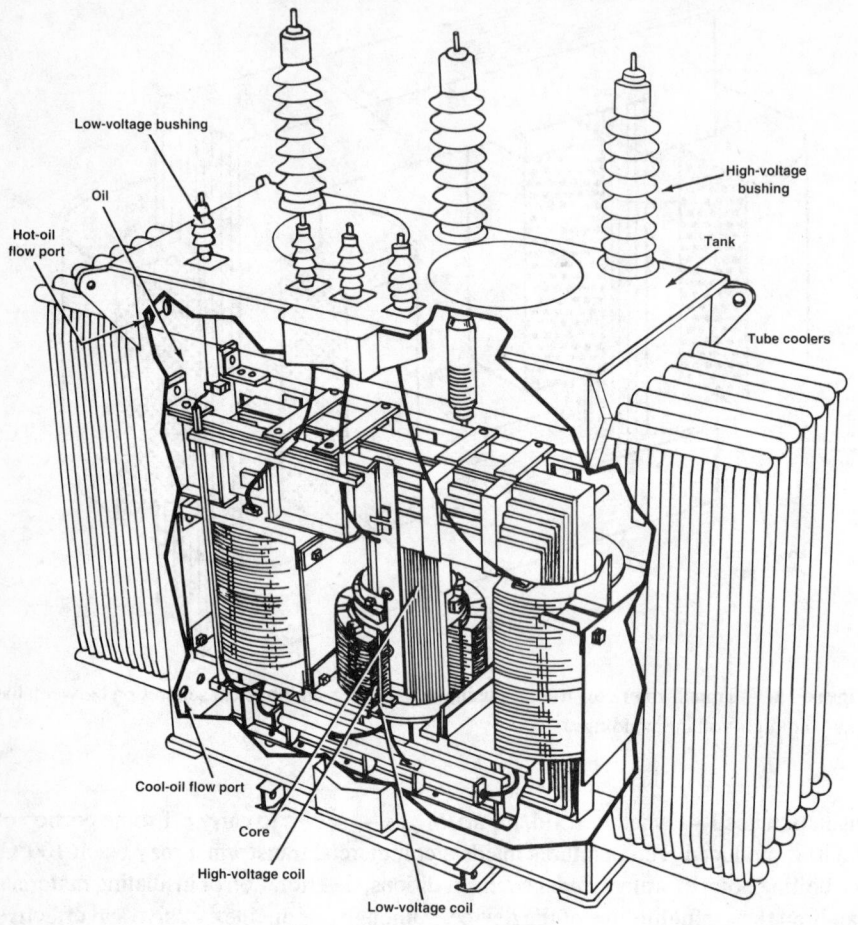

Low-voltage bushing

Oil

High-voltage
bushing

Hot-oil
flow port

Tank

Tube coolers

Cool-oil flow port

Core

High-voltage coil

Low-voltage coil

Figure 1.7   Construction of an oil-filled three-phase power transformer used for commercial power distribution.

that is produced when a magnetic field is developed around a coil of wire. Inductance presents an opposition to the change in current flow in a circuit. This opposition is evident in the diagram shown in Figure 1.8. In a purely inductive circuit (containing no resistance), the voltage will lead the current by 90°. However, because all practical circuits have resistance, the offset will vary from one circuit to the next. Figure 1.9 illustrates a circuit in which voltage leads current by 30°. The angular separation between voltage and current is referred to as the *phase angle*. The phase angle increases as the inductance of the circuit increases. Any inductive circuit exhibits the property of inductance, including electrical power-transmission and -distribution lines. The *henry* (H) is the unit of measurement for inductance. A circuit has a 1 H

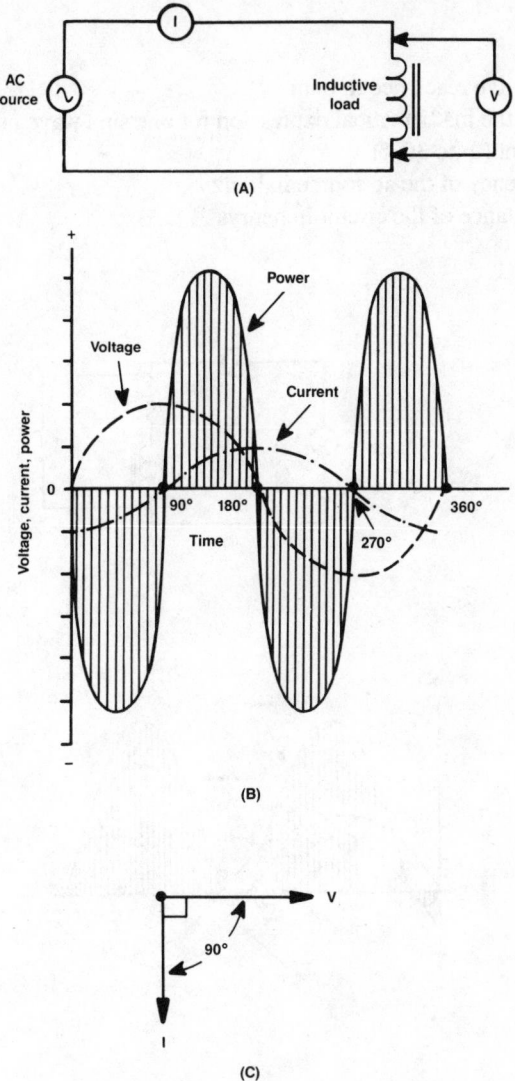

Figure 1.8 Purely inductive circuit: (a) circuit diagram; (b) representative waveforms; (c) vector representation.

inductance if a current changing at a rate of 1 A/s produces an induced counter-emf of 1 V.

In an inductive circuit with ac applied, an opposition to current flow is created by the inductance. This opposition is known as *inductive reactance* (X_l). The inductive reactance of a given ac circuit is determined by the inductance of the circuit and the rate of current change. Inductive reactance may be expressed as:

$X_l = 2 \pi fL$

Where:

X_l = inductive reactance in ohms

2π = 6.28, the mathematical expression for one sine wave of alternating current (0° to 360°)

f = frequency of the ac source in hertz

L = inductance of the circuit in henrys

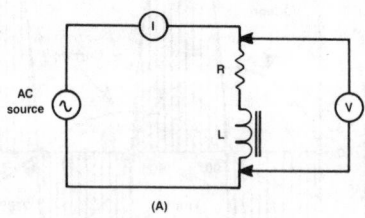

(A)

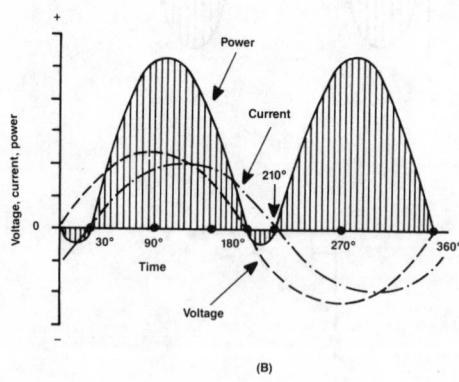

(B)

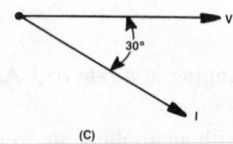

(C)

Figure 1.9 Resistive-inductive circuit: (a) circuit diagram; (b) representative waveforms; (c) vector representation.

Full Load Percent Impedance.   The full load percent impedance (FLPI) of a transformer is an important parameter in power-supply system design. FLPI is determined by the construction of the core and physical spacing between the primary and secondary windings. Typical FLPI values range from 1 percent to 5 percent. FLPI is a measure of the ability of a transformer to maintain rated voltage with a varying load. The lower the FLPI, the better the regulation. FLPI also determines the maximum fault current that the transformer can deliver. For example, if a 5 percent FLPI transformer supplying 5 A nominal at the secondary is short-circuited, the device can, theoretically, supply 100 A at full voltage. A similar transformer with a 10 percent FLPI can supply only 50 A when short-circuited. Typical short-circuit currents for a selection of small three-phase transformers are listed in Table 1.1.

Table 1.1   Full Load Percent Impedance Short-Circuit Currents for a Selection of Three-Phase Transformers

DC amps (kVA/kV)	Full load percent impedance	Symmetrical short-circuit current
1 A	1	57.7
1 A	2	28.8
1 A	3	19.3
1 A	4	14.4
1 A	5	11.5
2 A	1	115.5
2 A	2	57.7
2 A	3	38.5
2 A	4	28.8
2 A	5	23.1
3 A	1	173.2
3 A	2	86.6
3 A	3	57.7
3 A	4	43.3
3 A	5	34.6
4 A	1	230.9
4 A	2	115.5
4 A	3	77.0
4 A	4	57.7
4 A	5	46.2
5 A	1	288.6
5 A	2	144.3
5 A	3	96.2
5 A	4	72.2
5 A	5	57.7

1.2.2   Power Generators

Any ac power system begins with a generating source. A simplified diagram of a three-phase generator is shown in Figure 1.10. Poles A', B', and C' represent the start of each of the phase windings, while poles A, B, and C represent the ends of each of the windings. The windings of the generator may be connected in either of two ways:

- *Wye configuration.* A circuit arrangement in which the schematic diagram of the windings form a "Y."
- *Delta configuration.* A circuit arrangement in which the schematic diagram of the windings form a "delta."

Figure 1.11 illustrates the connection arrangements.

The generator shown in Figure 1.10 is a rotating-field type of device. A magnetic field is developed by an external dc voltage. Through electromagnetic induction, a current is induced into each of the stationary (stator) coils of the generator. Because each of the phase windings is separated by 120°, the output voltage of the generator also is offset for each phase by 120°. This concept is illustrated in Figure 1.12. Three-phase power is used almost exclusively for power distribution because it is an efficient method of transporting electrical energy. Most industrial plants use three-

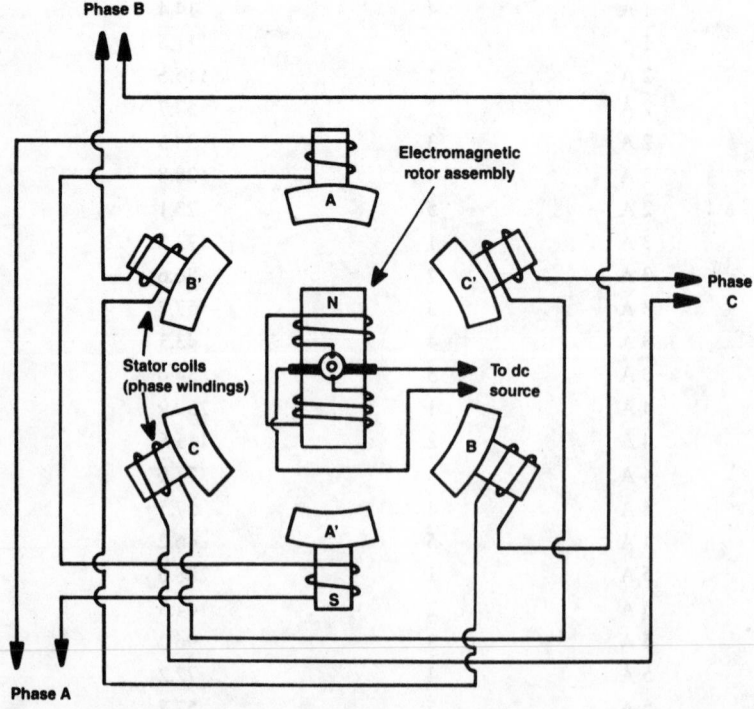

Figure 1.10   Simplified diagram of a three-phase ac generator.

phase power, and for commercial and residential use, three separate single-phase voltages can be derived from a three-phase transmission line.

Electrical power can be produced in many ways, including chemical reactions, heat, light, or mechanical energy. Most electrical power produced today is through hydro-electric plants, nuclear energy, and by burning coal, oil, or natural gas. Fossil-fuel and nuclear-fission plants use steam turbines to deliver the mechanical energy required to

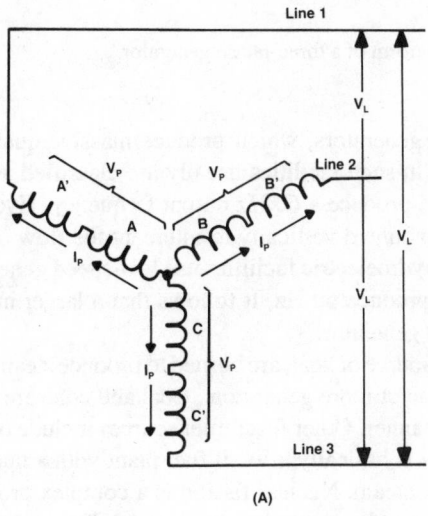

(A)

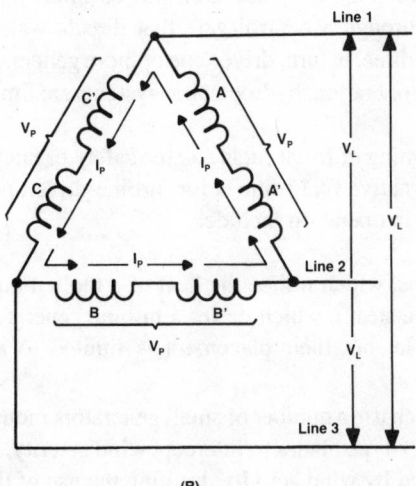

(B)

Figure 1.11 Generator circuit configurations: (a) wye; (b) delta.

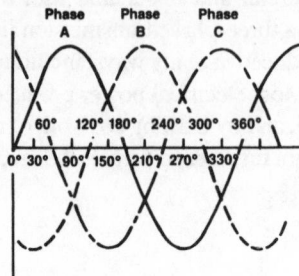

Figure 1.12   Output waveform of a three-phase generator.

rotate large three-phase generators, which produce massive quantities of electrical power. Generators used in such facilities usually are classified as high-speed units, operating at 3600 rpm to produce a 60 Hz output frequency. Hydroelectric systems use hydraulic turbines, mounted vertically to intercept the flow of water to produce electrical energy. Most hydroelectric facilities use low-speed generators, operating at 600 rpm or 900 rpm to produce 60 Hz. It follows that a larger number of poles are required for a low-speed generator.

Fossil fuels, used as a source of heat, are burned to produce steam in a boiler system. The steam then drives one or more generators. Coal and coke are used commonly to produce energy in this manner. Other fossil fuel sources include oil and natural gas.

A nuclear power plant is basically a fossil fuel plant with a nuclear power source to produce heat and then steam. Nuclear fission is a complex process that results in the division of the nucleus of an atom into two nuclei. This splitting of the atom is initiated by bombardment of the nucleus with neutrons, gamma rays, or other charged particles.

A hydroelectric system is the simplest of all power plants. Flowing water from a reservoir is channeled through a control gate that directs water to the blades of a hydraulic turbine. The turbine, in turn, drives one or more generators. Although simple in design and efficient in operation, hydroelectric systems are limited by the availability of a water reservoir.

Concern about the burning of fossil fuels and the safety of nuclear power has led to the development of alternative fuel sources for turbine-driven power plants. Power-generating systems now in operation include:

- Geothermal systems, which utilize the heat of a molten mass in the interior of the earth to produce steam, which drives a turbine generator. Such systems are efficient and simple, but their placement is limited to areas of geothermal activity.
- Wind systems, which use a number of small generators mounted on supports and attached to propeller-type blades to intercept wind activity. Naturally, generator output is determined by wind activity, limiting the use of these systems on any large scale.

Significant variations in load requirements must be satisfied at different times by a generating plant. Because of wide variations in load demands, much of the generating capability of a facility may be unused during low-demand periods. Two mathematical ratios commonly are used to measure utility service:

- *Load factor.* The average load for a given period of time divided by the peak load for that same period.
- *Capacity factor.* The average load for a given period of time divided by the output capacity of the power plant.

Under ideal conditions, both the load factor and the capacity factor are unity (100 percent). Commercial power systems use a number of three-phase generators connected in parallel, and synchronized in phase, to supply the load requirements.

1.2.3 Capacitors

A capacitor consists, basically, of two conductors separated by a dielectric. The operation of a capacitor in a circuit is dependent upon its ability to charge and discharge. When a capacitor charges, an excess of electrons is accumulated on one plate, and a deficiency of electrons is created on the other plate. Capacitance is determined by the size of the conductive material (the plates), and their separation (determined by the thickness of the dielectric material). Capacitance is directly proportional to plate size and inversely proportional to the distance between the plates. The unit of capacitance is the *farad* (F). A capacitance of 1 F results when a potential of 1 V causes an electric charge of 1 coulomb to accumulate on a capacitor. Capacitance is expressed as a ratio of electric charge to the voltage applied:

$$C = Q/V$$

Where:

C = capacitance
Q = charge
V = voltage

If a direct current is applied to a capacitor, the device will charge to the value of the applied voltage. After the capacitor is fully charged, it will block the flow of direct current. However, if an ac voltage is applied, the changing value of current will cause the device to alternately charge and discharge. In a purely capacitive circuit, the situation shown in Figure 1.13 will exist. The greatest amount of current will flow when the voltage changes most rapidly. The most rapid change in voltage occurs at the 0° and 180° positions in the sine wave where the polarity changes. At these positions, maximum current is developed in the circuit, as shown. It is evident by studying the waveform that, in a purely capacitive circuit, voltage will lag current by 90°. Because all practical circuits contain some resistance, a lag of 0° to 90° may be

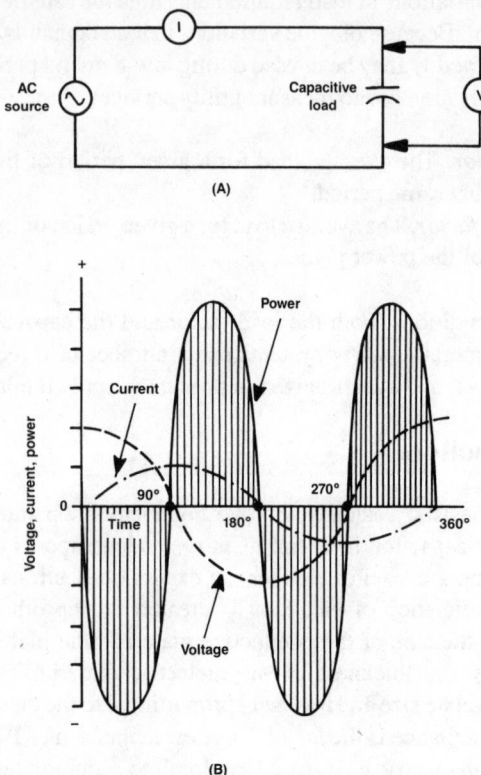

(A)

(B)

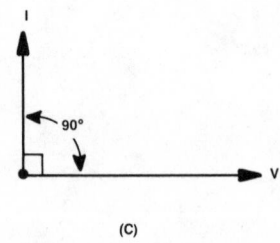

(C)

Figure 1.13 Purely capacitive circuit: (a) circuit diagram; (b) representative waveforms; (c) vector representation.

experienced in practice. Figure 1.14 illustrates a case in which voltage lags current by 30°.

Because of the electrostatic field developed around a capacitor, an opposition to the flow of alternating current exists. This opposition is known as *capacitive reactance*:

$$X_c = \frac{1}{2\pi fC}$$

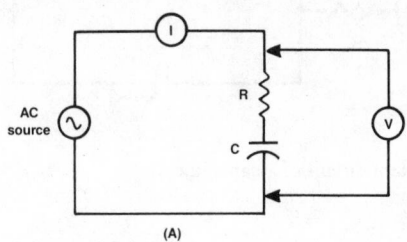

(A)

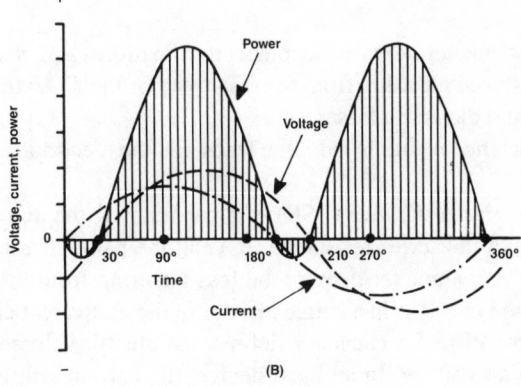

(B)

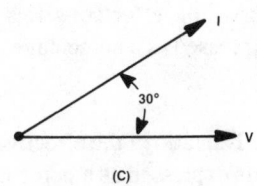

(C)

Figure 1.14   Resistive-capacitive circuit: (a) circuit diagram; (b) representative waveforms; (c) vector representation.

Where:

X_C	= capacitive reactance in ohms Ω
2π	= the mathematical expression of one sine wave
f	= frequency in hertz
C	= capacitance in farads

The dielectric used for a given capacitor varies, depending upon the application. Common dielectrics include air, gas, mica, glass, and ceramic. Each has a different

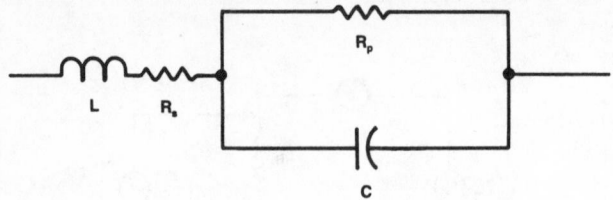

Figure 1.15   Equivalent circuit of a capacitor.

dielectric constant, temperature range, and thickness. In addition to capacitance, a practical capacitor has inductance and resistance components, as shown in Figure 1.15. The stray components are identified as follows:

R_S = series resistance of wire leads, contact terminations, and electrodes
R_p = shunt resistance resulting from the resistivity of the dielectric and case material, and dielectric losses
L = stray inductance resulting from the leads and the electrodes

The *equivalent series resistance* (ESR) of a capacitor is the ac resistance of the device reflecting both the series resistance (R_s) and the parallel resistance (R_p) at a given frequency. This parameter permits the loss resulting from the foregoing elements to be expressed as a loss in a single resistor in the equivalent circuit.

The power factor (PF) of a capacitor defines the electrical losses in the device operating under an ac voltage. In an ideal device, the current will lead the applied voltage by 90°. A practical capacitor, because of its dielectric, electrode, and contact termination losses, exhibits a phase angle of less than 90°. The power factor of a capacitor is defined as the ratio of the effective series resistance to the impedance of the capacitor. PF usually is expressed as a percentage. Other important specifications for capacitors include:

- *Dissipation factor* (DF). The ratio of the effective series resistance to capacitive reactance. DF normally is expressed as a percentage.
- *Leakage current*. The current flowing through the capacitor when a dc voltage is applied.
- *Insulation resistance*. The ratio of the applied voltage to the leakage current. Insulation resistance is normally expressed in megohms.
- *Ripple current/voltage*. The rms value of the maximum allowable alternating current or voltage (superimposed on any dc level) at a specific frequency at which the capacitor may be operated continuously at a specified temperature.
- *Surge voltage*. The maximum operating voltage of the capacitor at any temperature.

1.2.4 Transmission Circuits

The heart of any utility power-distribution system is the cable used to tie distant parts of the network together. Conductors are rated by the American Wire Gauge (AWG) scale. The smallest is no. 36, and the largest is no. 0000. There are 40 sizes in between. Sizes larger than no. 0000 AWG are specified in *thousand circular mil* units, referred to as "MCM" units (M is the Roman numeral expression for 1000). The cross-sectional area of a conductor doubles with each increase of three AWG sizes. The diameter doubles with every six AWG sizes.

Most conductors used for power transmission are made of copper or aluminum. Copper is the most common. Stranded conductors are used where flexibility is required. Stranded cables usually are more durable than solid conductor cables of the same AWG size. For long distances, utilities typically use uninsulated aluminum conductors or aluminum conductor steel-reinforced cables. For shorter distances, insulated copper wire normally is used.

Ampacity is the measure of the ability of a conductor to carry electric current. Although all metals will conduct current to some extent, certain metals are more efficient than others. The three most common high-conductivity conductors are:

- Silver, with a resistivity of 9.8 Ω/circular mil-foot
- Copper, with a resistivity of 10.4 Ω/circular mil-ft
- Aluminum, with a resistivity of 17.0 Ω/circular mil-ft

The ampacity of a conductor is determined by the type of material used, the cross-sectional area, and the heat-dissipation effects of the operating environment. Conductors operating in free air will dissipate heat more readily than conductors placed in a larger cable or in a raceway with other conductors.

1.2.5   Control and Switching Systems

Specialized hardware is necessary to interconnect the elements of a power-distribution system. Utility control and switching systems operate under demanding conditions, including high voltage and current levels, exposure to lightning discharges, and 24-hour-a-day use. For reliable performance, large margins of safety must be built into each element of the system. The primary control and switching elements are high-voltage switches and protection devices.

High-voltage switches are used to manage the distribution network. Most disconnect switches function to isolate failures or otherwise reconfigure the network. Air-type switches are typically larger versions of the common "knife switch" device. To prevent arcing, air switches are changed only when power is removed from the circuit. These types of switches may be motor-driven or manually operated.

Oil-filled circuit breakers are used at substations to interrupt current when the line is hot. The contacts usually are immersed in oil to minimize arcing. Oil-filled circuit breakers are available for operation at up to 500 kV. Magnetic air breakers are used primarily for low-voltage indoor applications.

Protection devices include fuses and lightning arresters. Depending upon the operating voltage, various types of fuses may be used. Arc suppression is a key consideration in the design and operation of a high-voltage fuse. A method must be provided to extinguish the arc that develops when the fuse element begins to open.

Lightning arresters are placed at numerous points in a power-distribution system. Connected between power-carrying conductors and ground, they are designed to operate rapidly and repeatedly if necessary. Arresters prevent flashover faults between power lines and surge-induced transformer and capacitor failures. The devices are designed to extinguish rapidly, after the lightning discharge has been dissipated, to prevent power follow-on damage to system components.

High-voltage insulators permit all of the foregoing hardware to be reliably interconnected. Most insulators are made of porcelain. The mechanical and electrical demands placed on high-voltage insulators are stringent. When exposed to rain or snow, the devices must hold off high voltages. They also must support the weight of heavy conductors and other components.

1.3 UTILITY AC POWER SYSTEM

The details of power distribution vary from one city or country to another, and from one utility company to another, but the basics are the same. Figure 1.16 shows a simplified distribution network. Power from a generating station or distribution grid comes into an area substation at 115 kV or higher. The substation consists of switching systems, step-down transformers, fuses, circuit breakers, reclosers, monitors, and control equipment. The substation delivers output voltages of approximately 60 kV to subtransmission circuits, which feed distribution substations. The substations convert the energy to approximately 12 kV and provide voltage regulation and switching provisions that permit *patching around* a problem. The 12 kV lines power pole- and surface-mounted transformers, which supply various voltages to individual loads. Typical voltage configurations include:

- 120/208 V wye
- 277/480 V wye
- 120/240 V single phase
- 480 V delta

Fuses and circuit breakers are included at a number of points in the 12 kV distribution system to minimize fault-caused interruptions of service. *Ground-fault interrupters* (GFIs) also are included at various points in the 12 kV system to open the circuit if excessive ground currents begin to flow on the monitored line. *Reclosers* may be included as part of overcurrent protection of the 12 kV lines. They will open the circuit if excessive currents are detected, and reclose after a preset length of time. The recloser will perform this trip-off/reset action a preset number of times before being locked out.

In some areas, the actions of circuit breakers, pole-mounted switches, and reclosers are controlled by 2-way radio systems that allow status interrogation and switching of the remotely located devices from a control center. Some utilities use this method sparingly; others make extensive use of it.

Depending on the geographic location, varying levels of lightning protection are included as part of the ac power-system design. Most service drop transformers (12

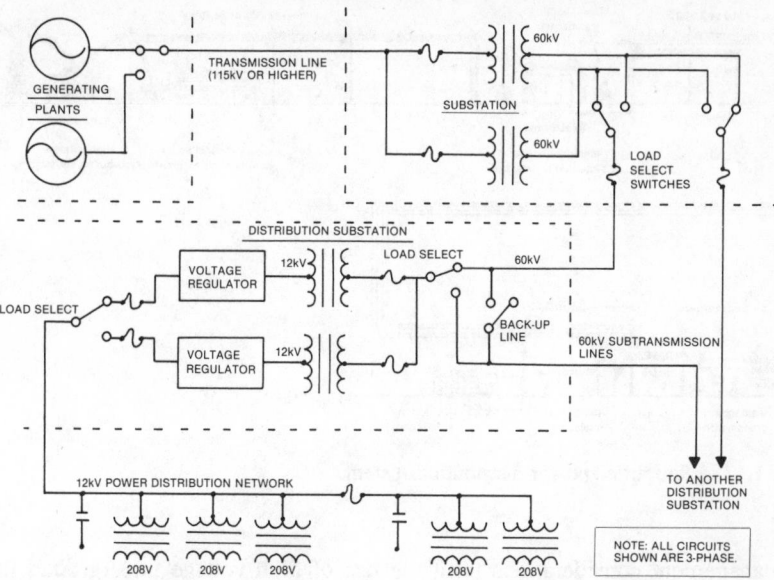

Figure 1.16 Simplified block diagram of a basic utility company power-distribution system. The devices shown as fuses could be circuit breakers or reclosers, which function as automatic-resetting circuit breakers. All circuits shown are three-phase. The capacitors perform power factor correction duty.

kV to 208 V) have integral lightning arresters. In areas of severe lightning, a ground (or *shield*) wire will be strung between the top insulators of each pole, attracting lightning to the ground wire, and away from the hot leads.

1.3.1 Power Distribution

The distribution of power over a utility company network is a complex process involving a number of power-generating plants, transmission lines, and substations. The physical size of a metropolitan power-distribution and control system is immense. Substations use massive transformers, oil-filled circuit breakers, huge strings of insulators, and high-tension conductors in distributing power to customers. Power-distribution and -transmission networks interconnect generating plants into an *area grid*, to which *area loads* are attached. Most utility systems in the United States are interconnected to one extent or another. In this way, power-generating resources can be shared as needed. The potential for single-point failure also is reduced in a distributed system.

A typical power-distribution network is shown in Figure 1.17. Power-transmission lines operate at voltage levels from 2.3 kV for local distribution to 500 kV or more for distribution between cities or generating plants. Long-distance direct-current transmission lines also are used, with potentials of 500 kV to 600 kV. Underground power lines are limited to short runs in urban areas. Increased installation costs and cable

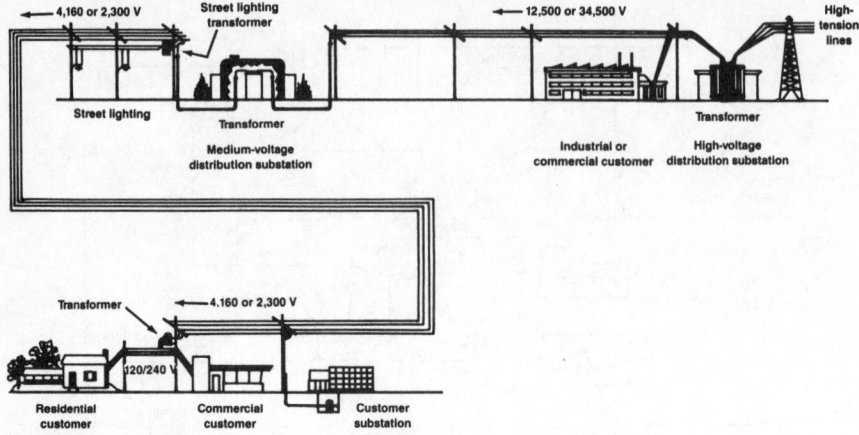

Figure 1.17   Simplified power-distribution system.

heat-management considerations limit the use of high-voltage underground lines. Wide variations in standard voltage levels can be found within any given system. Each link in the network is designed to transfer energy with the least I^2R loss, thereby increasing overall system efficiency. The following general classifications of power-distribution systems can be found in common use:

- *Radial system.* The simplest of all distribution networks, a single substation supplies power to all loads in the system. See Figure 1.18.
- *Ring system.* Distribution lines encircle the service area, with power being delivered from one or more sources into substations near the service area. Power is then distributed from the substations through the radial transmission lines. See Figure 1.19.
- *Network system.* A combination of the radial and ring distribution systems. Although such a system is more complex than either of the previous configurations, reliability is improved significantly. The network system, illustrated in Figure 1.20, is one of the most common power-distribution configurations.

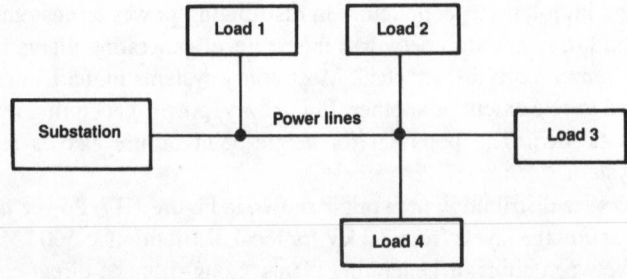

Figure 1.18   Radial power-transmission system.

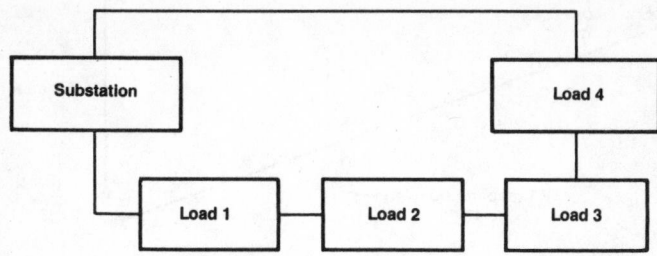

Figure 1.19   Ring power-transmission system.

1.3.2   Power Factor

Power factor is an important, but often misunderstood, characteristic of ac power systems. It is defined as the ratio of *true power* to *apparent power*, generally expressed as a percentage. (PF also may be expressed as a decimal value.) Reactive loads (inductive or capacitive) act on power systems to shift the current out of phase with the voltage. The cosine of the resulting angle between the current and voltage is the power factor.

A utility line that is feeding an inductive load (which is most often the case) is said to have a *lagging* power factor, while a line feeding a capacitive load has a *leading* power factor. See Figure 1.21. A poor power factor will result in excessive losses along utility company feeder lines because more current is required to supply a given load with a low power factor than the same load with a power factor close to unity.

For example, a motor requiring 5 kW from the line is connected to the utility service entrance. If it has a power factor of 86 percent, the apparent power demanded by the load will be 5 kW divided by 86 percent, or more than 5.8 kW. The true power is 5 kW, and the apparent power is 5.8 kW. The same amount of work is being done by the motor, but the closer the power factor is to unity, the more efficient the system. To

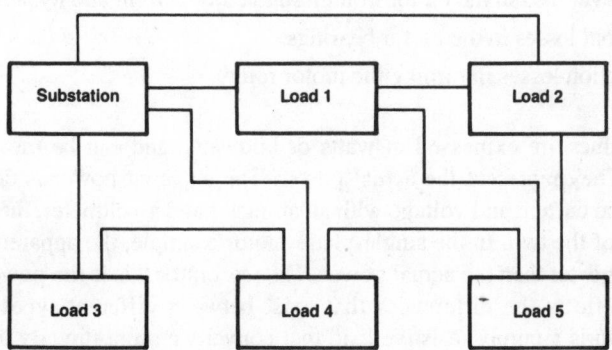

Figure 1.20   Network power-transmission system.

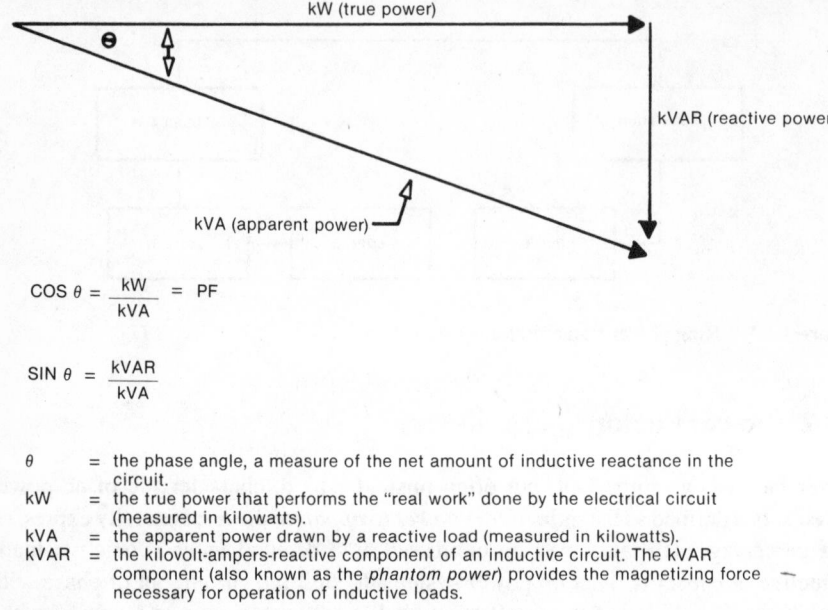

$$\text{COS } \theta = \frac{kW}{kVA} = PF$$

$$\text{SIN } \theta = \frac{kVAR}{kVA}$$

θ = the phase angle, a measure of the net amount of inductive reactance in the circuit.
kW = the true power that performs the "real work" done by the electrical circuit (measured in kilowatts).
kVA = the apparent power drawn by a reactive load (measured in kilowatts).
kVAR = the kilovolt-ampers-reactive component of an inductive circuit. The kVAR component (also known as the *phantom power*) provides the magnetizing force necessary for operation of inductive loads.

Figure 1.21   The mathematical relationships of an inductive circuit as they apply to power factor (PF) measurements. Reducing the kVAR component of the circuit causes Ø to diminish, improving the PF. When kW is equal to kVA, the phase angle is zero and the power factor is unity (100 percent).

expand upon this example, for a single-phase electric motor the actual power is the sum of several components, namely:

- The work performed by the system, specifically lifting, moving, or otherwise controlling an object.
- Heat developed by the power lost in the motor winding resistance.
- Heat developed in the motor iron through eddy-current and hysteresis losses.
- Frictional losses in the motor bearings.
- Air friction losses in turning the motor rotor.

All these values are expressed in watts or kilowatts, and can be measured with a wattmeter. They represent the actual power. The apparent power is determined by measuring the current and voltage with an ammeter and a voltmeter, then calculating the product of the two. In the single-phase motor example, the apparent power thus obtained is greater than the actual power. The reason for this is the power factor.

The PF reflects the differences that exist between different types of loads. A soldering iron is a purely resistive load that converts current directly into heat. The current is called *actual current* because it contributes directly to the production of actual power. On the other hand, the single-phase electric motor represents a partially

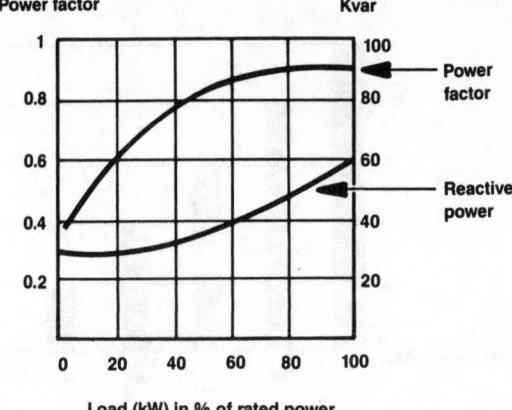

Figure 1.25   The relationship of reactive power and power factor to percentage of load for a 100 kW, three-phase induction motor.

from 230 V to 13.8 kV for pole-mounted applications. The PF correction capacitors are connected in parallel with the utility lines as close as practical to the low-PF loads. The primary disadvantage of static PF correction capacitors is that they cannot be adjusted for changing power factor conditions. Remotely operated relays may be used, however, to switch capacitor banks in and out of the circuit as required. Power factor correction also can be accomplished by using *synchronous capacitors* connected across utility lines. Synchronous capacitors may be adjusted to provide varying capacitance to correct for varying PF loads. The capacitive effect of a synchronous capacitor is changed by varying the dc excitation voltage applied to the rotor of the device.

Utilities usually pass on to customers the costs of operating low-PF loads. Power factor may be billed as one, or a combination, of the following:

- A penalty for PF below a predetermined value, or a credit for PF above a predetermined value.
- An increasing penalty for decreasing PF.
- A charge on monthly kVAR (kilovolt-ampere reactive) hours.
- A straight charge for the maximum value of kVA used during the month.

Aside from direct costs from utility companies for low-PF operation, the end-user experiences a number of indirect costs as well. When a facility operates with a low overall PF, the amount of useful electrical energy available inside the plant at the distribution transformer is reduced considerably because of the amount of reactive energy that the transformer(s) must carry. Figure 1.26 illustrates the reduction in available power from a distribution transformer when presented with varying PF loads. Figure 1.27 illustrates the increase in I^2R losses in feeder and branch circuits with varying PF loads. These conditions result in the need for oversized cables, transformers, switchgear, and protection circuits.

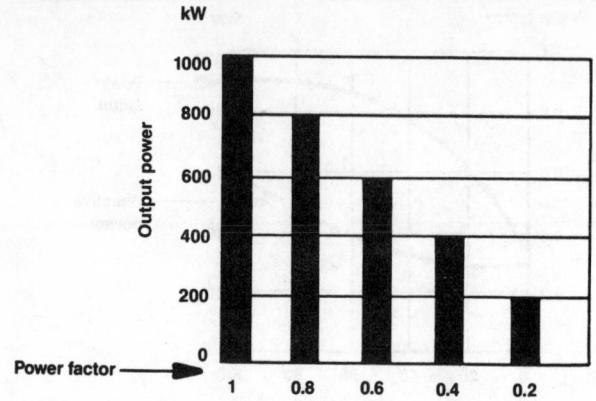

Figure 1.26   The effects of power factor on output of a transformer.

On-Site Power Factor Correction.   The first step in correcting for low PF is to determine the current situation at a given facility. Clamp-on power factor meters are available to users to prevent shutdown time at a facility. Power factor can be improved in two ways:

- Reduce the amount of reactive energy by eliminating unloaded motors and transformers.
- Apply external compensation capacitors to correct the low-PF condition.

PF correction capacitors perform the function of an energy-storage device. Instead of transferring reactive energy back and forth between the load and the power source, the magnetizing current reactive energy is stored in a capacitor at the load. Capacitors are rated in kVARs, and are available for single- and multiphase loads. Usually more

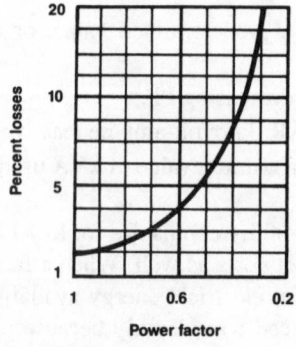

Figure 1.27   The relationship between power factor and percentage losses in system feeder and branch circuits.

than one capacitor is required to yield the desired degree of PF correction. The capacitor rating required in a given application may be determined by using lookup tables provided by PF capacitor manufacturers. Installation options include:

- Individual capacitors placed at each machine.
- A group or bank installation for an entire area of the plant.
- A combination of the two approaches.

When rectifier loads creating harmonic load current are the cause of a low-PF condition, the addition of PF correcting capacitors will not necessarily provide the desired improvement. The capacitors, in some cases, may actually raise the line current and fail to improve the power factor. Harmonic currents generally are most apparent in the neutral of three-phase circuits. Conductors supplying three-phase rectifiers using a neutral conductor require a neutral conductor that is as large as the phase conductors. A reduced neutral should not be permitted. When adding capacitors for PF correction, be careful to avoid any unwanted voltage resonances that might be excited by harmonic load currents.

If a delta/wye connected power transformer is installed between the power source and the load, the power factor at the transformer input generally will reflect the average PF of the loads on the secondary. This conclusion works on the assumption that the low PF is caused by inductive and capacitive reactances in the loads. However, if the load current is rich in harmonics from rectifiers and switching regulators, some of the harmonic currents will flow no farther toward the power source than the transformer delta winding. The third harmonic and multiples of three will flow in the delta winding and will be significantly reduced in amplitude. By this means, the transformer will provide some improvement in the PF of the total load.

An economic evaluation of the cost vs. benefits, plus a review of any mandatory utility company limits that must be observed for PF correction, will determine how much power factor correction, if any, may be advisable at a given facility. Correction to 85 percent will satisfy most requirements. No economic advantage is likely to result from correcting to 95 percent or greater. Overcorrecting a load by placing too many PF correction capacitors may reduce the power factor after reaching unity, and may cause uncontrollable overvoltages in low-kVA-capacity power sources.

PF correcting capacitors usually offer some benefits in absorbing line-voltage impulse-type noise spikes. However, if the capacitors are switched on and off, they will create significant impulses of their own. Switching may be accomplished with acceptably low disturbance through the use of soft-start or preinsertion resistors. Such resistors are connected momentarily in series with the capacitors. After a brief delay (0.5 s or less) the resistors are short-circuited, connecting the capacitors directly across the line.

Installation of PF correction capacitors at a facility is a complicated process that requires a knowledgeable consultant and licensed electrician. The local utility company should be contacted before any effort is made to improve the PF of a facility.

1.3.3   Utility Company Interfacing

Most utility company-to-customer connections are the *delta-wye* type shown in Figure 1.28. This transformer arrangement usually is connected with the delta side facing the high voltage and the wye side facing the load. This configuration provides good isolation of the load from the utility and somewhat retards the transmission of transients from the primary to the secondary. The individual three-phase loads are denoted in the figure by Z1, Z2, and Z3. They carry load currents as shown. For a wye-connected system, it is important that the building neutral lead is connected to the midpoint of the transformer windings, as shown. The neutral line provides a path for the removal of harmonic currents that may be generated in the system as a result of rectification of the secondary voltages.

In some areas, an *open-delta* arrangement is used by the utility company to supply power to customers. The open-delta configuration is shown in Figure 1.29. Users often encounter problems when operating sensitive three-phase loads from such a connection because of poor voltage-regulation characteristics during varying load conditions. The open-delta configuration is also subject to high third-harmonic content and transient propagation. The three loads and their respective load currents are shown in the diagram.

Other primary power connection arrangements are possible, such as *wye-to-wye* or *delta-to-delta*. Like the delta-to-wye configuration, they are not susceptible to the problems that may be experienced with the open-delta (or *V-V*) service.

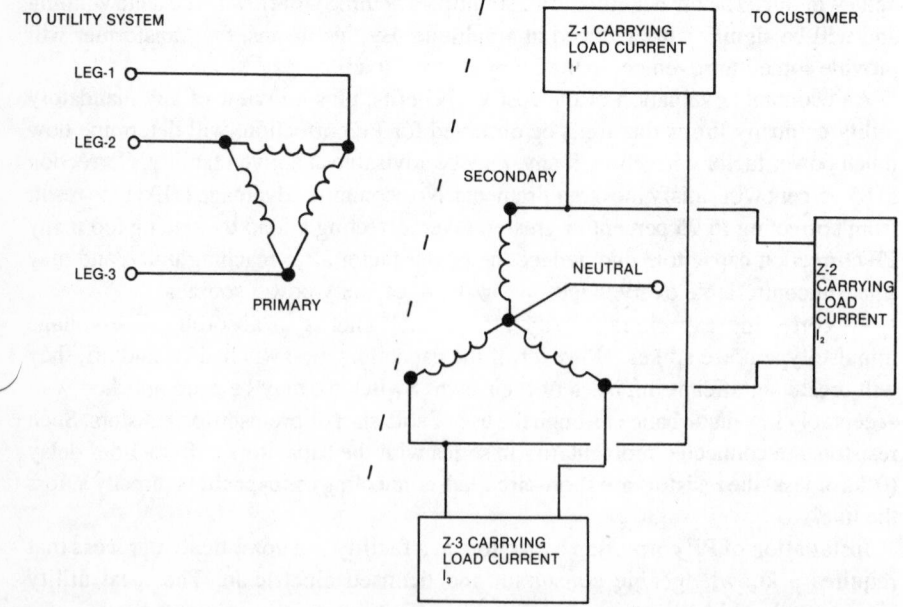

Figure 1.28   Delta-wye transformer configuration for utility company power distribution. This common type of service connection transformer provides good isolation of the load from the high-voltage distribution-system line.

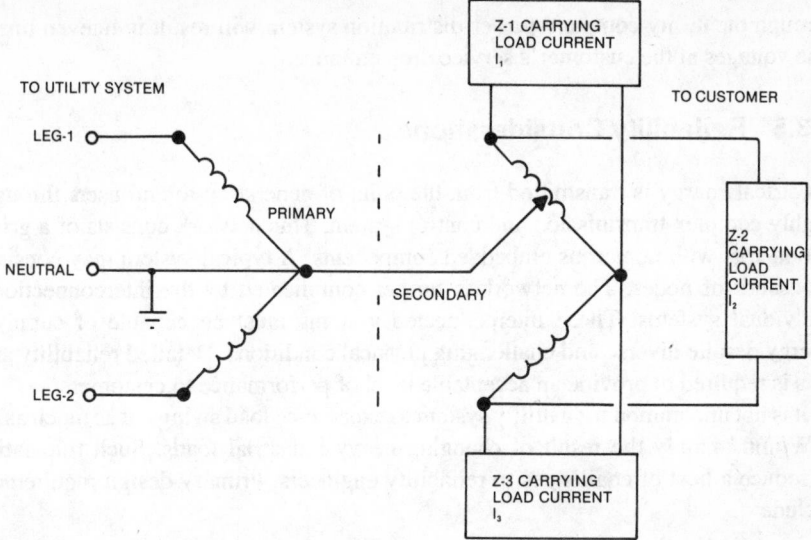

Figure 1.29   The open-delta (or V-V) utility company service connection transformer. This configuration is not recommended because of its poor voltage regulation under varying load, high third-harmonic content, and transient-disturbance propagation.

The open-delta system can develop a considerable imbalance among the individual phases in either voltage or phase or both. This can introduce a strong 120 Hz ripple frequency in three-phase power supplies, which are designed to filter out a 360 Hz ripple. The effects of this 120 Hz ripple may be increased noise in the supply and possible damage to protection devices across the power-supply chokes. Depending on the loading of an open-delta transformer, high third-harmonic energy can be transferred to the load, producing transients of up to 300 percent of the normal voltage. The result could be severely strained rectifiers, capacitors, and inductors in the power supply as well as additional output noise of the supply.

1.3.4 Phase-to-Phase Balance

The phase-to-phase voltage balance of a utility company line is important at most types of facilities, not only because of the increased power-supply ripple that it may cause, but also because of the possible heating effects. Even simple three-phase devices such as motors should be operated from a power line that is well-balanced, preferably within 1 percent. Studies have shown that a line imbalance of only 3.5 percent can produce a 25 percent increase in the heat generated by a three-phase motor. A 5 percent imbalance can cause a 50 percent increase in heat, which is potentially destructive. Similar heating also can occur in the windings of three-phase power transformers used in industrial equipment.

Phase-to-phase voltage balance can be measured accurately over a period of several days with a slow-speed chart recorder. The causes of imbalanced operation are generally large, single-phase power users on local distribution lines. Uneven currents through the utility company power-distribution system will result in uneven line-to-line voltages at the customer's service drop entrance.

1.3.5 Reliability Considerations

Electrical energy is transmitted from the point of generation to end-users through a highly complex transmission and control system. This network consists of a grid of conductors with numerous embedded components. A typical system may consist of thousands of nodes. The network is further complicated by the interconnection of individual systems. These interconnected systems must be capable of supplying energy despite diverse and challenging physical conditions. Detailed reliability analysis is required to provide an acceptable level of performance to customers.

It is not uncommon for a utility system to experience load swings of as much as 150 MW/min, mainly the result of changing heavy industrial loads. Such fluctuations introduce a host of challenges to reliability engineers. Primary design requirements include:

- The ability to handle large, constantly changing load demands from customers.
- A high degree of power-supply reliability.
- Tight voltage regulation across the system.
- A strong power source to provide for high inrush currents typically experienced at industrial plants.
- Rapid and effective fault isolation. Failures in one part of the system should have minimal effect on other portions of the network.

Utility companies are meeting these goals with improved reliability analysis, more operating reserve, and computerized control systems. Rapid response to load changes and fault conditions is necessary to ensure reliable service to customers. Improved telemetry systems provide system controllers with more accurate information on the state of the network, and advanced computer-control systems enable split-second decisions that minimize service disruptions.

1.4 BIBLIOGRAPHY

Acard, Christian: "Power Factor Measurement and Correction," *Plant Electrical Systems*, Intertec Publishing, Overland Park, KS, March 1981.

Fardo, S., and D. Patrick: *Electrical Power Systems Technology*, Prentice-Hall, Englewood Cliffs, NJ, 1985.

Federal Information Processing Standards Publication no. 94, *Guideline on Electrical Power for ADP Installations*, U.S. Department of Commerce, National Bureau of Standards, Washington, DC, 1983.

Fink, D., and D. Christiansen: *Electronics Engineers' Handbook*, 3rd ed., McGraw-Hill, New York, 1989.

Jay, Frank: *IEEE Standard Directory of Electrical and Electronics Terms*, 3rd ed., IEEE, New York, 1984.

Kramer, Robert A.: "Analytic and Operational Consideration of Electric System Reliability," *Proceedings of the Reliability and Maintainability Symposium*, IEEE, New York, 1991.

Nott, Ron: "The Sources of Atmospheric Energy," *Proceedings* of the SBE National Convention and Broadcast Engineering Conference, Society of Broadcast Engineers, Indianapolis, 1987.

Technical staff, "The Susceptibility of the Open Delta Connection to Third Harmonic Disturbances," Harris Corporation, technical paper, Quincy, IL.

Tyrrell, Peter A.: "Transformer Impedance Versus Your Rectifiers," *Broadcast Engineering* magazine, Intertec Publishing, Overland Park, KS, September 1980.

2

THE ORIGINS OF AC LINE DISTURBANCES

2.1 INTRODUCTION

Transient overvoltages come in a wide variety of forms, from a wide variety of sources. They can, however, be broken down into two basic categories: (1) those generated through natural occurrences, and (2) those generated through the use of equipment, either on-site or elsewhere.

2.2 NATURALLY OCCURRING DISTURBANCES

Natural phenomena of interest to facility managers consist mainly of lightning and related disturbances. The *lightning effect* can be compared to that of a capacitor, as shown in Figure 2.1. A charged cloud above the earth will create an oppositely charged area below it of about the same size and shape. When the voltage difference is sufficient to break down the dielectric (air), the two "plates" of the "capacitor" will arc over and neutralize their respective charges. If the dielectric spacing is reduced, as in the case of a conductive steel structure (such as a transmitting tower), the arc-over will occur at a lower-than-normal potential, and will travel through the conductive structure.

The typical duration of a lightning flash is approximately 0.5 s. A single flash is made up of various discharge components, among which are typically three or four high-current pulses called *strokes*. Each stroke lasts about one 1 ms; the separation between strokes is typically several tens of milliseconds. Lightning often appears to flicker because the human eye can just resolve the individual light pulses that are produced by each stroke.

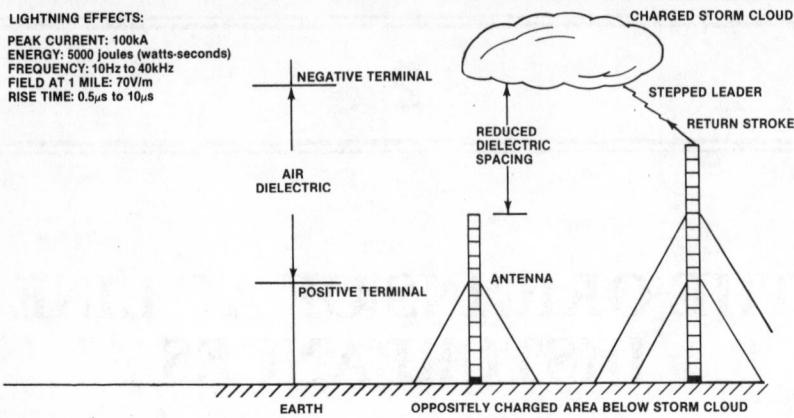

Figure 2.1   The lightning effect and how it can be compared to a more familiar mechanism, the capacitor principle. Also shown are the parameters of a typical lightning strike.

2.2.1   Sources of Atmospheric Energy

Lightning is one of the more visible effects of atmospheric electricity. Stellar events that occurred light-years ago spray the earth and its atmosphere with atoms that have been stripped of most or all of their electrons. In the process of entering the atmosphere, these particles collide with air molecules, which are knocked apart, creating billions more ion pairs each second. Even though these ions may exist for only about 100 s, they constantly are being replenished from deep space. The existence of ions in the atmosphere is the fundamental reason for atmospheric electricity. The primary sources of this energy are:

- *Cosmic rays.* Charged particles emitted by all radiating bodies in space. Most of these particles (ions) expend their energy in penetrating the envelope of air surrounding the earth. Through this process, they create more ions by colliding with air atoms and molecules. One high-energy particle may create up to a billion pairs of ions, many of which will become atmospheric electricity.
- *Solar wind.* Charged particles from the sun that continuously bombard the surface of the earth. Because about half of the earth's surface is always exposed to the sun, variations are experienced from day to night. Solar wind particles travel at only 200 to 500 miles per second, compared with cosmic particles that travel at near the speed of light. Because of their slower speed, solar wind particles have less of an effect on air atoms and molecules.
- *Natural radioactive decay.* The natural disintegration of radioactive elements. In the process of radioactive decay, air molecules are ionized near the surface of the earth. One of the results is radon gas.
- *Static electricity.* Energy generated by the interaction of moving air and the earth.
- *Electromagnetic generation.* Energy generated by the movement of air molecules through the magnetic field of the earth.

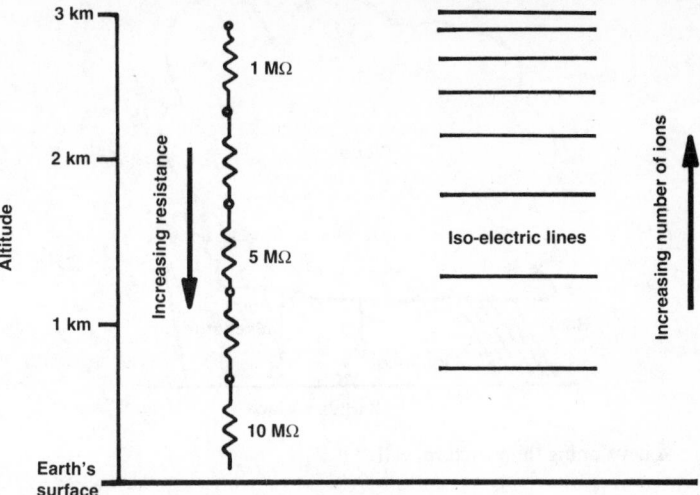

Figure 2.2   The effects of atmospheric conductivity.

The combined effects of cosmic rays and solar wind account for most atmospheric electrical energy.

Atmospheric energy is present at all times, even during clear weather conditions. This energy takes the form of a voltage differential between the surface of the earth and the ionosphere of 300 to 400 kV. The voltage gradient is nonlinear; near the surface it may be 150 V/m of elevation, but it diminishes significantly at higher altitudes. Under normal conditions, the earth is negative with respect to the ionosphere, and ions flow between the two entities. Because there are fewer free ions near the earth than the ionosphere, the volts/meter value is thought to be greater because of the lower effective conductivity of the air. This concept is illustrated in Figure 2.2.

Thermodynamic activity in a developing storm cloud causes it to become a powerfully charged cell, usually negatively charged on the bottom and positively charged on the top. See Figure 2.3. This voltage difference causes a distortion in the voltage gradient and, in fact, the polarity inverts, with the earth becoming positive with reference to the bottom of the cloud. This voltage gradient increases to a high value, sometimes exceeding 10 kV/m of elevation. The overall charge between the earth and the cloud may be on the order of 10 to 100 MV, or more. When sufficient potential difference exists, a lightning flash may occur.

Figure 2.4 shows the flash waveform for a typical lightning discharge. The rise time is very fast, in the microsecond range, as the lightning channel is established. The trailing edge exhibits a slow decay; the decay curve is known as a *reciprocal double exponential* waveform. The trailing edge is the result of the resistance of the ionized channel depleting energy from the cloud. The path length for a lightning discharge is measured in kilometers. The most common source of lightning is cumulonimbus cloud forms, although other types of clouds (such as nimbostratus) occasionally may produce activity.

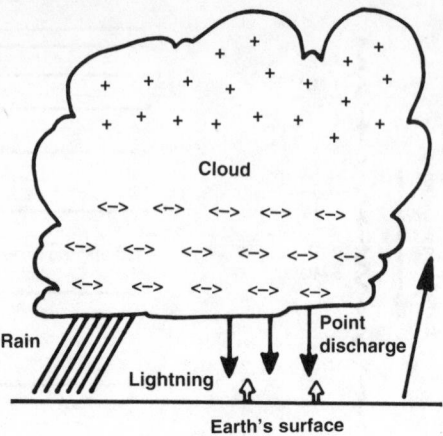

Figure 2.3   A developing thunderstorm cell.

Although most lightning strikes are *negative* (the bottom of the cloud is negative with respect to the earth), *positive strikes* also can occur. Such strikes have been estimated to carry as much as 10 times the current of a negative strike. A positive flash can carry 200 kiloamps (kA) or more of discharge current. Such "hot strikes," as they are called, can cause considerable damage. Hot strikes can occur in the winter, and are often the aftereffect of a particularly active storm. After a number of discharges, the lower negative portion of the cloud will become depleted. When charged, the lower portion may have functioned as a screen or shield between the earth and the upper, positively charged portion of the cloud. When depleted, the shield is removed,

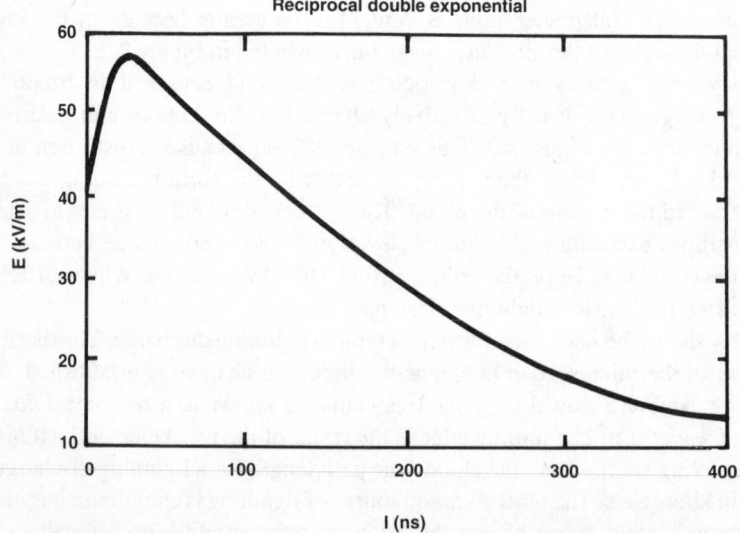

Figure 2.4   Discharge waveform for a typical lightning strike.

exposing the earth to the massive charge in the upper cloud containing potentials of perhaps 500 MV or more.

2.2.2   Characteristics of Lightning

A typical lightning flash consists of a stepped leader that progresses toward the ground at a velocity that may exceed 50 m/μs. When sufficient potential difference between the cloud and the ground exists, arcs move from the ground to the leader column, completing the ionized column from cloud to ground. A fast and bright return stroke then moves upward along the leader column at about one-third the speed of light. Peak currents from such a lightning flash may exceed 100 kA, with a total charge as high as 100 coulombs (C). Although averages are difficult to assess where lightning is concerned, a characteristic flash exhibits a 2 μs rise time, and a 10 to 40 μs decay to a 50 percent level. The peak current will average 18 kA for the first impulse, and about half that for the second and third impulses. Three to four strokes per flash are common.

A lightning flash is a constant current source. Once ionization occurs, the air becomes a conductive plasma reaching 60,000°F, and becomes luminous. The resistance of an object struck by lightning is of small consequence except for the power dissipation on that object, which is equivalent to I^2R. Fifty percent of all strikes will have a first discharge of at least 18 kA, 10 percent will exceed 60 kA, and only 1 percent will exceed 120 kA.

Four specific types of cloud-to-ground lightning have been identified. They are categorized in terms of the direction of motion (upward or downward) and the sign of the electric charge (positive or negative) of the initiating leader. The categories, illustrated in Figure 2.5, are defined as follows:

- *Category 1*. Negative leader cloud-to-ground discharge. By far the most common form of lightning, such discharges account for 90 percent or more of the cloud-to-ground flashes worldwide. Such events are initiated by a downward-moving negatively charged leader.
- *Category 2*. Positive leader ground-to-cloud discharge. This event is initiated by an upward-initiated flash from earth, generally from a mountaintop or tall steel structure. Category 2 discharges are relatively rare.
- *Category 3*. Positive leader cloud-to-ground discharge. Less than 10 percent of cloud-to-ground lightning worldwide is of this type. Positive discharges are initiated by leaders that do not exhibit the distinct steps of their negative counterparts. The largest recorded peak currents are in the 200–300 kA range.
- *Category 4*. Negative leader ground-to-cloud discharge. Relatively rare, this form of lightning begins with an upward leader that exhibits a negative charge. Similar to Category 2 discharges, Category 4 discharges occur primarily from a mountaintop or tall steel structure.

An idealized lightning flash is shown in Figure 2.6. The stepped leader initiates the first return stroke in a negative cloud-to-ground flash by propagating downward in a series of discrete steps, as shown. The breakdown process sets the stage for a negative charge to be lowered to the ground. A fully developed leader lowers 10 C or more of

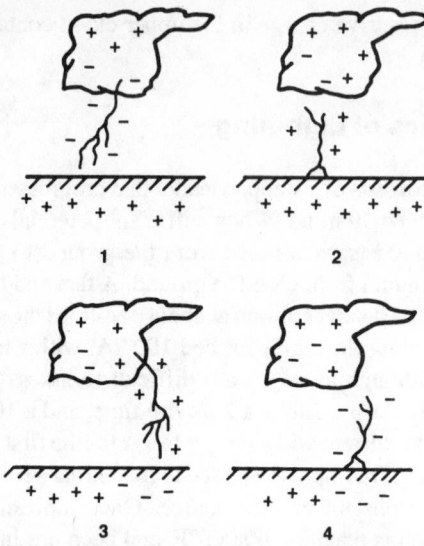

Figure 2.5 Four general types of lightning activity. (Adapted from: K. Berger, "Blitzstorm-Parameter von Aufwärtsblitzen," *Bull. Schweitz. Electrotech.* Ver., vol. 69, pp. 353–360, 1978.)

negative cloud charge to near the ground within a few tens of milliseconds. The average return leader current measures from 100 A to 1 kA. During its trip toward earth, the stepped leader branches in a downward direction, producing the characteristic lightning discharge.

The electrical potential difference between the bottom of the negatively charged leader channel and the earth may exhibit a magnitude in excess of 100 MV. As the leader tip nears ground level, the electric field at sharp objects on the ground increases until the breakdown strength of the atmosphere is exceeded. At that point, one or more upward-moving discharges are initiated, and the *attachment process* begins. The leader channel is discharged when the first return stroke propagates up the previously ionized and charged leader path. This process will repeat if sufficient potential exists after the initial stroke. The time between successive strokes in a flash is usually several tens of milliseconds.

Cloud-to-Cloud Activity. A cloud discharge may be defined as any lightning event that does not connect with the earth. Most lightning discharges occur within the confines of the cloud. Cloud discharges can be subdivided into intra-cloud, inter-cloud, and cloud-to-air flashes. Although relatively insignificant insofar as earthbound equipment is concerned, current movement between clouds can create a corresponding *earth current*.

It is estimated that only about 10 to 25 percent of lightning occurs from cloud to ground; most discharges consist of intra-cloud activity. The reason is the enormous voltage difference that builds up between the top and bottom of a storm cloud. Furthermore, the region between the top and bottom of the cloud may be more highly

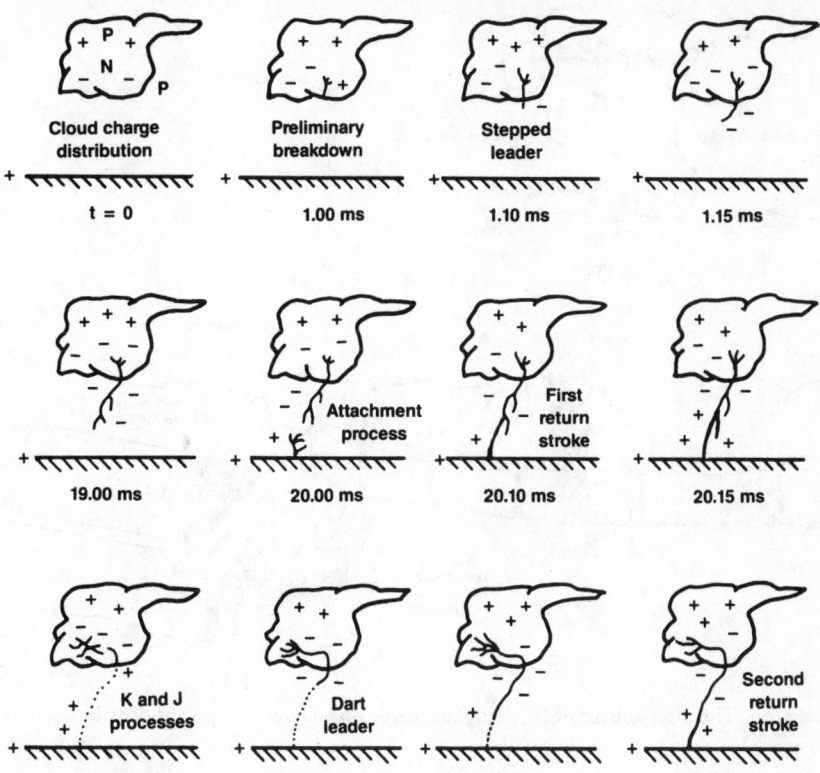

Figure 2.6   The mechanics of a lightning flash. (Adapted from: Martin A. Uman, "Natural and Artificially Initiated Lightning and Lightning Test Standards," *Proceedings of the IEEE*, vol. 76, no. 12, IEEE, New York, December 1988.)

ionized than the region between the bottom of the cloud and the earth. Currents developed by cloud-to-cloud discharges can induce significant voltages in conductors buried in-line with the charge movement. The *windstorm effect* also can induce voltages in above- or below-ground conductors as a result of rapid changes in the electrical potential of the atmosphere.

It is unnecessary, therefore, for atmospheric charge energy to actually strike a conductor of concern, such as a transmitting tower or utility company pole. In many cases, significant voltage transients can be generated solely by induction. Cloud-to-cloud charge movements generate horizontally polarized radiation, and cloud-to-ground discharges generate vertically polarized radiation. Field strengths exceeding 70 V/m can be induced in conductors a mile or so from a large strike.

Figure 2.7 illustrates the mechanisms of lightning damage. Traveling waves of voltage and current follow all conductive paths until the flash energy has been dissipated. Reflections occur at discontinuities, such as lightning arresters (points 1, 2, 3, and 5) and transformers (points 4 and 6).

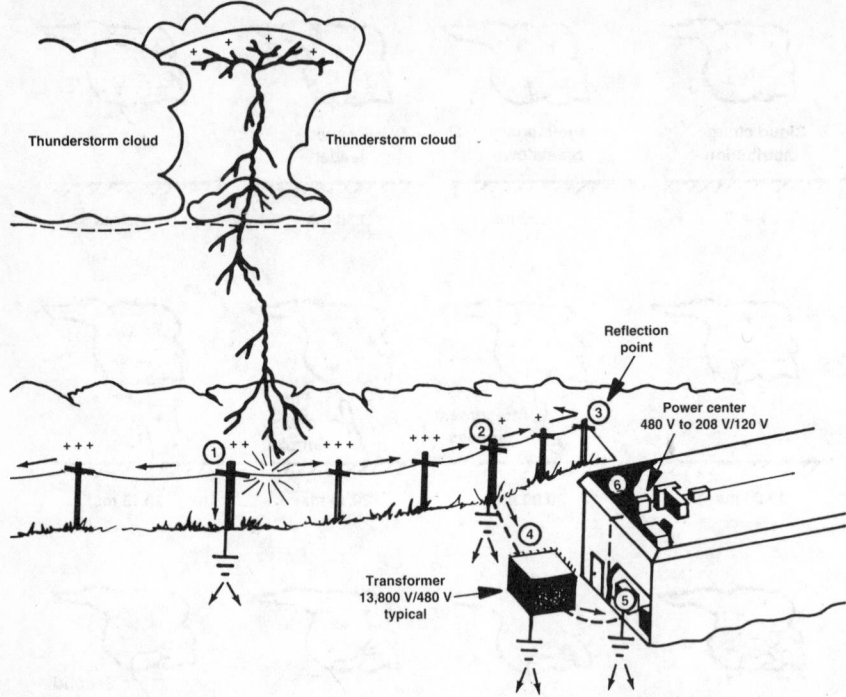

Figure 2.7   The mechanisms of lightning discharge on an overhead power-distribution line.

2.2.3 Lightning Protection

Research into the physical properties of lightning and related phenomena has two basic goals: (1) identify the character and severity of the threat, and (2) devise methods to prevent damage resulting from atmospheric activity. Many different approaches have been taken over the years for controlling the damaging potential of lightning; some have become widely accepted, others remain controversial. The issue of lightning *prevention* clearly falls into the second category.

Application of the *point discharge* theory, the basis of lightning prevention schemes, is controversial to begin with. Still, it offers the promise of a solution to a serious problem faced by nearly all telecommunications operators. The goal is to dissipate static charges around a given structure at a rate sufficient to maintain the charge below the value at which a lightning flash will occur. The theory holds that discharge from the point of an electrode to a surrounding medium will follow predictable rules of behavior. The sharper the point, the greater the discharge. The greater the number of discharge points, the more efficient the dissipation system. Several static dissipators based on this theory are shown in Figure 2.8. Key design elements for such dissipators include:

- Radius of the dissipator electrode. The goal of the dissipator is to create a high field intensity surrounding the device. Theory states that the electric field

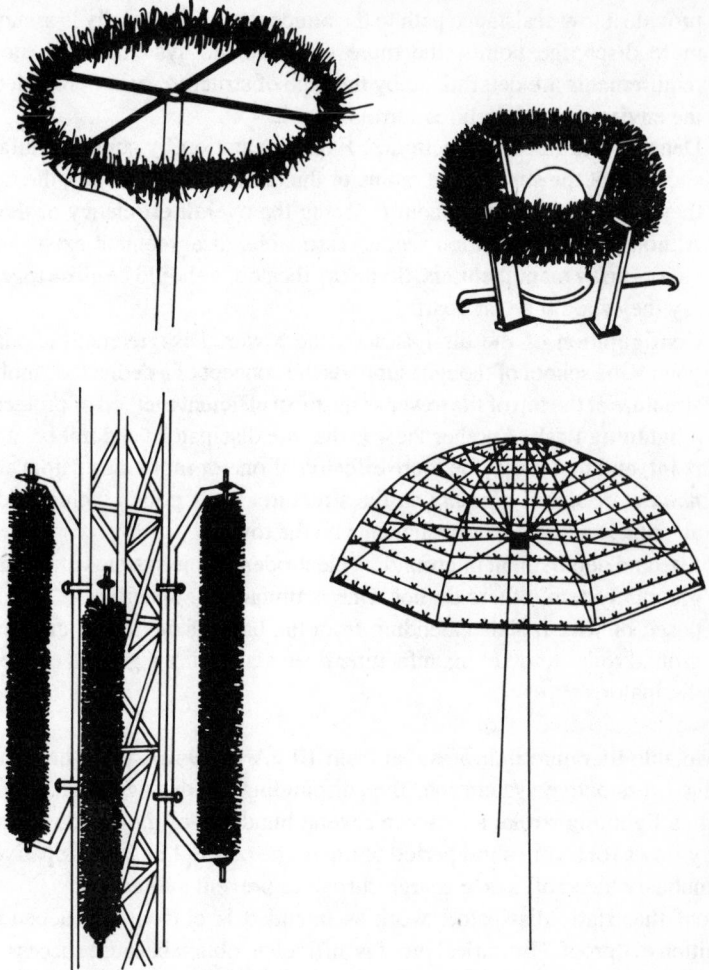

Figure 2.8   Various types of static dissipation arrays.

intensity will increase as the electrode radius is reduced. Dissipators, therefore, generally use the smallest-radius electrodes possible, consistent with structural integrity. There is, however, disagreement among certain dissipation-array manufacturers on this point. The "optimum wire size," according to available literature, varies from 0.005-in- to 1/8-in-thick tapered spikes.

- Dissipator construction material. Important qualities include conductivity and durability. The dissipator should be a good conductor to provide: (1) the maximum discharge of current during normal operation, and (2) an efficient path for current flow in the event of a direct strike.
- Number of dissipator electrodes. Calculating the number of dissipator points is, again, the subject of some debate. However, because the goal of the system is to

provide a low-resistance path to the atmosphere, it generally is assumed that the more discharge points, the more effective the system. Dissipator electrode requirements are determined by the type of structure being protected as well as the environmental features surrounding it.

- Density of dissipator electrodes. Experimentation by some manufacturers has shown that the smaller the radius of the dissipator electrodes, the more closely they can be arranged without reducing the overall efficiency of the dissipator. Although this convention seems reasonable, disagreement exists among dissipation-array manufacturers. Some say the points should be close together; others say they should be far apart.
- Configuration of the dissipator on the tower. Disagreement abounds on this point. One school of thought supports the concept of a dedicated "umbrella-type" structure at the top of the tower as the most efficient method of protecting against a lightning flash. Another view is that the dissipator need not be at the highest point, and that it may be more effective if one or more dissipators are placed at *natural dissipation points* on the structure. Such points include side-mounted antennas and other sharp elements on the tower.
- Size and deployment of grounding electrodes. Some systems utilize an extensive ground system, others do not. One manufacturer specifies a "collector" composed of wire radials extending from the base of the tower and terminated by ground rods. Another manufacturer does not require a ground connection to the dissipator.

Available literature indicates that from 10 μA to 10 mA flow through a properly designed dissipative system into the surrounding air during a lightning storm. Although a lightning stroke may reach several hundreds of thousands of amperes, this energy flows for a very short period of time. The concept of the dissipative array is to continuously bleed off space charge current to prevent a direct hit.

Proof that static dissipators work as intended is elusive and depends upon the definition of "proof." Empirical proof is difficult to obtain because successful performance of a static dissipator is evidenced by the absence of any results. Supporting evidence, both pro and con, is available from end-users of static dissipators, and from those who have studied this method of reducing the incidence of lightning strikes to a structure. The situation is doubly confusing because at least six companies manufacture lightning prevention equipment, and few of them agree on any major points.

Protection Area.   The placement of a tall structure over low-profile structures tends to protect the facilities near the ground from lightning flashes. The tall structure, typically a communications tower, is assumed to shield the facility below it from hits. This cone of protection is determined by the following:

- Height of the tall structure
- Height of the storm cloud above the earth

The higher the cloud, the larger the radius of the base of the protecting cone. The ratio of radius to base to height varies approximately from one to two.

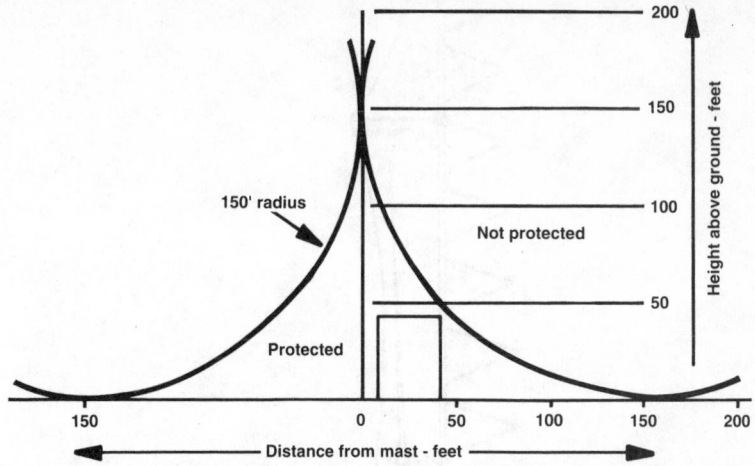

Figure 2.9   Protection zone for a tall tower under the "rolling sphere" theory.

Conventional wisdom has held that a tower, whether protected with a static dissipation array or simply a tower-top lightning rod, provided a cone of protection stretching out on all sides of the structure at an angle of about 45°. Although this theory held favor for many years, modifications have been proposed. One school of thought suggests that a smaller cone of perhaps 30° from the tower is more realistic. Another suggests that the cone theory is basically flawed and, instead, proposes a "rolling sphere" approach. This theory states that areas enclosed below a 150-ft rolling sphere will enjoy protection against lightning strikes. The concept is illustrated in Figure 2.9. Note that the top of the tower shown in the figure is assumed to experience limited protection. The concept, as it applies to side-mounted antennas, is shown in Figure 2.10. The antenna is protected through the addition of two horizontally mounted lightning rods, one above the antenna and one below.

2.2.4 Electrostatic Discharge

A static charge is the result of an excess or deficiency of electrons on a given surface. The relative level of electron imbalance determines the static charge. Simply stated, a charge is generated by physical contact between, and then separation of, two materials. One surface loses electrons to the other. The types of materials involved, and the speed and duration of motion between the materials, determine the charge level. Electrostatic energy is a stationary charge phenomenon that can build up in either a nonconductive material or in an ungrounded conductive material. The charge can occur in one of two ways:

- *Polarization*. Charge buildup when a conductive material is exposed to a magnetic field.
- *Triboelectric effects*. Charge buildup that occurs when two surfaces contact and then separate, leaving one positively charged and one negatively charged.

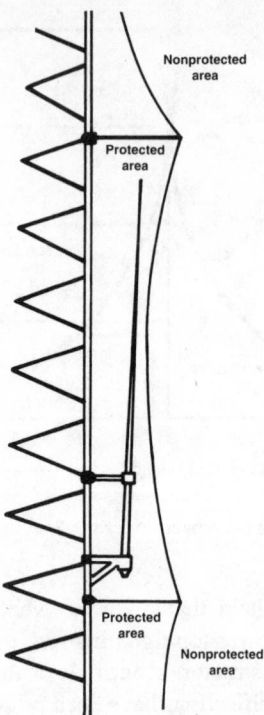

Figure 2.10   Protection zone for side-mounted antenna under the "rolling sphere" theory.

Friction between two materials increases the triboelectric charge by increasing the surface area that experiences contact. For example, a person accumulates charge by walking across a nylon carpet; discharge occurs when the person touches a conductive surface.

Triboelectric Effects.   Different materials have differing potentials for charge. Nylon, human and animal hair, wool, and asbestos have high positive triboelectric potential. Silicon, polyurethane, rayon, and polyester have negative triboelectric potentials. Cotton, steel, paper, and wood all tend to be relatively neutral materials. The intensity of the triboelectric charge is inversely proportional to the relative humidity (RH). As humidity increases, electrostatic discharge (ESD) problems decrease. For example, a person walking across a carpet can generate a 1.5 kV charge at 90 percent RH, but will generate as much as 35 kV at 10 percent RH.

When a charged object comes in contact with another object, the electrostatic charge will attempt to find a path to ground, discharging into the contacted object. The current level is very low (typically less than 0.1 nA), but the voltage can be high (25 to 50 kV).

ESD also can collect on metallic furnishings, such as chairs and equipment racks. Sharp corners and edges, however, encourage a corona that tends to bleed the charge

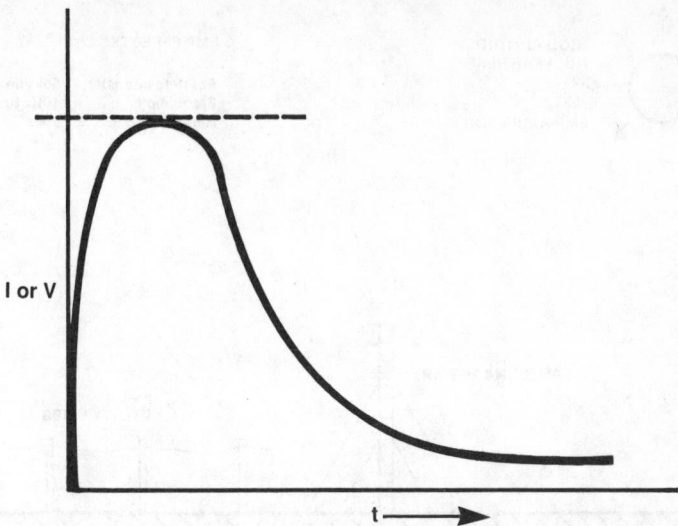

Figure 2.11   Discharge waveform for an ESD event. (Data from: Technical staff, Military/Aerospace Products Division, *The Reliability Handbook*, National Semiconductor, Santa Clara, CA, 1987.)

off such objects. The maximum voltage normally expected for furniture-related ESD is about 6 to 8 kV. Because metallic furniture is much more conductive than humans, however, furniture-related ESD generally will result in higher peak discharge currents. Figure 2.11 shows a waveform of a typical ESD event.

2.2.5   EMP Radiation

Electromagnetic pulse (EMP) radiation is the result of an intense release of electromagnetic waves that follows a nuclear explosion. See Figure 2.12. The amount of damaging energy is a function of the altitude of detonation and the size of the device. A low-altitude or surface burst would generate a strong EMP covering a few thousand square kilometers. However, the effects of the radiation would be meaningless, because the blast would destroy most structures in the area. A high-altitude burst, on the other hand, presents a real threat to all types of communications and electronic systems. Such an explosion would generate an EMP with a radius of more than 1000 km, a large portion of the United States.

The sudden release of gamma rays in a nuclear explosion would cause almost instant ionization (the removal of electrons from atoms) of the atmospheric gases that surround the detonation area. Free electrons are driven outward. In a high-altitude event, the gamma rays travel great distances before ionizing the upper atmosphere. The forced movement of these electrons, which will again recombine with atoms in the atmosphere, creates a pulsed electromagnetic field.

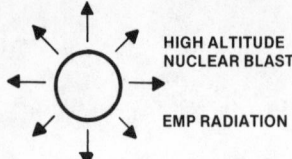

HIGH ALTITUDE
NUCLEAR BLAST

EMP RADIATION

EMP EFFECTS:

Far field intensity	50kV/m
Frequency	10Hz to 100MHz
Rise time	20ns

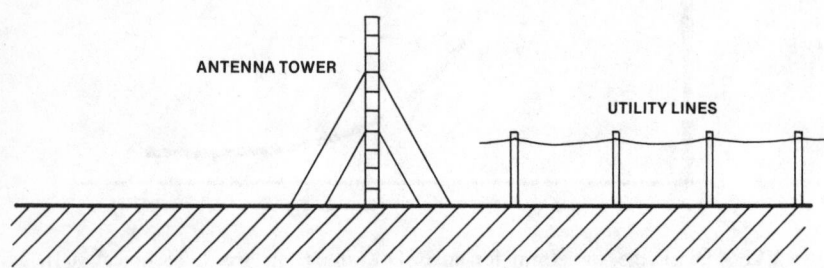

ANTENNA TOWER

UTILITY LINES

Figure 2.12   The EMP effect and how it can induce damaging voltages onto utility company lines and antenna structures. The expected parameters of an EMP also are shown.

The amplitude and polarization of the field produced by a high-altitude detonation depends on the altitude of the burst, the yield of the device, and the orientation of the burst with respect to the receiving point. The EMP field creates a short but intense broadband radio frequency pulse with significant energy up to 100 MHz. Most of the radiated energy, however, is concentrated below 10 MHz. Figure 2.13 shows the distribution of energy as a function of frequency. The electric field can be greater than 50 kV/m, with a rise time measured in the tens of nanoseconds. Figure 2.14 illustrates the field of a simulated EMP discharge.

Many times, lightning and other natural occurrences cause problems not because they strike a given site, but because they strike part of the utility power system and are brought into the facility via the ac lines. Likewise, damage that could result from EMP radiation would be most severe to equipment connected to the primary power source, because it is generally the most exposed part of any facility. Table 2.1 lists the response of various systems to an EMP event.

2.2.6 Coupling Transient Energy

The utility power-distribution system can couple transient overvoltages into a customer's load through induction or direct-charge injection. As stated previously, a lightning flash a mile away from a 12 kV line can create an electromagnetic field with a strength of 70 V/m or more. Given a sufficiently long line, substantial voltages can

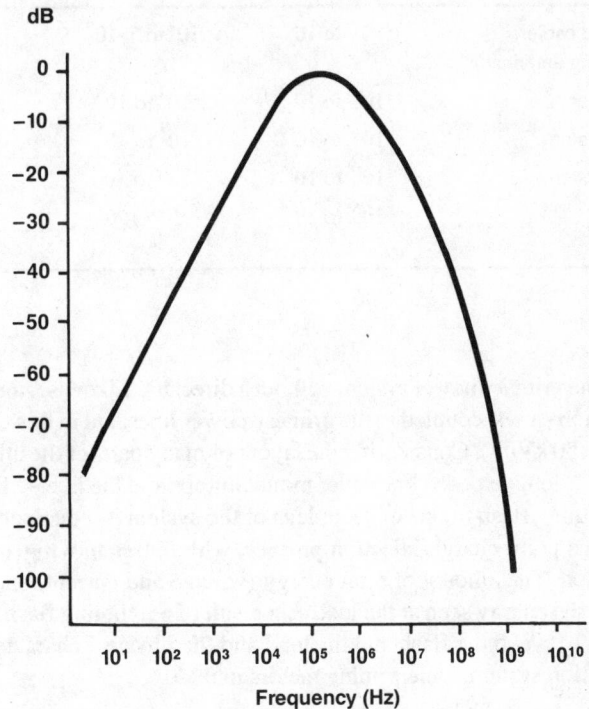

Figure 2.13 Normalized EMP spectrum. Approximately 99 percent of the radiated energy is concentrated between 10 kHz and 100 MHz.

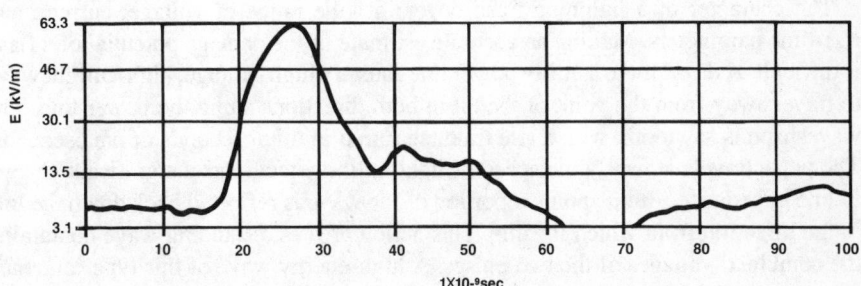

Figure 2.14 Discharge waveform for a nuclear EMP event.

Table 2.1   Effects of an EMP Event on Various Systems

Type of Conductor	Rise Time (sec)	Peak Voltage (volts)	Peak Current (amps)
Long unshielded cable (power lines, long antennas)	10^{-8} to 10^{-7}	10^5 to $5 \bullet 10^6$	10^3 to 10^4
HF antenna systems	10^{-8} to 10^{-7}	10^4 to 10^6	500 to 10^4
VHF antenna system	10^{-9} to 10^{-8}	10^3 to 10^5	100 to 10^3
UHF antenna system	10^{-9} to 10^{-8}	100 to 10^4	10 to 100
Shielded cable	10^{-6} to 10^{-4}	1 to 100	0.1 to 50

be coupled to the primary power system without a direct hit. Likewise, the field created by EMP radiation can be coupled to the primary power lines, but in this case at a much higher voltage (50 kV/m). Considering the layout of many parts of the utility company power system — long exposed lines over mountaintops and the like — the possibility of a direct lightning flash to one or more legs of the system is a distinct one.

Lightning is a point charge-injection process, with pulses moving away from the point of injection. The amount of total energy (voltage and current) and the rise and decay times of the energy seen at the load as a result of a lightning flash are functions of the distance between the flash and the load and the physical characteristics of the power-distribution system. Determining factors include:

- Wire size
- Number and sharpness of bends
- Types of transformers
- Types of insulators
- Placement of lightning suppressors

The character of a lightning flash covers a wide range of voltage, current, and rise-time parameters. Making an accurate estimate of the damage potential of a flash is difficult. A direct hit to a utility power line causes a high-voltage, high-current wave to travel away from the point of the hit in both directions along the power line. The waveshape is sawtooth, with a rise time measured in microseconds or nanoseconds. The pulse travels at nearly the speed of light until it encounters a significant change in line impedance. At this point, a portion of the wave is reflected back down the line in the direction from which it came. This action creates a standing wave containing the combined voltages of the two pulses. A high-energy wave of this type can reach a potential sufficient to arc over to another parallel line, a distance of about 8 ft on a local feeder (typically 12 kV) power pole.

2.3 EQUIPMENT-CAUSED TRANSIENT DISTURBANCES

Equipment-caused transients in the utility power system are a consequence of the basic nature of alternating current. A sudden change in an electric circuit will cause a transient voltage to be generated because of the stored energy contained in the circuit inductances (*L*) and capacitances (*C*). The size and duration of the transient depend on the values of *L* and *C* and the waveform applied.

A large step-down transformer, the building block of a power system, normally will generate transient waveforms when energized or deenergized. As illustrated in Figure 2.15, the stray capacitances and inductances of the secondary can generate a brief oscillating transient of up to twice the peak secondary voltage when the transformer is energized. The length of this oscillation is determined by the values of *L* and *C* in the circuit.

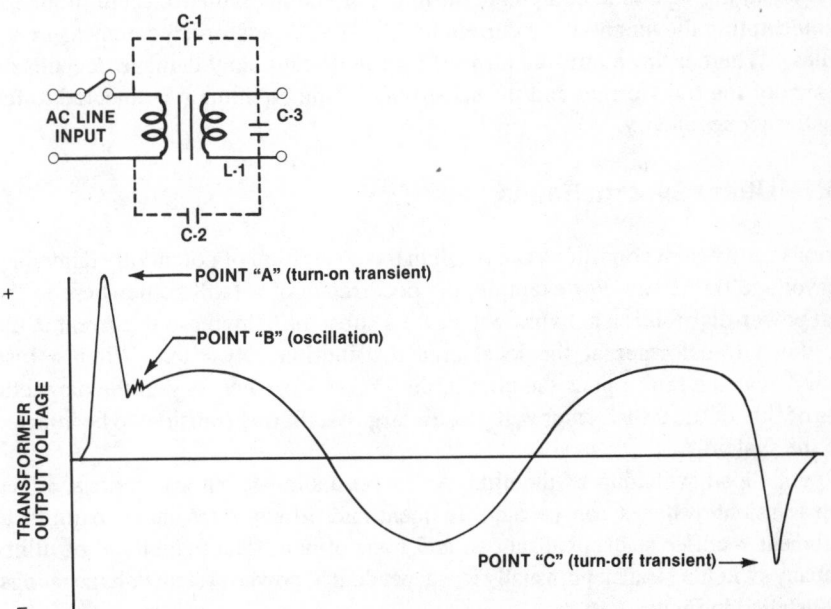

Figure 2.15 The causes of inductor turn-on and turn-off spikes. The waveforms are exaggerated to illustrate the transient effects. C1, C2, and C3 are stray capacitances that form a divider network between the primary and secondary, causing the turn-on spike shown at point *A*. The oscillation shown at point *B* is caused by the interaction of the inductance of the secondary (L1) and C3. The spike shown at point *C* is the result of power interruption to the transformer primary, which causes the collapsing lines of flux to couple a high-voltage transient into the secondary circuit.

A more serious problem can be encountered when energizing a step-down transformer. The load is, in effect, looking into a capacitive divider from the secondary into the primary. If the interwinding capacitance is high and the load capacitance is low, a spike of as much as the full primary voltage can be induced onto the secondary, and thus, onto the load. This spike does not carry much energy because of its short duration, but sensitive equipment on the load side could be damaged upon reapplication of power to a utility company pole transformer, for example, as would occur after a power outage. A typical transformer may have 0.002 µF series capacitance from input to output. A simple Faraday shielded isolation transformer can reduce this series capacitance to 30 pF, nearly two orders of magnitude lower.

Deenergizing a large power transformer also can cause high-voltage spikes to be generated. Interrupting the current to the primary windings of a transformer, unless switched off at or near the zero crossing, will cause the collapsing magnetic lines of flux in the core to couple a high-voltage transient into the secondary circuit. If a low-impedance discharge path is not present, this spike will be impressed upon the load. Transients in excess of 10 times the normal secondary voltage have been observed when this type of switching occurs. Such spikes can have damaging results to equipment on-line. For example, the transient produced by interrupting the magnetizing current to a 150 kVA transformer can measure 9 J (joules). Whether these turn-on, turn-off transients cause any damage depends on the size of the transformer and the sensitivity of the equipment connected to the. transformer secondary.

2.3.1   Utility System Faults

Various utility fault conditions can result in the generation of potentially damaging overvoltage transients. For example, the occurrence of a fault somewhere in the local power-distribution network will cause a substantial increase in current in the step-down transformer at the local area distribution substation. When a fuse located near the fault opens the circuit, the excess stored energy in the magnetic lines of flux of the transformer will cause a large oscillating transient to be injected into the system.

Routine load switching by the utility will have a similar, but less serious, effect. Such transient voltages can be quite frequent and, in some instances, harmful to equipment rectifier stacks, capacitors, and transformers. The magnitude of utility company switching transients usually is independent of power-system voltage ratings, as illustrated in Figure 2.16.

2.3.2   Switch Contact Arcing

Transients generated by contact bounce occur not only because of physical bouncing upon closing or opening, but also because of arcing between contacts, the result of transients that are generated when an inductive load is deenergized. The principle is illustrated in Figure 2.17. When current is interrupted to an inductor, the magnetic

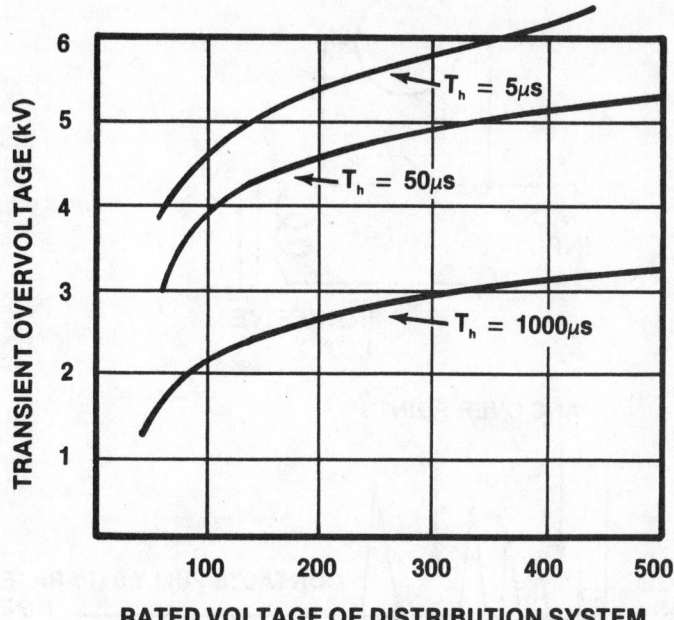

Figure 2.16 The relationship (computed and measured) between utility company system voltage and switching transient peaks. Switching transients are plotted as a function of normal operating voltages for three values of the transient tail time to half-potential. It can be seen that there is no direct linear increase in transient amplitude as the system voltage is increased.

lines of flux will try to maintain themselves by charging stray capacitances. The current will oscillate in the inductance and capacitance at a high frequency. If sufficiently high voltages are generated in this process, an arc will jump the contacts after they have opened, clearing the oscillating current. As the contacts separate further, the process is repeated until the voltage generated by the collapsing lines of flux is no longer sufficient to jump the widening gap between the contacts. This voltage then may look for another discharge path, such as interwinding arcing or other components in parallel with the inductor. Contact arcing also may occur when an inductor is switched on, if the contacts bounce open after first closing. Figure 2.18 illustrates typical contacting characteristics for energizing a relay mechanism; deenergizing of relay contacts is illustrated in Figure 2.19.

It should be noted that the ringing of an inductive circuit exposed to a transient disturbance produces not only the overshoot that most engineers are familiar with, but also a potentially damaging *undershoot*. This effect is illustrated in Figure 2.20. Sudden and severe polarity reversals such as the one illustrated can have damaging effects on semiconductor devices.

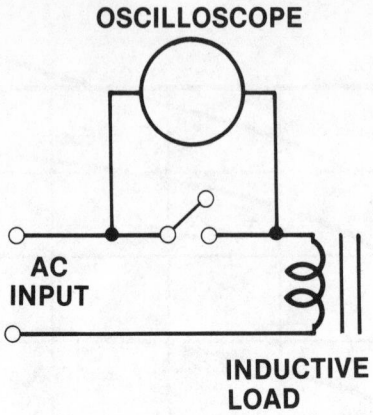

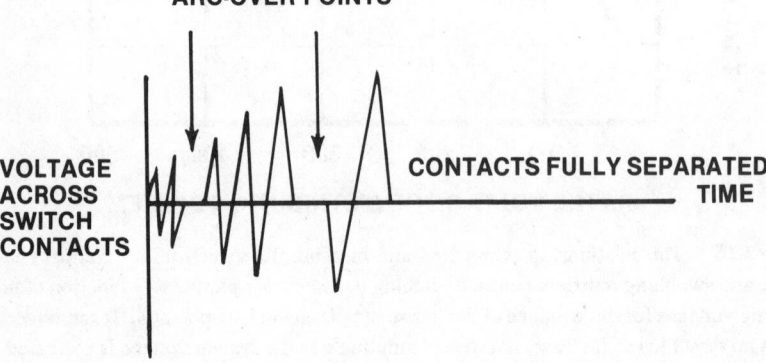

Figure 2.17   The mechanics of contact bounce: (a, *top*) measurement configuration; (b, *bottom*) representative waveforms. The waveforms shown are exaggerated to illustrate the transient effects. For clarity, the modulating effect of the ac line voltage is not shown.

2.3.3   Telephone System Transients

Overvoltages on telephone loops and data lines generally can be traced to the 60 Hz power system and lightning. Faults, crossed lines, and bad grounding can cause energy to be injected into or coupled onto telco circuits from utility company power lines when the cables share common poles or routing paths. Direct lightning hits to telco lines will generate huge transients on low-level audio, video, or data circuits. Buried phone company cables also are subject to damaging transients because of charge movements in the earth resulting from lightning flashes and cloud-to-cloud discharge activity. Voltages can be induced in cable shield material and in the lines themselves. EMP radiation can penetrate a buried telco line in a similar manner.

Lightning or other transient currents usually travel along a telephone cable shield until dissipated, either through ground connections (in the case of pole-mounted cables) or through cable-to-earth arc-overs (in the case of buried cables). This activity usually does not cause damage to the shield material itself, but it can induce transient

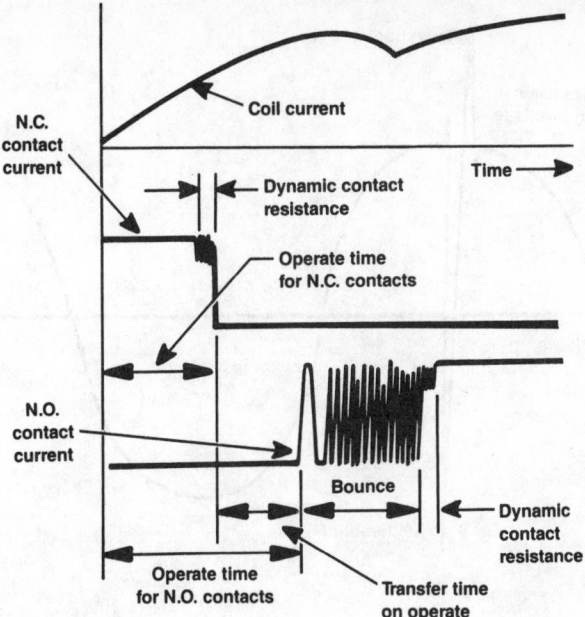

Figure 2.18   Mechanics of relay contact bounce during the energizing period.

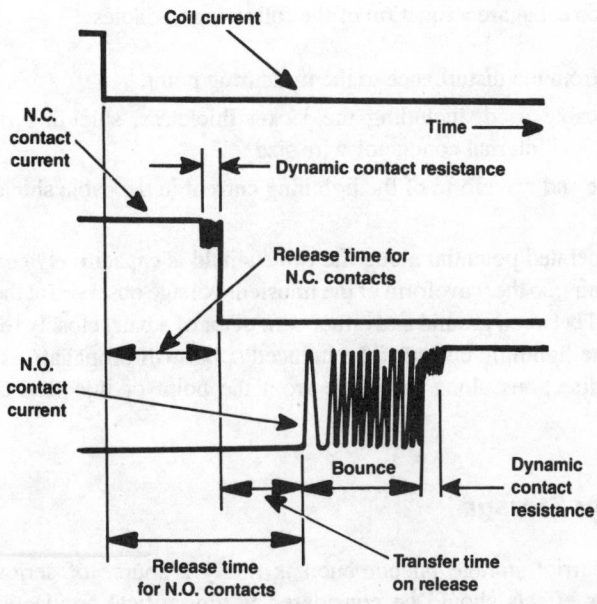

Figure 2.19   Mechanics of relay contact bounce during the deenergizing period.

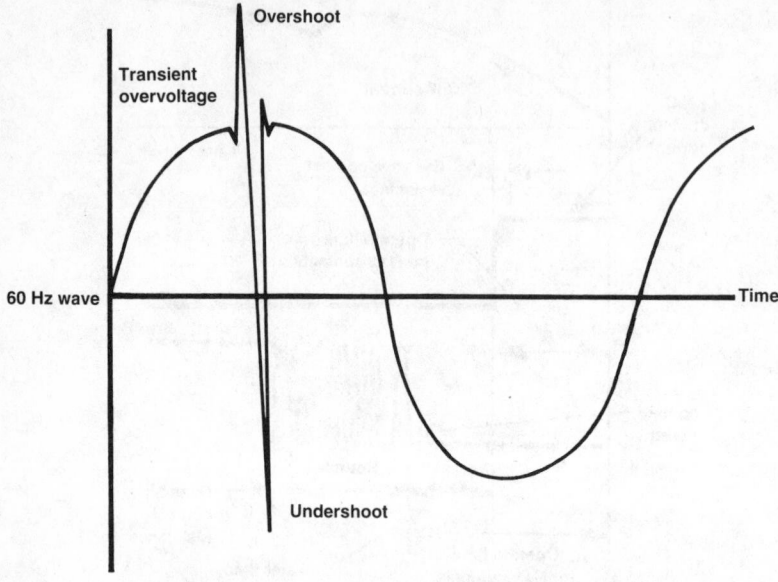

Figure 2.20   The "iceberg" effect of many transient waveforms. The overshoot often is followed by an equally damaging undershoot before leveling off.

voltages on internal conductors, which may be harmful to central office equipment or end-user devices and systems. The characteristics of the transient that can appear at the end of a telco cable are a function of the following variables:

- Distance from the disturbance to the measuring point.
- Type of cable used, including the jacket thickness, shielding material and thickness, and internal conductor wire size.
- Amplitude and waveform of the lightning current in the cable shield.

The current-generated potential along the cable shield is capacitively coupled to the internal cable pairs, so the waveform of the transient voltage observed at the measuring point (measured between ground and either conductor of a pair) closely resembles the waveform of the lightning current. The induced spike will propagate as a traveling wave in both directions along the cable from the point of injection or region of induction.

2.3.4 Carrier Storage

Although the carrier storage phenomenon is rarely a source of serious transient disturbances, its effects should be considered in any critical application. When a silicon diode switches from a forward conduction mode to a reverse blocking state, the presence of stored carriers at the device junction can prevent an immediate cessation of current flow. These stored carriers have the effect of permitting current to flow in the reverse direction during a brief portion of the ac cycle.

The carrier storage current is limited only by the applied voltage and external circuit design. The current flow is brief, as carriers are removed rapidly from the junction by internal recombination and the sweeping effects of the reverse current. This removal of carriers causes the diode to revert to its blocking condition, and the sudden cessation of what can be a large reverse current may cause damaging voltage transients in the circuit if there is appreciable system inductance and if transient suppression has not been included in the system design.

The reverse current caused by carrier storage usually is not excessive in normal operation of power rectifier circuits. Carrier storage does not, in itself, constitute a hazard, especially at ac power line frequencies. The carrier storage effect can, however, lead to complications in certain switching arrangements. For example, current will tend to "free-wheel" through rectifier diodes after the supply voltage has been removed in an inductive circuit until the stored energy has been dissipated. If the supply voltage is reapplied while this free-wheeling process is under way, some of the diodes in the circuit will be required to conduct in a forward direction, but others will be required to block. While the latter diodes are recovering from the carrier storage free-wheeling current, a short-circuit effect will be experienced by the source, causing potentially damaging surge currents.

2.4   TRANSIENT-GENERATED NOISE

Problems can be caused in a facility by transient overvoltages not only through device failure, but also because of logic state upsets. Studies have shown that an upset in the logic of typical digital circuitry can occur with transient energy levels as low as $1 \bullet 10^{-9}$ J. Such logic-state upsets can result in microcomputer latchup, lost or incorrect data, program errors, and control-system shutdown. Transient-induced noise is a major contributor to such problems.

2.4.1   Noise Sources

Utility company transients, lightning, and ESD are the primary sources of random, unpredictable noise. Unless properly controlled, these sources represent a significant threat to equipment reliability. Although more predictable, other sources, such as switch contact arcing and SCR switching, also present problems for equipment users.

ESD Noise.   The upper frequency limit for an ESD discharge can exceed 1 GHz. Determining factors include the voltage level, relative humidity, speed of approach, and shape of the charged object. At such frequencies, circuit board traces and cables function as fairly efficient receiving antennas.

Electrical noise associated with ESD may enter electronic equipment by either conduction or radiation. In the near field of an ESD (within a few tens of centimeters) the primary type of radiated coupling may be either inductive or capacitive, depending on the impedances of the ESD source and the receiver. In the far field, electromagnetic

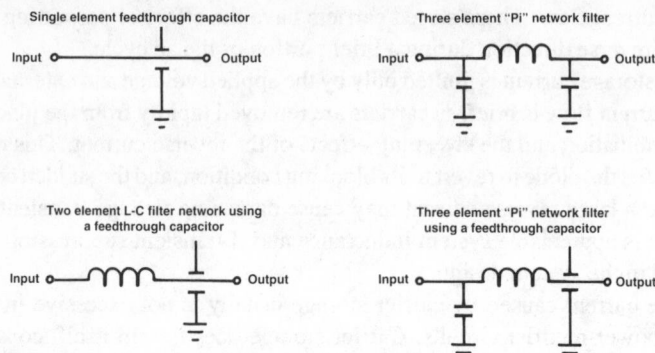

Figure 2.21   Common radio frequency interference (RFI) filtering networks. These filters, combined with tight mechanical assembly construction and printed wiring board ground plane shielding, will reduce significantly the possibility of noise-caused equipment disturbances.

field coupling predominates. Circuit operation is upset if the ESD-induced voltages and/or currents exceed typical signal levels in the system. Coupling of an ESD voltage in the near field is determined by the impedance of the circuit:

- High-impedance circuit. Capacitive coupling will dominate; ESD-induced voltages will be the major problem.
- Low-impedance circuit. Inductive coupling will dominate; ESD-induced currents will be the major problem.

Contact Arcing.   Switch contact arcing and similar repetitive transient-generating operations can induce significant broadband noise into an electrical system. Noise generated in this fashion is best controlled at its source, almost always an inductive load. Noise can travel through power lines and create problems for microcomputer equipment either through direct injection into the system power supply, or through coupling from adjacent cables or printed wiring board (PWB) traces.

SCR Switching.   SCR power controllers are a potential source of noise-induced microcomputer problems. Each time an SCR is triggered into its active state in a resistive circuit, the load current goes from zero to the load-limited current value in less than a few microseconds. This step action generates a broadband spectrum of energy, with an amplitude inversely proportional to frequency. Electronic equipment using full-wave SCR control in a 60 Hz circuit can experience such noise bursts 120 times a second.

In an industrial environment, where various control systems may be spaced closely, electrical noise can cause latchup problems or incorrect data in microcomputer equipment, or interaction between SCR firing units in machine controllers. Power-line cables within a facility can couple noise from one area of a plant to another, further complicating the problem. The solution to the SCR noise problem is found by looking at both the source of the interference and the susceptible hardware. The use of good transient-suppression techniques in the application of SCR power controllers will eliminate noise generation in all but the most critical of applications.

As a further measure of insurance, sensitive electronic equipment should be shielded adequately against noise pickup, including metal cabinet shields, ac power line filters, and input/output line feedthrough RF filters. Fortunately, most professional/industrial equipment is designed with RF shielding as a concern. Figure 2.21 shows some of the more common RFI shielding techniques.

2.5 POWER-DISTURBANCE CLASSIFICATIONS

Short-term ac voltage disturbances can be classified into four major categories, as illustrated in Figure 2.22. The categories are defined according to peak value and duration:

- *Voltage surge.* An increase of 10 to 35 percent above the normal line voltage for a period of 16 ms to 30 s.
- *Voltage sag.* A decrease of 10 to 35 percent below the normal line voltage for a period of 16 ms to 30 s.
- *Transient disturbance.* A voltage pulse of high energy and short duration impressed upon the ac waveform. The overvoltage pulse may be 1 to 100 times the normal ac potential and may last up to 15 ms. Rise times measure in the nanosecond range.
- *Momentary power interruption.* A decrease to zero voltage of the ac power line potential, lasting from 33 to 133 ms. (Longer-duration interruptions are considered power outages.)

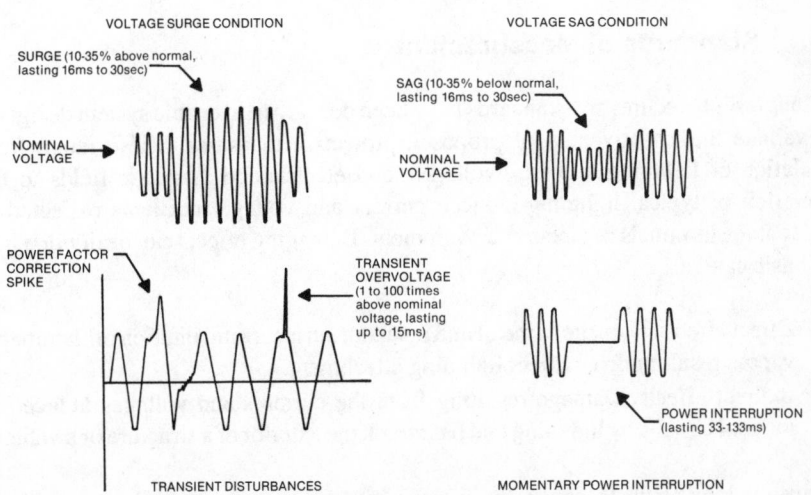

Figure 2.22   The four basic classifications of short-term power-line disturbances.

Voltage surges and sags occasionally result in operation problems for equipment on-line, but automatic protection or correction circuits generally take appropriate actions to ensure that there is no equipment damage. Such disturbances can, however, garble computer system data if the disturbance *transition time* (the rise or fall time of the disturbance) is sufficiently fast. System hardware also may be stressed if there is only a marginal power-supply reserve or if the disturbances are frequent.

Momentary power interruptions can cause a loss of volatile memory in computer-driven systems and place severe stress on hardware components, especially if the ac supply is allowed to surge back automatically without *soft-start* provisions. Successful system reset may not be accomplished if the interruption is sufficiently brief.

Although voltage sags, surges, and momentary interruptions can cause operational problems for equipment used today, the possibility of complete system failure because of one of these mechanisms is relatively small. The greatest threat to the proper operation of electronic equipment rests with transient overvoltage disturbances on the ac line. Transients are difficult to identify and difficult to eliminate. Many devices commonly used to correct sag and surge conditions, such as ferroresonant transformers or motor-driven autotransformers, are of limited value in protecting a load from high-energy, fast-rise-time disturbances.

In the computer industry, research has shown that most unexplained problems resulting in disallowed states of operation actually are caused by transient disturbances on the utility feed. With the increased use of microcomputers in industry, this consideration cannot be ignored. Because of the high potential that transient disturbances typically exhibit, they not only cause data and program errors, but also can damage or destroy electric components. This threat to electronic equipment involves sensitive integrated circuits and many other common devices, such as capacitors, transformers, rectifiers, and power semiconductors. Figure 2.23 illustrates the vulnerability of common components to high-energy pulses. The effects of transient disturbances on electronic devices are often cumulative, resulting in gradual deterioration and, ultimately, catastrophic failure.

2.5.1   Standards of Measurement

Various test procedures and standards have been developed to enable system designers to evaluate the effectiveness of proposed protective measures. These range from simulation of lightning currents, voltages, and electric and magnetic fields to the generation of typical lightning-induced current and voltage transients expected to appear at the terminals of electronic equipment. Damaging effects can be divided into two basic categories:

- Direct effects. Damage to metal and insulator surfaces, and ignition of flammable vapors resulting from direct lightning attachment.
- Indirect effects. Damage resulting from the currents and voltages induced in internal circuits by lightning that has struck the exterior of a structure or a vehicle.

Because it is difficult to assess the threat posed by transient disturbances without guidelines on the nature of transients in ac power systems, a number of separate

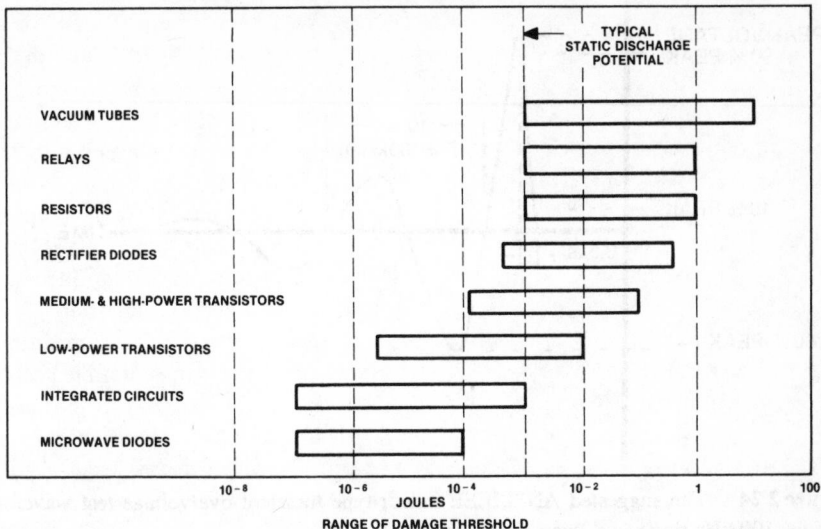

Figure 2.23   An estimation of the susceptibility of common electric devices to damage from transient disturbances. The vertical line marked "static discharge" represents the energy level of a discharge that typically can be generated by a person who touches a piece of equipment after walking across a carpeted floor.

standards have been developed for individual component groups. The best known was developed by a working group of the Institute of Electrical and Electronic Engineers (IEEE) to simulate indirect lightning effects. IEEE suggested two waveforms, one unidirectional and the other oscillatory, for measuring and testing transient-suppression components and systems in ac power circuits with rated voltages of up to 1 kV rms line-to-ground. The guidelines also recommend specific source impedance or short-circuit current values for transient analysis.

The voltage and current amplitudes, waveshapes, and source impedance values suggested in the American National Standards Institute/IEEE guide (ANSI/IEEE standard C62.41-1980) are designed to approximate the vast majority of high-level transient disturbances, but are not intended to represent worst-case conditions — a difficult parameter to predict. The timing of a transient overvoltage with respect to the power line waveform is also an important parameter in the examination of ac disturbances. Certain types of semiconductors exhibit failure modes that are dependent on the position of a transient on the sine wave.

Figure 2.24 shows the ANSI/IEEE representative waveform for an indoor-type spike (for 120 to 240 V ac systems). Field measurements, laboratory observations, and theoretical calculations have shown that most transient disturbances in low-voltage indoor ac power systems have oscillatory waveshapes, instead of the unidirectional wave most often thought to represent a transient overvoltage. The oscillatory nature of the indoor waveform is the result of the natural resonant frequencies of the ac distribution system. Studies by the IEEE show that the oscillatory frequency range of

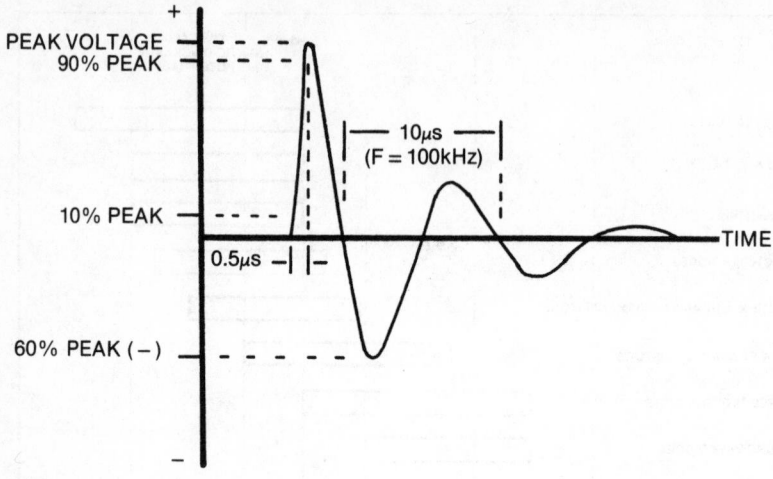

Figure 2.24   The suggested ANSI/IEEE indoor-type transient overvoltage test waveform (0.5 μs–100 kHz ring wave, open-circuit voltage).

such disturbances extends from 30 Hz to 100 kHz, and that the waveform changes, depending upon where it is measured in the power-distribution system.

The waveform shown in Figure 2.24 is the result of extensive study by the IEEE and other independent organizations of various ac power circuits. The representative waveshape for 120 and 240 V systems is described as a *0.5 μs–100 kHz ring wave*. This standard indoor spike has a rise time of 0.5 μs, then decays while oscillating at 100 kHz. The amplitude of each peak is approximately 60 percent of the preceding peak.

Figure 2.25 shows the ANSI/IEEE representative waveforms for an outdoor-type spike. The classic lightning overvoltage pulse has been established at a 1.2/50 μs waveshape for a voltage wave and an 8/20 μs waveshape for a current wave. Accordingly, the ANSI/IEEE standard waveshape is defined as *1.2/50 μs open-circuit voltage* (voltage applied to a high-impedance device), and *8/20 μs discharge current* (current in a low-impedance device).

The test waveshapes, although useful in the analysis of components and systems, are not intended to represent all transient patterns seen in low-voltage ac circuits. Lightning discharges can cause oscillations, reflections, and other disturbances in the utility company power system that can appear at the service drop entrance as decaying oscillations.

Unfortunately, most lightning disturbance standard waveforms address one or two characteristics of a discharge. Furthermore, standard waveforms are intended to represent typical events, not worst-case events. In testing to determine the immunity of a system to a direct lightning strike, a conservative approach is generally taken, in that the characteristics of a relatively severe flash are adopted. Figure 2.26 shows a test waveform specified for aerospace vehicles (MIL-STD-1757A). The current

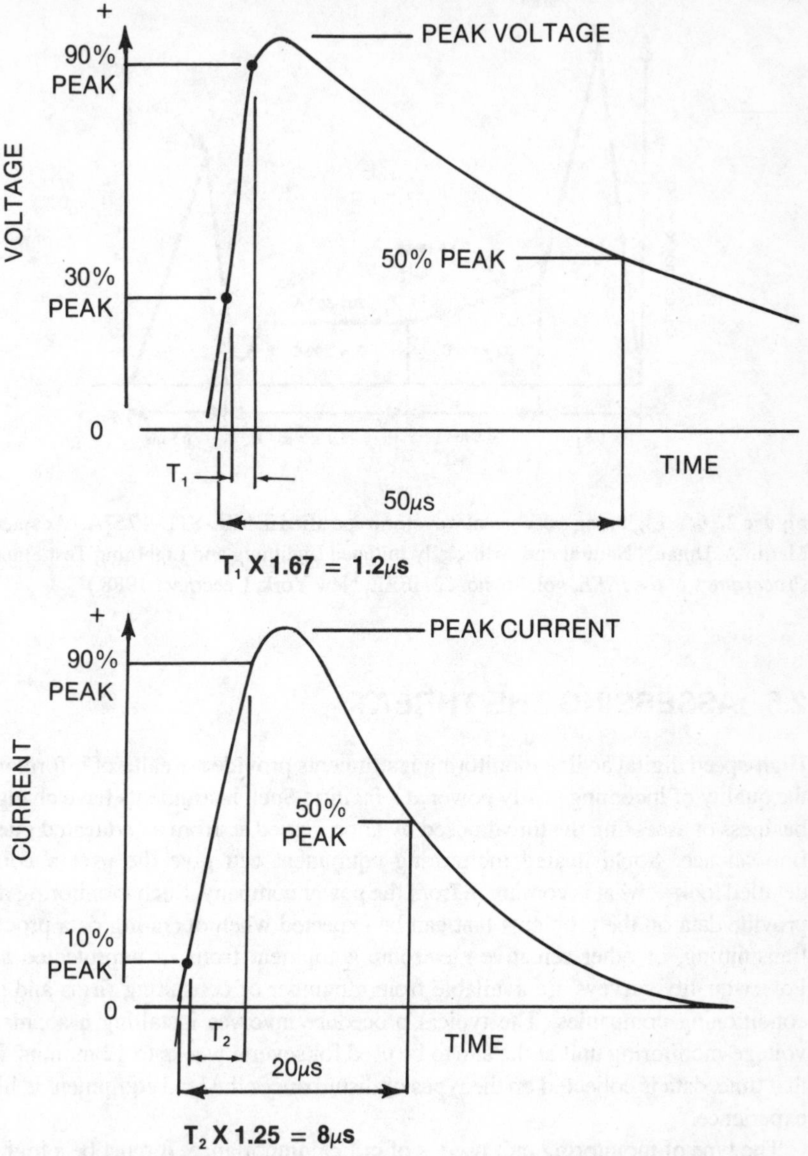

Figure 2.25   The unidirectional waveshape for outdoor-type transient overvoltage test analysis based on ANSI Standard C62.1: (a, *top*) open-circuit waveform; (b, *bottom*) discharge current waveform.

waveform includes one initial and one subsequent stroke, between which flows continuing current. Peak currents are 200 kA for the first stroke and 100 kA for the subsequent stroke. The total charge transfer is in excess of 200 C.

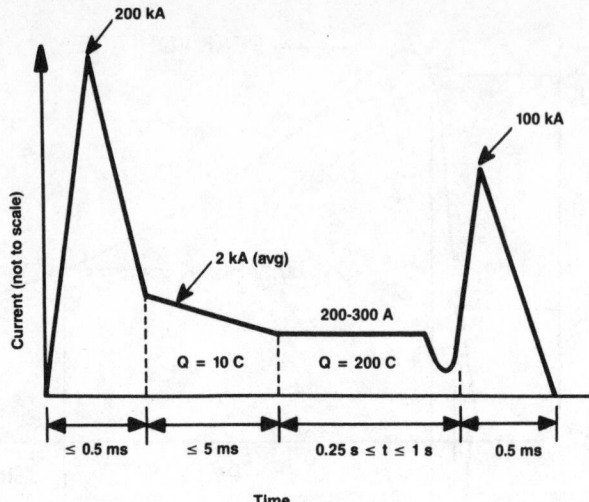

Figure 2.26   Lightning test current waveform specified in MIL-STD-1757A. (Adapted from: Martin A. Uman, "Natural and Artificially Initiated Lightning and Lightning Test Standards," *Proceedings of the IEEE*, vol. 76, no. 12, IEEE, New York, December 1988.)

2.6   ASSESSING THE THREAT

High-speed digital ac line monitoring instruments provide a wealth of information on the quality of incoming utility power at a facility. Such instruments have changed the business of assessing the threat posed by unprocessed ac from an educated guess to a fine science. Sophisticated monitoring equipment can give the user a complete, detailed look at what is coming in from the power company. Such monitoring devices provide data on the problems that can be expected when operating data processing, transmitting, or other sensitive electronic equipment from an unprotected ac line. Power-quality surveys are available from a number of consulting firms and power-conditioning companies. The typical procedure involves installing a sophisticated voltage-monitoring unit at the site to be used for several weeks to 12 months. During that time, data is collected on the types of disturbances the load equipment is likely to experience.

The type of monitoring unit used is of critical importance. It must be a high-speed system that stores disturbance data in memory and delivers a printout of the data on demand. Conventional chart recorders are too slow and lack sufficient sensitivity to accurately show short-duration voltage disturbances. Chart recorders can confirm the presence of long-term surge and sag conditions, but provide little useful data on transients.

2.6.1   Digital Measurement Instruments

Today's power monitors are sophisticated instruments that can identify not only ac power problems, but also provide the following useful supplementary data:

- Ambient temperature
- Relative humidity
- Radio frequency interference
- Sample voltage monitoring for ac and dc points
- Harmonic distortion content of the ac input signal

Digital technology offers a number of significant advantages beyond the capabilities of analog instruments. Digital ac line monitors can store in memory the signal being observed, permitting in-depth analysis impossible with previous technology. Because the waveform resides in memory, the data associated with the waveform can be transferred to a remote computer for real-time processing, or for processing at a later time.

The digital storage oscilloscope (DSO) is the forerunner of most digital ac line monitors. DSO technology forms the basis for most monitoring instruments. Before the introduction of the DSO, the term *storage oscilloscope* referred to a scope that used a *storage* CRT (cathode-ray tube). This type of instrument stored the display waveform as a trace on the scope face. The digital storage oscilloscope operates on a completely different premise. Figure 2.27 shows a block diagram of a DSO. Instead of being amplified and applied directly to the deflection plates of a CRT, the waveform first is converted into a digital representation and stored in memory. To reproduce the waveform on the CRT, the data is sequentially read and converted back into an analog signal for display.

The analog-to-digital (A/D) converter transforms the input analog signal into a sequence of digital bits. The amplitude of the analog signal, which varies continuously in time, is sampled at preset intervals. The analog value of the signal at each sample point is converted into a binary number. This *quantization* process is illustrated in Figure 2.28. The sample rate of a digital oscilloscope must be greater than 2 times the highest frequency to be measured. The higher the sampling rate relative to the input signal, the greater the measurement accuracy. A sample rate 10 times the input signal is sufficient for most applications. This rule of thumb applies for single-shot signals, or signals that are changing constantly. The sample rate also may be expressed as the *sampling interval*, or the period of time between samples. The sample interval is the inverse of the sampling frequency.

Although a DSO is specified by its maximum sampling rate, the actual rate used in acquiring a given waveform usually is dependent on the *time-per-division* setting of the oscilloscope. The *record length* (samples recorded over a given period of time) defines a finite number of sample points available for a given acquisition. The DSO must, therefore, adjust its sampling rate to fill a given record over the period set by the sweep control. To determine the sampling rate for a given sweep speed, the number

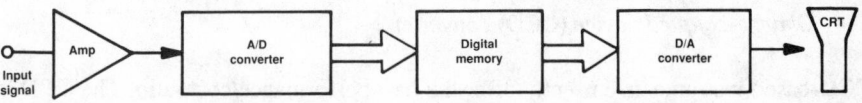

Figure 2.27   Simplified block diagram of a digital storage oscilloscope.

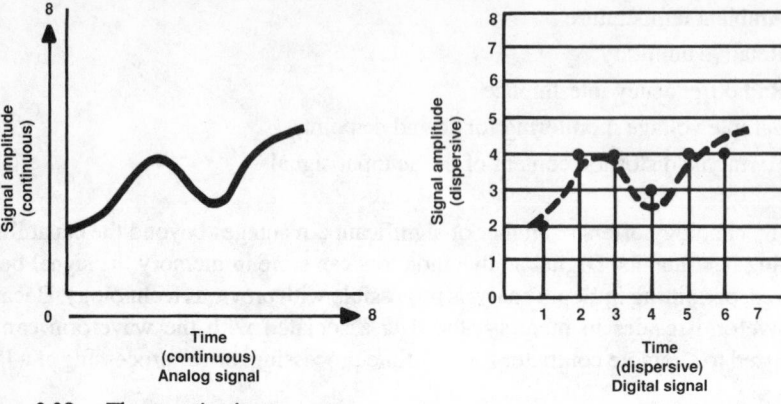

Figure 2.28   The quantization process.

of displayed points per division is divided into the sweep rate per division. Two additional features can modify the actual sampling rate:

- Use of an external clock for pacing the digitizing rate. With the internal digitizing clock disabled, the digitizer will be paced at a rate defined by the operator.
- Use of a *peak detection* (or *glitch capture*) mode. Peak detection allows the digitizer to sample at the full digitizing rate of the DSO, regardless of the time base setting. The minimum and maximum values found between each normal sample interval are retained in memory. These minimum and maximum values are used to reconstruct the waveform display with the help of an algorithm that recreates a smooth trace along with any captured glitches. Peak detection allows the DSO to capture glitches even at its slowest sweep speed. For higher performance, a technique known as *peak-accumulation* (or *envelope mode*) may be used. With this approach, the instrument accumulates and displays the maximum and minimum excursions of a waveform for a given point in time. This builds an envelope of activity that can reveal infrequent noise spikes, long-term amplitude or time drift, and pulse jitter extremes.

2.6.2 A/D Conversion

The analog-to-digital conversion process is critical to the overall accuracy of a digital instrument. Three types of converters commonly are used:

- *Flash converter*
- *Successive approximation* converter
- *Charge-coupled device* (CCD) converter

CCD-based instruments currently offer the best performance/cost ratio. The CCD is an analog storage array. The monitor can capture high-speed events in real time using a CCD, then route the captured information to low-cost A/D converters to be digitized.

Conversion resolution is determined by the number of bits into which the analog signal is transformed. The higher the number of bits available to the A/D converter, the more discrete levels the digital signal is able to describe. For example, an 8-bit digitizer has 256 levels, while a 10-bit digitizer has 1024 levels. Although a higher number of bits allows greater discrimination between voltage values, the overall accuracy of the instrument may be limited by other factors, including the accuracy of the analog amplifier(s) feeding the digitizer(s).

2.6.3   Digital Monitor Features

Advanced components and construction techniques have led to lower costs for digital instruments as well as higher performance. Digital monitors can capture and analyze transient signals from any number of sources. Automated features reduce testing and troubleshooting costs through the use of recallable instrument setups, direct parameter readout, and unattended monitoring. Digital monitors offer the following features:

- High resolution (determined by the quality of the analog-to-digital converter).
- Memory storage of digitized waveforms.
- Automatic setup for repetitive signal analysis. For complex multichannel configurations that are used often, front-panel storage/recall can save dozens of manual selections and adjustments. When multiple memory locations are available, multiple front-panel setups can be stored to save even more time.
- Auto-ranging. Many instruments will adjust automatically for optimum sweep, input sensitivity, and triggering. The instrument's microprocessor automatically configures the front panel for optimum display. Such features permit the operator to concentrate on making measurements, rather than adjusting the instrument.
- Instant hardcopy output from printers and plotters.
- Remote programmability via GPIB for automated test applications.
- Trigger flexibility. Single-shot digitizing monitors capture transient signals and allow the user to view the waveform that *preceded* the trigger point. Figure 2.29 illustrates the use of pre/post trigger for waveform analysis.
- Signal analysis. Intelligent systems can make key measurements and comparisons. Display capabilities include voltage peak, mean voltage, rms value, rise time, fall time, and frequency.

Digital memory storage offers a number of benefits, including:

- Reference memory. A previously acquired waveform can be stored in memory and compared with a sampled waveform. This feature is especially useful for identifying the source of a transient disturbance. Because certain transients have a characteristic "signature" waveform, it is possible to identify the source of a transient by examining the disturbance on a CRT. Nonvolatile battery-backed memory permits reference waveforms to be transported to field sites.
- Simple data transfers to a host computer for analysis or archive.
- Local data analysis through the use of a built-in microprocessor.
- Cursors capable of providing a readout of delta and absolute voltage and time.

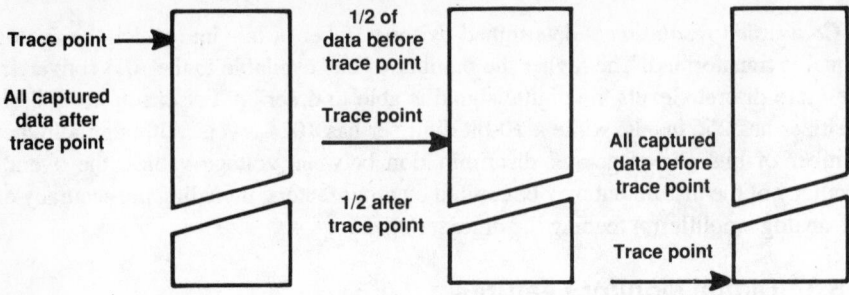

Figure 2.29   Use of pre/post trigger function for waveform analysis.

* No CRT blooming for display of fast transients.
* Full bandwidth capture of long-duration waveforms, thus storage of all the signal details. The waveform can be expanded after capture to expose the details of a particular section.

2.6.4   Capturing Transient Waveforms

Single-shot digitizing makes it possible to capture and clearly display transient and intermittent signals. With single-shot digitizing, the waveform is captured the first time it occurs, on the first trigger. It then can be displayed immediately or held in memory for analysis at a later date. Figure 2.30 illustrates the benefits of digital storage in capturing a transient waveform. An analog instrument often will fail to detect a transient pulse that a digital monitor can display clearly.

Basic triggering modes available on digital monitors permit the user to select the desired source, its coupling, level, and slope. More advanced instruments contain triggering circuitry that permits the user to trigger on elusive conditions, such as pulse widths less than or greater than expected, intervals less than or greater than expected, and specified external conditions. Many trigger modes are further enhanced by a feature that allows the user to hold off the trigger by a selectable time or number of events.

2.6.5   Case in Point

A recent study for a San Francisco Bay area company planning to install a new data processing center graphically demonstrates the scope of the transient problem. The firm wanted to determine the extent of transient activity that could be expected at the new site so that an informed decision could be made on the type of power conditioning needed. An ac line monitor was connected to the 480 V dedicated drop at the new facility for a period of 6 days. During that time, the monitor recorded thousands of spikes, many exceeding 2 kV, on one or more of three-phase inputs. The transient-

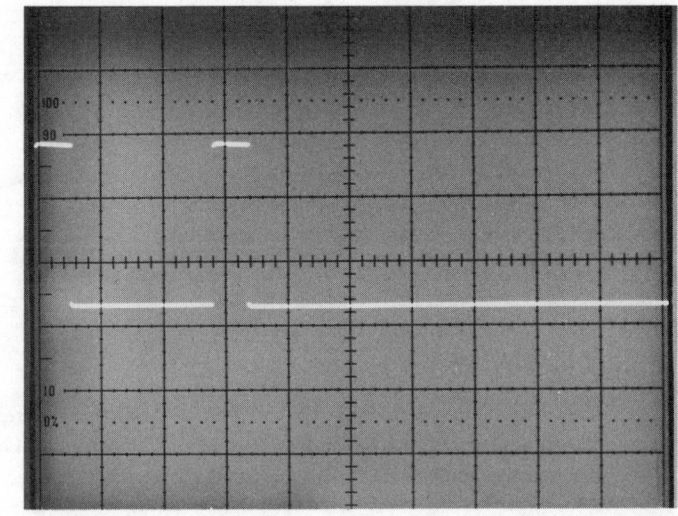

(a)

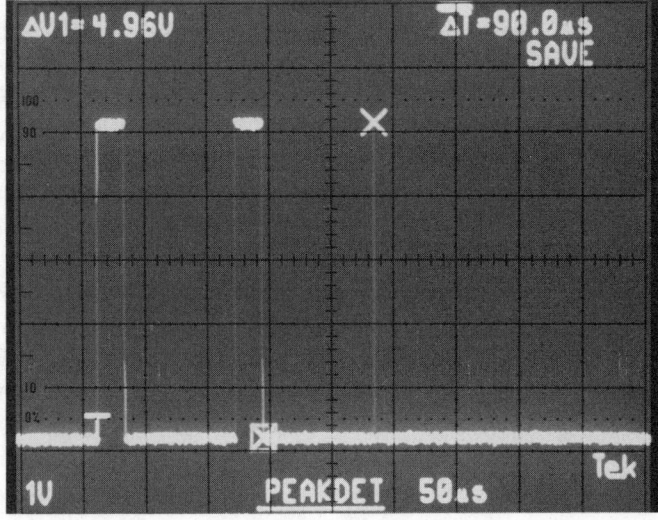

(b)

Figure 2.30 The benefits of a DSO in capturing transient signals: (a, *top*) analog display of a pulsed signal with transient present, but not visible; (b, *bottom*) display of the same signal clearly showing the transient. (Courtesy of Tektronix)

```
C  1792U   IMPULSE
A  0544U   IMPULSE
C  1984U   IMPULSE
A  0592U   IMPULSE
C  1856U   IMPULSE
A  0544U   IMPULSE
C  1824U   IMPULSE
A  0560U   IMPULSE
C  1488U   IMPULSE
A  0496U   IMPULSE
C  1664U   IMPULSE
A  0528U   IMPULSE
C  1600U   IMPULSE
A  0544U   IMPULSE
B  2480U   IMPULSE
   11:19:11
```

Figure 2.31 A portion of the ac power monitor readout from a San Francisco area power-quality study. The first column indicates on which phase (A, B, or C) the transient occurred. The second column is an actual readout of the transient (impulse) magnitude in volts.

recording threshold was 460 V above the nominal ac voltage level of 480 V, phase-to-phase.

An expert from the report summary stated that, on one particular day, the facility was plagued by many high-level transient periods, stretching from 8:30 A.M. until 3:00 P.M. The highest voltage recorded during this period was 4.08 kV. This transient activity occurred during periods of good weather. Figure 2.31 is part of the printout from the study. The data covers transients exceeding more than twice the normal line voltage that occurred within a period of just 30 s. Even though these transients were of brief duration, any sensitive equipment connected to the power line would suffer damage in a short period of time.

Although this is certainly a worst-case example of poor-quality ac power, it points out the need for a minimal amount of transient protection on all incoming power lines. It should be noted that few power drops are as bad as the one analyzed in this study. Furthermore, most transient activity on ac lines is generated by power customers, not utility companies. The environment in which a facility is located, therefore, can greatly affect the power quality at the site.

2.7 BIBLIOGRAPHY

Allen, George, and Donald Segall: "Monitoring of Computer Installations for Power Line Disturbances," *Proceedings of the IEEE Power Engineering Society*, IEEE, New York, 1974.

Berger, K.: "Blitzstorm-Parameter von Aufwärtsblitzen," *Bull. Schweitz. Electrotech. Ver.*, vol. 69, pp. 353–360, 1978.

Bishop, Don: "Lightning Sparks Debate: Prevention or Protection?," *Mobile Radio Technology* magazine, Intertec Publishing, Overland Park, KS, January 1989.

————: "Lightning Devices Undergo Tests at Florida Airports," *Mobile Radio Technology* magazine, Intertec Publishing, Overland Park, KS, May 1990.

Block, Roger: "The Grounds for Lightning and EMP Protection," PolyPhaser Corporation, Gardnerville, NV, 1987.

————: "Dissipation Arrays: Do They Work?," *Mobile Radio Technology* magazine, Intertec Publishing, Overland Park, KS, April 1988.

Breya, Marge: "New Scopes Make Faster Measurements," *Mobile Radio Technology* magazine, Intertec Publishing, Overland Park, KS, November 1988.

Defense Civil Preparedness Agency, *EMP and Electric Power Systems*. Publication TR-61-D, July 1973.

Drabkin, Dr. Mark, and Roy Carpenter, Jr.: "Lightning Protection Devices: How Do They Compare?," *Mobile Radio Technology* magazine, Intertec Publishing, Overland Park, KS, October 1988.

Defense Civil Preparedness Agency, *EMP Protection for Emergency Operating Centers*.

Federal Information Processing Standards Publication no. 94, *Guideline on Electrical Power for ADP Installations*, U.S. Department of Commerce, National Bureau of Standards, Washington, DC, 1983.

Fink, D., and D. Christiansen: *Electronics Engineers' Handbook*, 3rd ed., McGraw-Hill, New York, 1989.

Goldstein, M., and P. D. Speranza: "The Quality of U.S. Commercial AC Power," *Proceedings of the IEEE*, IEEE, New York, 1982.

Harris, Brad: "Understanding DSO Accuracy and Measurement Performance," *Electronic Servicing & Technology* magazine, Intertec Publishing, Overland Park, KS, April 1989.

————: "The Digital Storage Oscilloscope: Providing the Competitive Edge," *Electronic Servicing & Technology* magazine, Intertec Publishing, Overland Park, KS, June 1988.

Hoyer, Mike: "Bandwidth and Rise Time: Two Keys to Selecting the Right Oscilloscope," *Electronic Servicing & Technology* magazine, Intertec Publishing, Overland Park, KS, April 1990.

Jay, Frank: *IEEE Standard Directory of Electrical and Electronics Terms*, 3rd ed., IEEE, New York, 1984.

Jordan, Edward C.: *Reference Data for Engineers: Radio, Electronics, Computer, and Communications*, 7th ed., Howard W. Sams Company, Indianapolis, 1985.

Kaiser, Bruce A.: "Straight Talk on Static Dissipation," *Proceedings of ENTELEC 1988*, Energy Telecommunications and Electrical Association, Dallas, 1988.

————: "Can You Really Fool Mother Nature?," *Cellular Business* magazine, Intertec Publishing, Overland Park, KS, March 1989.

Key, Lt. Thomas: "The Effects of Power Disturbances on Computer Operation," IEEE Industrial and Commercial Power Systems Conference paper, Cincinnati, June 7, 1978.

Martzloff, F. D.: "The Development of a Guide on Surge Voltages in Low-Voltage AC Power Circuits," 14th Electrical/Electronics Insulation Conference, IEEE, Boston, October 1979.

Montgomery, Steve: "Advances in Digital Oscilloscopes," *Broadcast Engineering* magazine, Intertec Publishing, Overland Park, KS, November 1989.

Nott, Ron: "The Sources of Atmospheric Energy," *Proceedings of the SBE National Convention and Broadcast Engineering Conference*, Society of Broadcast Engineers, Indianapolis, 1987.

SCR Applications Handbook, International Rectifier Corporation, El Segundo, CA, 1977.

SCR Manual, 5th ed., General Electric Company, Auburn, NY.

Smeltzer, Dennis: "Getting Organized About Power," *Microservice Management* magazine, Intertec Publishing, Overland Park, Kan., March 1990.

Technical staff, Military/Aerospace Products Division, *The Reliability Handbook*, National Semiconductor, Santa Clara, Calif., 1987.

Technical staff, "Voltage Transients and the Semiconductor," *The Electronic Field Engineer*, pp. 37–40, March/April 1979.

Uman, Martin A.: "Natural and Artificially Initiated Lightning and Lightning Test Standards," *Proceedings of the IEEE*, vol. 76, no. 12, IEEE, New York, December 1988.

————: *The Lightning Discharge*, Academic Press, Orlando, FL, 1987.

Whitaker, Jerry: *Radio Frequency Transmission Systems: Design and Operation*, McGraw-Hill, New York, 1990.

————: *Maintaining Electronic Systems*, CRC Press, Boca Raton, FL, 1991.

3

THE EFFECTS OF TRANSIENT DISTURBANCES

3.1 INTRODUCTION

Protection against transient disturbances is a science that demands attention to detail. This work is not inexpensive. It is not something that can be accomplished overnight. Facility managers will, however, wind up paying for transient protection one way or another, either *before* problems occur or *after* problems occur. In the world of transient protection, there is truly no such thing as a free lunch.

Unfortunately, the power-quality problems affecting many regions are becoming worse, not better. Users cannot depend upon power suppliers to solve the transient problems that exist. Utility companies rarely are interested in discussing ac disturbances that are measured in the microseconds or nanoseconds. Such problems are usually beyond the control of the power provider. The problem must be solved, instead, at the input point of sensitive loads. Utilities traditionally have checked the quality of a customer's service drop by connecting a chart recorder to the line for a period of several days. The response time of such recorders, however, is far too slow to document transient spikes. Slow-speed analog recorders will show only long-term surge and sag conditions, which generally can be managed by the regulated power supplies or high-voltage protection systems normally used in high-quality equipment.

The degree of protection afforded a facility is generally a compromise between the line abnormalities that will account for more than 90 percent of the expected problems, and the amount of money available to spend on that protection. Each installation is unique and requires an assessment of not only the importance of keeping the system up and running at all times, but also the threat of transient disturbances posed by the ac line feed to the plant.

The first line of defense in the protection of electronic equipment from damaging transient overvoltages is the ac-to-dc power supply. Semiconductor power-supply components are particularly vulnerable to failure from ac line disturbances. Devices occasionally will fail from one large transient, but many more fail because of smaller, more frequent spikes that punch through the device junction. Such occurrences explain why otherwise reliable systems fail "without apparent reason."

3.2 SEMICONDUCTOR FAILURE MODES

Semiconductor devices may be destroyed or damaged by transient disturbances in one of several ways. The primary failure mechanisms include:

- Avalanche-related failure
- Thermal runaway
- Thermal second breakdown
- Metallization failure
- Polarity reversals

When a semiconductor junction fails because of overstress, a low-resistance path is formed that shunts the junction. This path is not a true short, but it is a close approximation. The shunting resistance may be less than 10 Ω in a junction that has been heavily overstressed. By comparison, the shunting resistance of a junction that has been only mildly overstressed may be as high as 10 MΩ. The formation of low-resistance shunting paths is the result of a junction's electrothermal response to overstress.

3.2.1 Device Ruggedness

The best-constructed device will fail if exposed to stress exceeding its design limits. The *safe operating area* (SOA) of a power transistor is the single most important parameter in the design of high-power semiconductor-based systems. Fortunately, advances in diffusion technology, masking, and device geometry have enhanced the power-handling capabilities of semiconductor devices.

A bipolar transistor exhibits two regions of operation that must be avoided:

- *Dissipation region.* Where the voltage-current product remains unchanged over any combination of voltage (V) and current (I). Gradually, as the collector-to-emitter voltage increases, the electric field through the base region causes hot spots to form. The carriers actually can punch a hole in the junction by melting silicon. The result is a dead (short-circuited) transistor.
- *Second breakdown* ($I_{s/b}$) *region.* Where power transistor dissipation varies in a nonlinear inverse relationship with the applied collector-to-emitter voltage when the transistor is forward-biased.

To get SOA data into some type of useful format, a family of curves at various operating temperatures must be developed and plotted. This exercise gives a clear

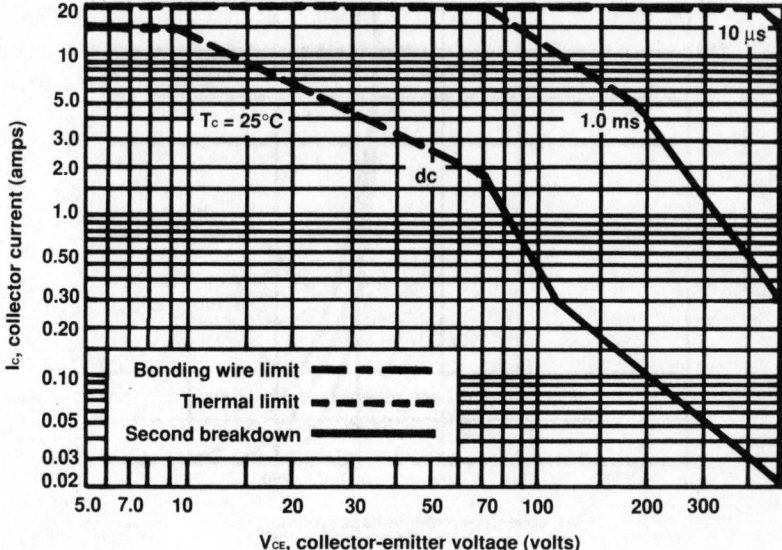

Figure 3.1 Forward bias safe operating area curve for a bipolar transistor (MJH16010A). (Courtesy of Motorola)

picture of what the data sheet indicates, compared with what happens in actual practice.

3.2.2 Forward Bias Safe Operating Area

The forward bias safe operating area (FBSOA) describes the ability of a transistor to handle stress when the base is forward-biased. Manufacturer FBSOA curves detail maximum limits for both steady-state dissipation and turn-on load lines. Because it is possible to have a positive base-emitter voltage and negative base current during the device storage time, forward bias is defined in terms of base current.

Bipolar transistors are particularly sensitive to voltage stress; more so than with stress induced by high currents. This situation is particularly true of switching transistors, and it shows up on the FBSOA curve. Figure 3.1 shows a typical curve for a common power transistor. In the case of the dc trace, the following observations can be made:

- The power limit established by the *bonding wire limit* portion of the curve permits 135 W maximum dissipation (15 A • 9 V).
- The power limit established by the *thermal limit* portion of the curve permits (at the maximum voltage point) 135 W maximum dissipation (2 A • 67.5 V). There is no change in maximum power dissipation.
- The power limit established by the *second breakdown* portion of the curve decreases dramatically from the previous two conditions. At 100 V, the maximum current is 0.42 A, for a maximum power dissipation of 42 W.

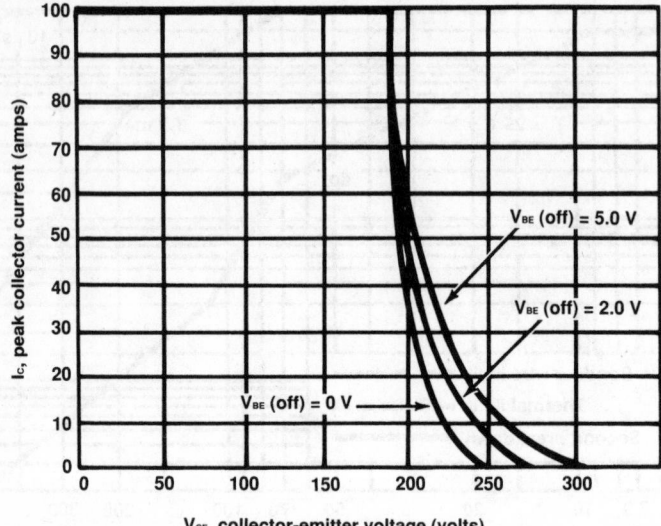

V_{CE}, collector-emitter voltage (volts)

Figure 3.2   Reverse bias safe operating area curve for a bipolar transistor (MJH10610A). (Courtesy of Motorola)

3.2.3   Reverse Bias Safe Operating Area

The reverse bias safe operating area (RBSOA) describes the ability of a transistor to handle stress with its base reverse-biased. As with FBSOA, RBSOA is defined in terms of current. In many respects, RBSOA and FBSOA are analogous. First among these is voltage sensitivity. Bipolar transistors exhibit the same sensitivity to voltage stress in the reverse bias mode as in the forward bias mode. A typical RBSOA curve is shown in Figure 3.2. Note that maximum allowable peak instantaneous power decreases significantly as voltage is increased.

3.2.4   Power-Handling Capability

The primary factor in determining the amount of power a given device can handle is the size of the active junction(s) on the chip. The same power output from a device may be achieved through the use of several smaller chips in parallel. This approach, however, may result in unequal currents and uneven distribution of heat. At high power levels, heat management becomes a significant factor in chip design.

Specialized layout geometries have been developed to ensure even current distribution throughout the device. One approach involves the use of a matrix of emitter resistances constructed so that the overall distribution of power among the parallel emitter elements results in even thermal dissipation. Figure 3.3 illustrates this *interdigited* geometry technique.

With improvements in semiconductor fabrication processes, output device SOA is primarily a function of the size of the silicon slab inside the package. Package type,

RESISTORS

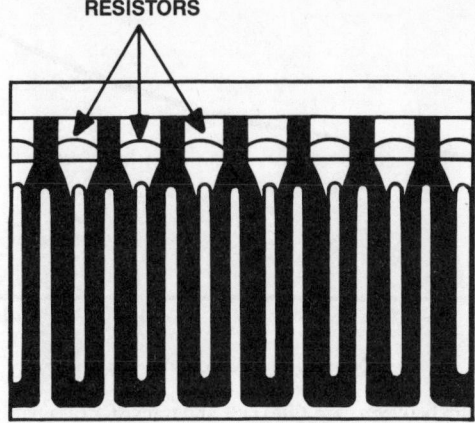

Figure 3.3   Interdigited geometry of emitter resistors used to balance currents throughout a power device chip.

of course, determines the ultimate dissipation because of thermal saturation with temperature rise. A good TO-3 or a 2-screw-mounted plastic package will dissipate approximately 350 to 375 W if properly mounted. Figure 3.4 demonstrates the relationships between case size and power dissipation for a TO-3 package.

3.2.5   Semiconductor Derating

Good design calls for a measure of caution in the selection and application of active devices. Unexpected operating conditions, or variations in the manufacturing process, may result in field failures unless a margin of safety is allowed. Derating is a common method of achieving such a margin. The primary derating considerations are:

- *Power derating.* Designed to hold the worst-case junction temperature to a value below the normal permissible rating.
- *Junction-temperature derating.* An allowance for the worst-case ambient temperature or case temperature that the device is likely to experience in service.
- *Voltage derating.* An allowance intended to compensate for temperature-dependent voltage sensitivity and other threats to device reliability as a result of instantaneous peak-voltage excursions caused by transient disturbances.

3.2.6   Failure Mechanisms

It is estimated that as much as 95 percent of all transistor failures in the field are directly or indirectly the result of excessive dissipation or applied voltages in excess of the maximum design limits of the device. There are at least four types of voltage breakdown that must be considered in a reliability analysis of discrete power

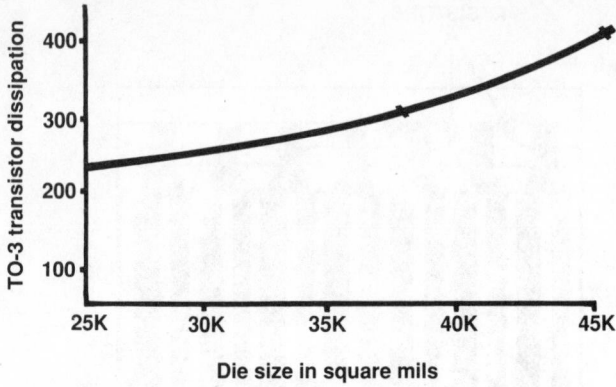

Figure 3.4   Relationship between case (die) size and transistor dissipation.

transistors. Although they are not strictly independent, each type can be treated separately. Keep in mind, however, that each is related to the others.

Avalanche Breakdown.   Avalanche is a voltage breakdown that occurs in the collector-base junction, similar to the *Townsend effect* in gas tubes. This effect is caused by the high dielectric field strength that occurs across the collector-base junction as the collector voltage is increased. This high-intensity field accelerates the free charge carriers so that they collide with other atoms, knocking loose additional free charge carriers that, in turn, are accelerated and have more collisions.

This multiplication process occurs at an increasing rate as the collector voltage increases, until at some voltage, V^a (avalanche voltage), the current suddenly tries to go to infinity. If enough heat is generated in this process, the junction will be damaged or destroyed. A damaged junction will result in higher-than-normal leakage currents, increasing the steady-state heat generation of the device, which ultimately may destroy the semiconductor junction.

Alpha Multiplication.   Alpha multiplication is produced by the same physical phenomenon that produces avalanche breakdown, but differs in circuit configuration. This effect occurs at a lower potential than the avalanche voltage and generally is responsible for collector-emitter breakdown when base current is equal to zero.

Punch-Through.   Punch-through is a voltage breakdown occurring at the collector-base junction because of high collector voltage. As collector voltage is increased, the *space charge region* (collector junction width) gradually increases until it penetrates completely through the base region, touching the emitter. At this point, the emitter and collector are effectively short-circuited together.

Although this type of breakdown occurs in some PNP junction transistors, alpha multiplication breakdown generally occurs at a lower voltage than punch-through. Because this breakdown occurs between the collector and emitter, punch-through is more serious in the common-emitter or common-collector configuration.

Thermal Runaway.   Thermal runaway is a regenerative process by which a rise in temperature causes an increase in the leakage current; in turn, the resulting increased collector current causes higher power dissipation. This action raises the junction temperature, further increasing leakage current.

If the leakage current is sufficiently high (resulting from high temperature or high voltage), and the current is not adequately stabilized to counteract increased collector current because of increased leakage current, this process can regenerate to a point that the temperature of the transistor rapidly rises, destroying the device. This type of effect is more prominent in power transistors, where the junction normally is operated at elevated temperatures and where high leakage currents are present because of the large junction area. Thermal runaway is related to the avalanche effect and is dependent upon circuit stability, ambient temperature, and transistor power dissipation.

Breakdown Effects.   The effects of the breakdown modes outlined manifest themselves in various ways on the transistor:

- Avalanche breakdown usually results in destruction of the collector-base junction because of excessive currents. This, in turn, results in an open circuit between the collector and base.
- Breakdown due to alpha multiplication and thermal runaway most often results in destruction of the transistor because of excessive heat dissipation that shows up electrically as a short circuit between the collector and emitter. This condition, which is most common in transistors that have suffered catastrophic failure, is not always detected easily. In many cases, an ohmmeter check may indicate a good transistor. Only after operating voltages are applied will the failure mode be exhibited.
- Punch-through breakdown generally does not cause permanent damage to the transistor; it can be a self-healing type of breakdown. After the overvoltage is removed, the transistor usually will operate satisfactorily.

3.2.7 Avalanche-Related Failure

A high reverse voltage applied to a nonconducting PN junction can cause *avalanche* currents to flow. Avalanche is the process resulting from high fields in a semiconductor device in which an electron, accelerated by the field, strikes an atom and releases more electrons, which continue the sequence. If enough heat is generated in this cycle, the junction can be damaged or destroyed.

If such a process occurs at the base and emitter junction of a transistor, the effect may be either minor or catastrophic. With a minor failure, the gain of the transistor can be reduced through the creation of *trapping centers*, which restrict the free flow of carriers. With a catastrophic failure, the transistor will cease to function.

Thermal Runaway.   A thermal runaway condition can be triggered by a sudden increase in gain resulting from the heating effect of a transient on a transistor. The transient can bring the device (operating in the active region) out of its safe operating area and into an unpredictable operating mode.

Thermal Second Breakdown.   Junction burnout is a significant failure mechanism for bipolar devices, particularly JFET (junction field-effect transistor) and Schottky devices. The junction between a P-type diffusion and an N-type diffusion normally has a positive temperature coefficient at low temperatures. Increased temperature will result in increased resistance. When a reverse-biased pulse is applied, the junction dissipates heat in a narrow *depletion region*, and the temperature in that area increases rapidly. If enough energy is applied in this process, the junction will reach a point at which the temperature coefficient of the silicon will turn negative. In other words, increased temperature will result in decreased resistance. A thermal runaway condition may then ensue, resulting in localized melting of the junction. If sustaining energy is available after the initial melt, the hot spot can grow into a *filament short*. The longer the energy pulse, the wider the resulting filament short. *Current filamentation* is a concentration of current flow in one or more narrow regions, which leads to localized heating.

After the transient has passed, the silicon will resolidify. The effect on the device may be catastrophic, or it may simply degrade the performance of the component. With a relatively short pulse, a hot spot may form, but not grow completely across the junction. As a result, the damage may not appear immediately as a short circuit, but manifest itself at a later time as a result of *electromigration* or another failure mechanism.

Metallization Failure.   The smaller device geometry required by high-density integrated circuits has increased the possibility of metallization failure resulting from transient overvoltages. Metallization melt is a power-dependent failure mechanism. It is more likely to occur during a short-duration, high-current pulse. Heat generated by a long pulse tends to be dissipated in the surrounding chip die.

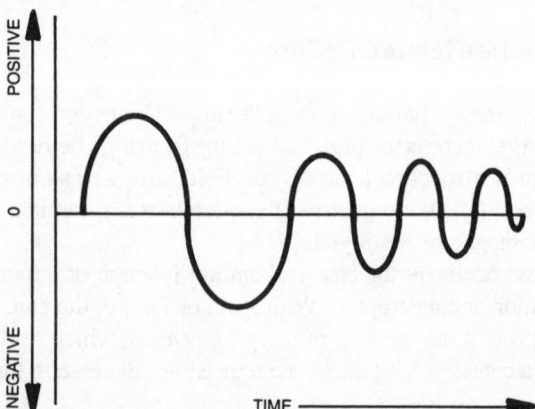

Figure 3.5   Waveshape of a typical transient disturbance. Note how the tail of the transient oscillates as it decays.

Metallization failure also may occur as a side effect of junction melt. The junction usually breaks down first, opening the way for high currents to flow. The metallization then heats until it reaches the melting point. Metallization failure results in an open circuit. A junction short circuit can, therefore, lead to an open-circuit failure.

Polarity Reversal.   Transient disturbances typically build rapidly to a peak voltage and then decay slowly. If enough inductance and/or capacitance is present in the circuit, the tail will oscillate as it decays. This concept is illustrated in Figure 3.5. The oscillating tail can subject semiconductor devices to severe voltage polarity reversals, forcing the components into or out of a conducting state. This action can damage the semiconductor junction or result in catastrophic failure.

3.3 MOSFET DEVICES

Power MOSFETs (metal-oxide semiconductor field-effect transistors) have found numerous applications because of their unique performance attributes. A variety of specifications can be used to indicate the maximum operating voltages a specific device can withstand. The most common specifications include:

- Gate-to-source breakdown voltage
- Drain-to-gate breakdown voltage
- Drain-to-source breakdown voltage

These limits mark the maximum voltage excursions possible with a given device before failure. Excessive voltages cause carriers within the depletion region of the reverse-biased PN junction to acquire sufficient kinetic energy to result in ionization. Voltage breakdown also can occur when a *critical electric field* is reached. The magnitude of this voltage is determined primarily by the characteristics of the die itself.

3.3.1 Safe Operating Area

The safe dc operating area of a MOSFET is determined by the rated power dissipation of the device over the entire drain-to-source voltage range (up to the rated maximum voltage). The maximum drain-source voltage is a critical parameter. If exceeded even momentarily, the device can be damaged permanently.

Figure 3.6 shows a representative SOA curve for a MOSFET. Notice that limits are plotted for several parameters, including drain-source voltage, thermal dissipation (a time-dependent function), package capability, and drain-source on-resistance. The capability of the package to withstand high voltages is determined by the construction of the die itself, including bonding wire diameter, size of the bonding pad, and internal thermal resistances. The drain-source on-resistance limit is simply a manifestation of Ohm's law; with a given on-resistance, current is limited by the applied voltage.

To a large extent, the thermal limitations described in the SOA chart determine the boundaries for MOSFET use in linear applications. The maximum permissible junction temperature also affects the pulsed current rating when the device is used as a

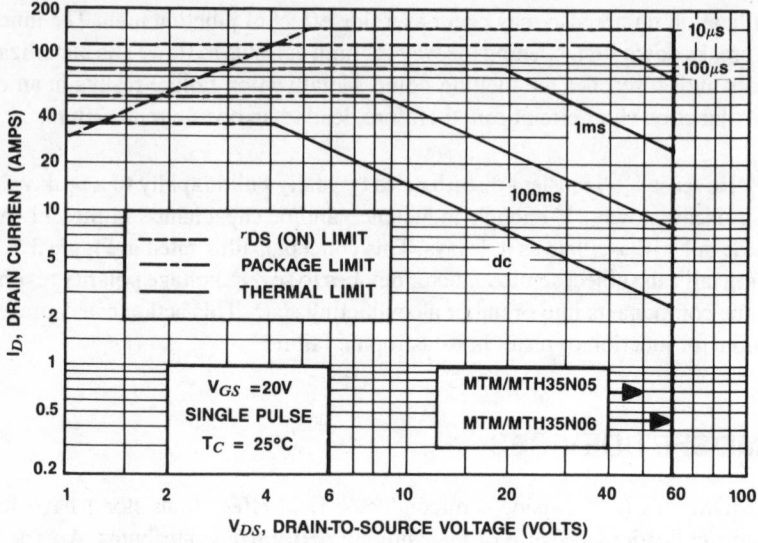

Figure 3.6 Safe operating area (SOA) curve for a power FET device.

switch. MOSFETs are, in fact, more like rectifiers than bipolar transistors with respect to current ratings; their peak current ratings are not gain-limited, but thermally limited.

In switching applications, total power dissipation comprises both switching losses and on-state losses. At low frequencies, switching losses are small. As the operating frequency increases, however, switching losses become a significant factor in circuit design. The *switching safe operating area* (SSOA) defines the MOSFET voltage and current limitations during switching transitions. Although the SSOA chart outlines both turn-on and turn-off boundaries, it is used primarily as a source for turn-off SOA data. As such, it is the MOSFET equivalent of the reverse-biased SOA (RBSOA) curve of bipolar transistors. As with the RBSOA rating, turn-off SOA curves are generated by observing device performance as it switches a clamped inductive load. Figure 3.7 shows a typical SSOA chart for a family of MOSFET devices.

Figure 3.8 illustrates an FET device switching an inductive load in a circuit with no protection from *flyback* (back-emf) voltages. The waveform depicts the turn-off voltage transient resulting from the load and the parasitic lead and wiring inductance. The device experiences an avalanche condition for about 300 ns at its breakdown voltage of 122 V. Placing a clamping diode across the inductive load suppresses most (but not all) of the transient. See Figure 3.9. The drain-to-source (V_{ds}) voltage still will overshoot the supply rail by the sum of the effects of the diode's forward recovery characteristics, the diode lead inductance, and the parasitic series inductances. If the series resistance of the load is small in comparison with its inductance, a simple diode clamp may allow current to circulate through the load-diode loop for a significant period of time after the MOSFET is turned off. When this residual current is unacceptable, a resistance can be inserted in series with the diode at the expense of increasing the peak flyback voltage seen at the drain.

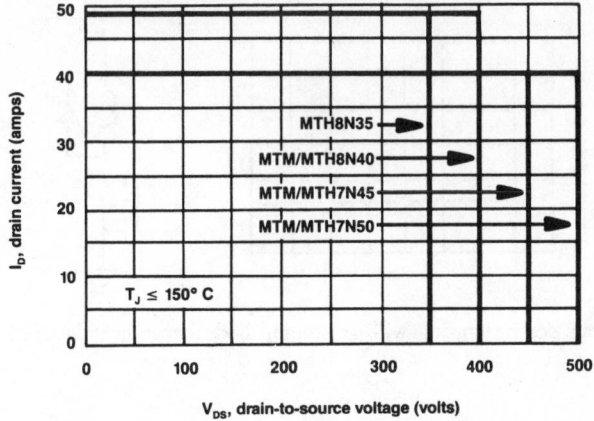

Figure 3.7 Maximum rated switching safe operating area of the MTM8N40 MOSFET. (Courtesy of Motorola)

Protecting the drain-source from voltage transients with a zener diode (a wideband device) is another simple and effective solution. Except for the effects of the lead and wiring inductances and the negligible time required to avalanche, the zener will clip the voltage transient at its breakdown voltage. A slow-rise-time transient will be clipped completely; a rapid-rise-time transient may momentarily exceed the zener breakdown. These effects are shown in Figure 3.10.

Figure 3.11 shows an RC clamp network that suppresses flyback voltages greater than the potential across the capacitor. Sized to sustain nearly constant voltage during the entire switch cycle, the capacitor absorbs energy only during transients and dumps that energy into the resistance during the remaining portion of the cycle.

A series RC snubber circuit is shown in Figure 3.12. Although the circuit effectively reduces the peak drain voltage, it is not as efficient as a true clamping scheme. Whereas a clamping network dissipates energy only during the transient, the RC snubber

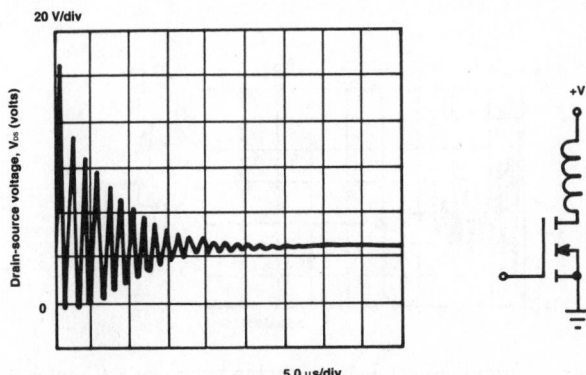

Figure 3.8 Drain-source transient resulting from switching off an unclamped inductive load. (Courtesy of Motorola)

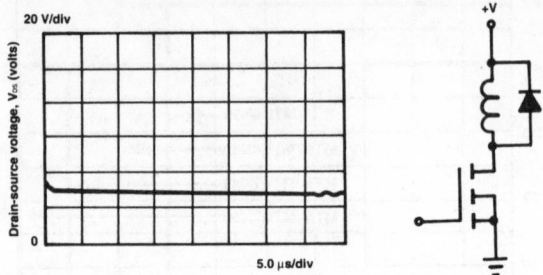

Figure 3.9   Drain-source transient with a clamping diode across the inductive load. (Courtesy of Motorola)

absorbs energy during portions of the switching cycle that are not overstressing the MOSFET. This configuration also slows turn-on times because of the additional drain-source capacitance that must be charged.

Until recently, a MOSFET's maximum drain-to-source voltage specification prohibited even instantaneous excursions beyond stated limits; the first power MOSFET devices were never intended to be operated in avalanche. As is still the case with most bipolar transistors, avalanche limitations simply were not specified. Some devices happened to be rugged, while others were not. Manufacturers now have designed power MOSFET devices that are able to sustain substantial currents in avalanche at elevated junction temperatures. As a result, newly designed "ruggedized" devices are replacing older MOSFETs in critical equipment.

3.3.2   MOSFET Failure Modes

The thermal and electrical stresses that a MOSFET device may experience during switching can be severe, particularly during turn-off when an inductive load is present.

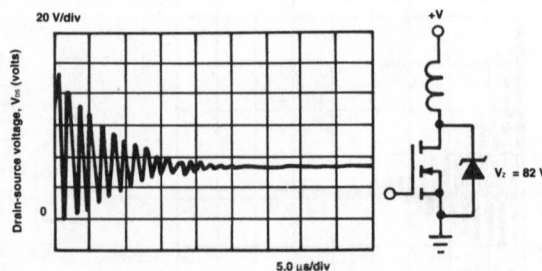

Figure 3.10   Drain-source transient with a clamping zener diode. (Courtesy of Motorola)

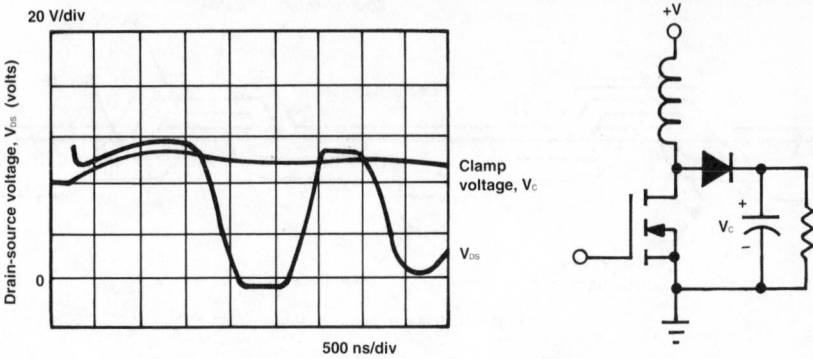

Figure 3.11   Transient waveforms for a gated RC clamp. (Courtesy of Motorola)

When power MOSFETs were introduced, it usually was stated that, because the MOSFET was a majority carrier device, it was immune to second breakdown as observed in bipolar transistors. It must be understood, however, that a parasitic bipolar transistor is inherent in the structure of a MOSFET. This phenomenon is illustrated in Figure 3.13. The parasitic bipolar transistor can allow a failure mechanism similar to second breakdown. Research has shown that if the parasitic transistor becomes active, the MOSFET may fail. This situation is particularly troublesome if the MOSFET drain-source breakdown voltage is approximately twice the collector-

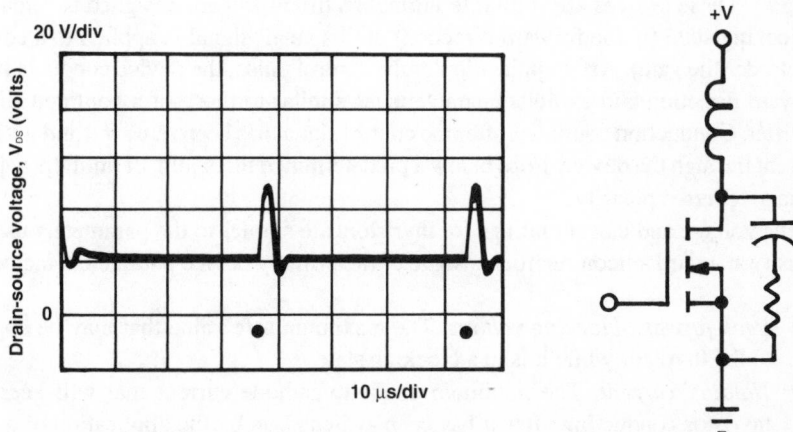

Figure 3.12   Drain-source transient with an RC snubber circuit. (Courtesy of Motorola)

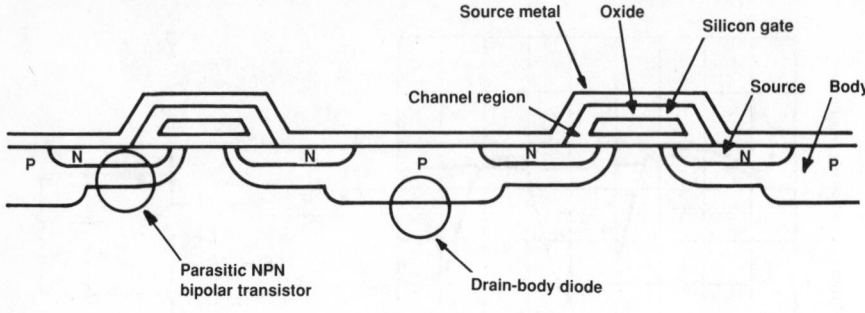

Figure 3.13   Cross section of a power MOSFET device showing the parasitic bipolar transistor and diode inherent in the structure.

emitter sustaining voltage of the parasitic bipolar transistor. This failure mechanism results, apparently, when the drain voltage snaps back to the sustaining voltage of the parasitic device. This *negative resistance characteristic* can cause the total device current to constrict to a small number of cells in the MOSFET structure, leading to device failure. The precipitous voltage drop synonymous with second breakdown is a result of avalanche injection and any mechanism, electric or thermal, that can cause the current density to become large enough for avalanche injection to occur.

3.4   THYRISTOR COMPONENTS

The term *thyristor* identifies a general class of solid-state silicon controlled rectifiers (SCRs). These devices are similar to normal rectifiers, but are designed to remain in a blocking state (in the forward direction) until a small signal is applied to a control electrode (the gate). After application of the control pulse, the device conducts in the forward direction and exhibits characteristics similar to those of a common silicon rectifier. Conduction continues after the control signal has been removed and until the current through the device drops below a predetermined threshold, or until the applied voltage reverses polarity.

The voltage and current ratings for thyristors are similar to the parameters used to classify standard silicon rectifiers. Some of the primary device parameters include:

- *Peak forward blocking voltage*. The maximum safe value that may be applied to the thyristor while it is in a blocking state.
- *Holding current*. The minimum anode-to-cathode current that will keep the thyristor conducting after it has been switched on by the application of a gate pulse.
- *Forward voltage drop*. The voltage loss across the anode-to-cathode current path for a specified load current. Because the ratio of rms-to-average forward current varies with the angle of conduction, power dissipation for any average current also varies with the device angle of conduction. The interaction of forward

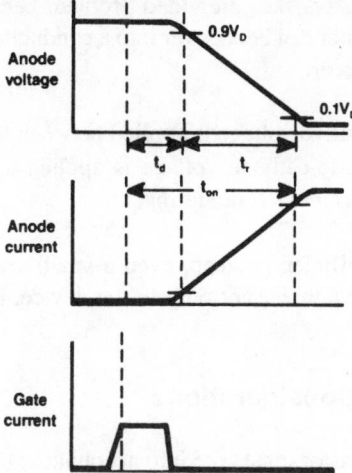

Figure 3.14   Turn-on waveforms for an SCR device. T_d = delay time interval between a specified point at the beginning of a gate pulse and the instant at which the principal voltage drops to a specified value. T_r = rise time between the principal voltage dropping from one value to a second lower value when the thyristor turns from off to on.

voltage drop, phase angle, and device case temperature generally are specified in the form of one or more graphs or charts.

- *Gate trigger sensitivity.* The minimum voltage and/or current that must be applied to the gate to trigger a specific type of thyristor into conduction. This value must take into consideration variations in production runs and operating temperature. The minimum trigger voltage is not normally temperature-sensitive, but the minimum trigger current can vary considerably with thyristor case temperature.
- *Turn-on time.* The length of time required for a thyristor to change from a nonconducting state to a conducting state. When a gate signal is applied to the thyristor, anode-to-cathode current begins to flow after a finite delay. A second switching interval occurs between the point at which current begins to flow and the point at which full anode current (determined by the instantaneous applied voltage and the load) is reached. The sum of these two times is the turn-on time. The turn-on interval is illustrated in Figure 3.14.
- *Turn-off time.* The length of time required for a thyristor to change from a conducting state to a nonconducting state. The turn-off time is composed of two individual periods: the *storage time* (similar to the storage interval of a saturated transistor) and the *recovery time.* If forward voltage is reapplied before the entire turn-off time has elapsed, the thyristor will conduct again.

3.4.1 Failure Modes

Thyristors, like diodes, are subject to damage from transient overvoltages because the peak inverse voltage or instantaneous forward voltage (or current) rating of the device may be exceeded. Thyristors face an added problem because of the possibility of device misfiring. A thyristor can break over into a conduction state, regardless of gate drive, if either of these occur:

1.   Too high of a positive voltage is applied between the anode and cathode.
2.   A positive anode-to-cathode voltage is applied too quickly, exceeding the dv/dt (delta voltage/delta time) rating.

If the leading edge is sufficiently steep, even a small voltage pulse can turn on a thyristor. This represents a threat not only to the device, but also to the load that it controls.

3.4.2   Application Considerations

Any application of a thyristor must take into account the device dv/dt rating and the electrical environment in which it will operate. A thyristor controlling an appreciable amount of energy should be protected against fast-rise-time transients that may cause the device to break over into a conduction state. The most basic method of softening the applied anode-to-cathode waveform is the resistor/capacitor snubber network shown in Figure 3.15. This standard technique of limiting the applied dv/dt relies on the integrating ability of the capacitor. In the figure, C1 snubs the excess transient energy, while R1 defines the applied dv/dt with L_t, the external system inductance.

An applied transient waveform (assuming an infinitely sharp wavefront) will be impressed across the entire protection network of C1, R1, and L_t. The total distributed and lumped system inductance, L_t, plays a significant role in determining the ability of C1 and R1 to effectively snub a transient waveform. Power sources that are *stiff* (having little series inductance or resistance) will present special problems to design engineers seeking to protect a thyristor from steep transient waveforms.

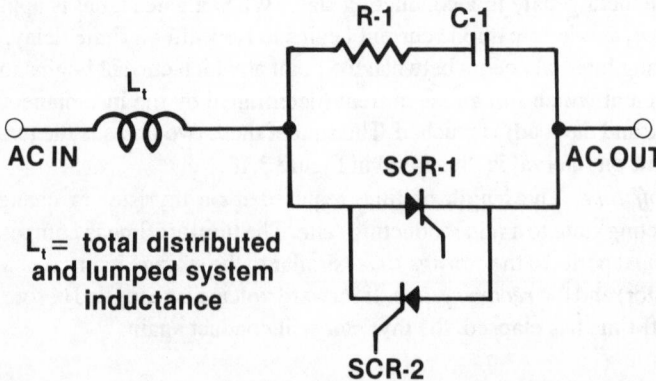

Figure 3.15   The basic RC snubber network commonly used to protect thyristors from fast-rise-time transients.

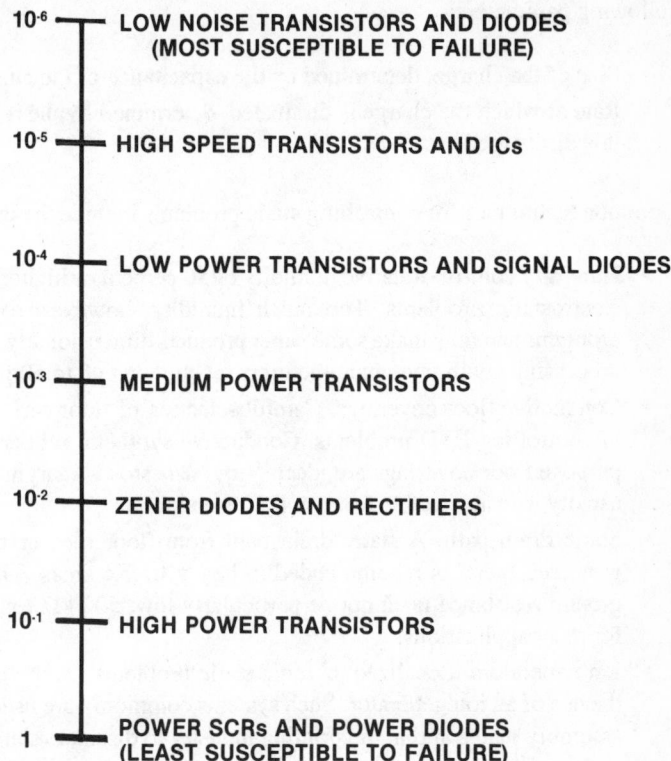

Figure 3.16 An estimate of the susceptibility of semiconductor devices to failure because of transient energy. The estimate assumes a transient duration of several microseconds.

Exposure of semiconductors to a high-transient environment can cause a degrading of the device, which eventually may result in total failure. Figure 3.16 shows the energy vs. survival scale for several types of semiconductors.

3.5 ESD FAILURE MODES

Low-power semiconductors are particularly vulnerable to damage from ESD discharges. MOS devices tend to be more vulnerable than other components. The gate of a MOS transistor is especially sensitive to electrical overstress. Application of excessive voltage may exceed the dielectric standoff voltage of the chip structure and punch through the oxide, forming a permanent path from the gate to the semiconductor below. An ESD pulse of 25 kV usually is sufficient to rupture the gate oxide. The scaling of device geometry that occurs with LSI (large-scale integrated) or VLSI (very large-scale integrated) components complicates this problem. The degree of damage caused by electrostatic discharge is a function of the following parameters:

- Size of the charge, determined by the capacitance of the charged object.
- Rate at which the charge is dissipated, determined by the resistance into which it is discharged.

Common techniques for controlling static problems include the following:

- Humidity control. Relative humidity of 50 percent or higher will greatly inhibit electrostatic problems. Too much humidity, however, can create corrosion problems and may make some paper products dimensionally unstable. Most data processing equipment manufacturers recommend 40 to 60 percent RH.
- Conductive floor coverings. Careful selection of floor surfaces will aid greatly in controlling ESD problems. Conductive synthetic rubber and other special-purpose floor coverings are ideal. Vinyl-asbestos is marginal. Nylon carpeting usually is unacceptable from an ESD standpoint.
- Static drain path. A static drain path from floor tiles or mats to the nearest grounded metal is recommended in heavy traffic areas. The floor surface-to-ground resistance need not be particularly low, 500 kΩ to 20 MΩ is adequate for most applications.
- Ion generators. Localized, chronic static problems can be neutralized through the use of an ion generator. Such systems commonly are used in semiconductor assembly plants and in the printing industry to dissipate static charges.

3.5.1   Failure Mechanisms

Destructive voltages and/or currents from an ESD event may result in device failure because of thermal fatigue and/or dielectric breakdown. MOS transistors normally are constructed with an oxide layer between the gate conductor and the source-drain channel region, as illustrated in Figure 3.17 for a metal gate device, and Figure 3.18 for a silicon gate device. Bipolar transistor construction, shown in Figure 3.19, is less susceptible to ESD damage because the oxide is used only for surface insulation.

Oxide thickness is the primary factor in MOS ruggedness. A thin oxide is more susceptible to electrostatic punch-through, which results in a permanent low-resistance short circuit through the oxide. Where pinholes or other weaknesses exist in the oxide, damage is possible at a lower charge level. Semiconductor manufacturers have reduced oxide thickness as they have reduced device size. This trend has resulted in a significant increase in sensitivity to ESD damage.

Detecting an ESD failure in a complex device may present a significant challenge for quality control engineers. For example, erasable programmable read-only memory (EPROM) chips use oxide layers as thin as 100 angstroms, making them susceptible to single-cell defects that can remain undetected until the damaged cell itself is addressed. An electrostatic charge small enough that it does not result in oxide breakdown still can cause lattice damage in the oxide, lowering its ability to withstand

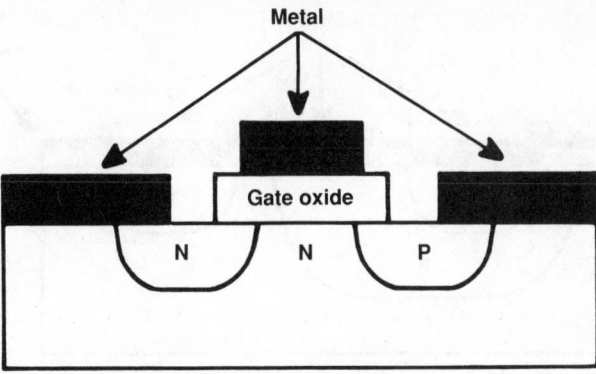

Figure 3.17   Construction of a metal gate NMOS transistor. (Data from: Technical staff, Military/Aerospace Products Division, *The Reliability Handbook*, National Semiconductor, Santa Clara, CA, 1987.)

subsequent ESD exposure. A weakened lattice will have a lower breakdown threshold voltage.

Table 3.1 lists the susceptibility of various semiconductor technologies to ESD-induced failure. Table 3.2 lists the ESD voltage levels that can result from common workbench operations.

Latent Failures.   Immediate failure resulting from ESD exposure is easily determined: The device no longer works. A failed component may be removed from the subassembly in which it is installed, representing no further reliability risk to the system. Not all devices exposed to ESD, however, fail immediately. Unfortunately, there is little data dealing with the long-term reliability of devices that have survived ESD exposure. Some experts, however, say that two to five devices are degraded for every one that fails. Available data indicates that latent failures can occur in both bipolar and MOS chips, and that there is no direct relationship

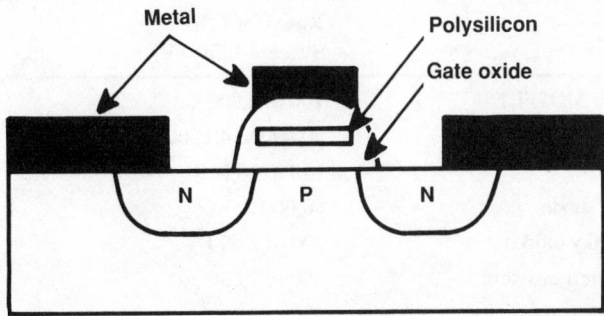

Figure 3.18   Construction of a silicon gate NMOS transistor. (Data from: Technical staff, Military/Aerospace Products Division, *The Reliability Handbook*, National Semiconductor, Santa Clara, CA, 1987.)

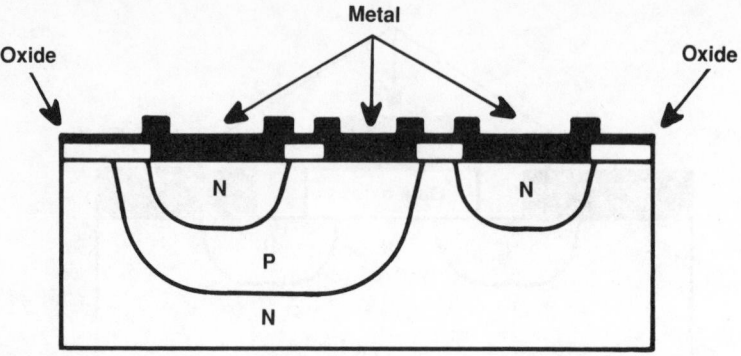

Figure 3.19   Construction of a bipolar transistor. (Data from: Technical staff, Military/Aerospace Products Division, *The Reliability Handbook*, National Semiconductor, Santa Clara, CA, 1987.)

between the susceptibility of a device to catastrophic failure and its susceptibility to latent failure. Damage can manifest itself in one of two primary mechanisms:

- Shortened lifetime (a possible cause of many infant mortality failures seen during burn-in).
- Electrical performance shifts, many of which may cause the device to fail electrical limit tests.

Case in Point.   Figure 3.20 shows an electron microscope photo of a chip that failed because of an overvoltage condition. An ESD to this MOSFET damaged one of the metallization connection points of the device, resulting in catastrophic failure. Note

Table 3.1   The Susceptibility of Various Technologies from ESD-Induced Damage (*Data courtesy of Motorola*)

Device Type	Range of ESD Susceptibility (V)
Power MOSFET	100 to 2,000
Power Darlington	20,000 to 40,000
JFET	140 to 10,000
Zener diode	40,000
Schottky diodes	300 to 2,500
Bipolar transistors	380 to 7,000
CMOS	250 to 2,000
ECL	500
TTL	300 to 2,500

Table 3.2   Electrostatic Voltages That Can Be Developed through Common Workbench
Activities (*Data courtesy of Motorola*)

| | Electrostatic Voltages | |
Means of Static Generation	10–20 % RH	65–90 % RH
Walking across carpet	35,000	1,500
Walking on vinyl floor	12,000	250
Worker at bench	6,000	100
Handling vinyl envelope	7,000	600
Handling common polybag	20,000	1,200

the spot where the damage occurred. The objects in the photo that look like bent nails
are actually gold lead wires with a diameter of 1 mil. By contrast, a typical human hair
is about 3 mils in diameter. The original photo was shot at X200 magnification. Figure
3.21 offers another view of the MOSFET damage point, but at X5000. The character
of the damage can be observed. Some of the aluminum metallization has melted and
can be seen along the bottom edge of the hole.

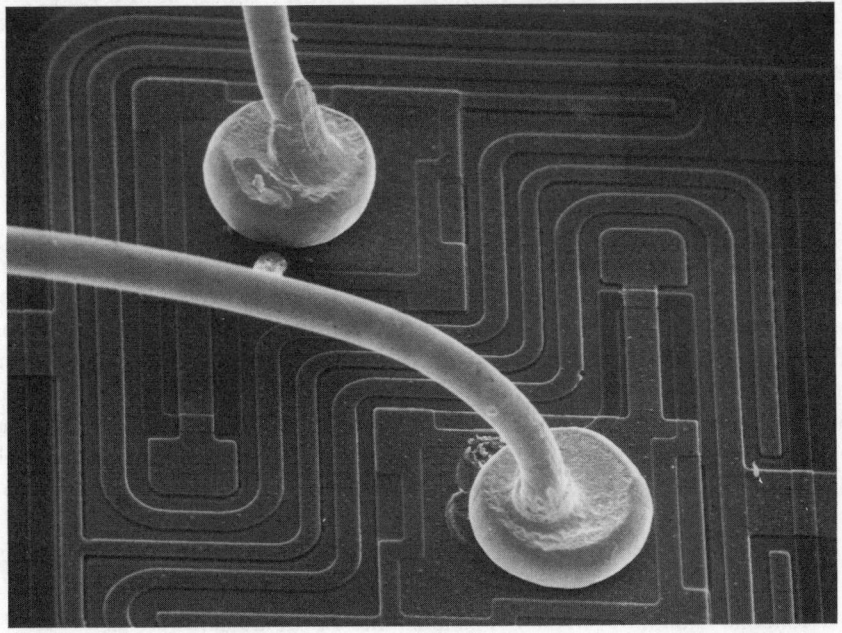

Figure 3.20   A scanning electron microscope photo illustrating ESD damage to the metallization of a MOSFET device.

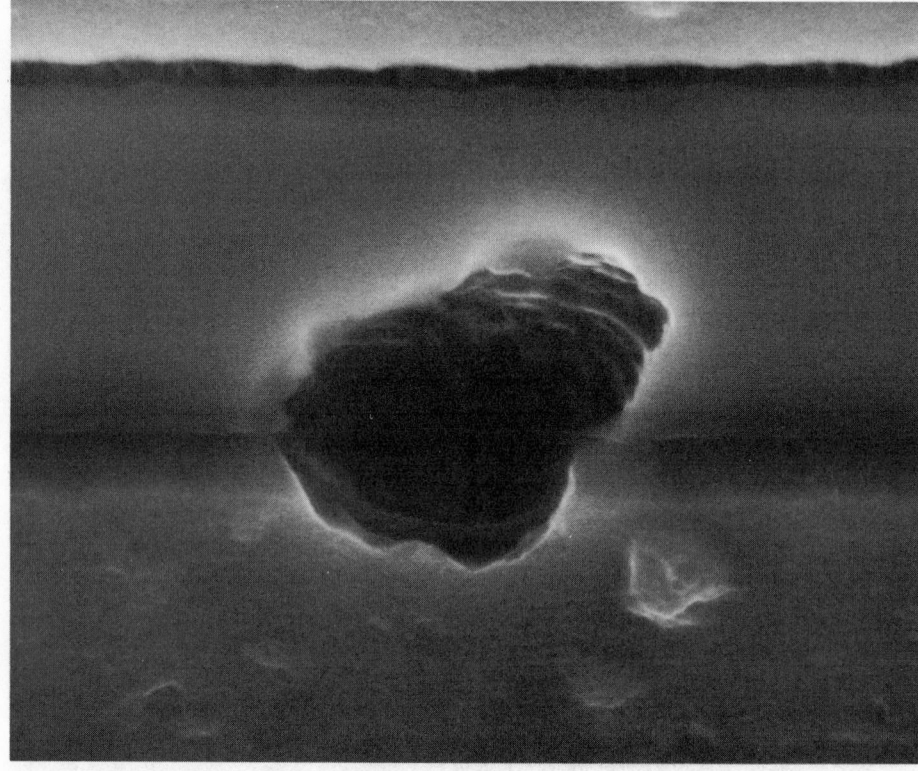

Figure 3.21   The device shown in the previous figure at X5000 magnification. The character of the damage can be observed.

3.6 SEMICONDUCTOR DEVELOPMENT

Semiconductor failures caused by high-voltage stresses are becoming a serious concern for engineers, operators, and technical managers as new, high-density integrated circuits are placed into service. Internal IC connection lines that were 2 microns a few years ago have been reduced to less than 1.0 micron today. Spacing between leads has been reduced by a factor of 4 or more. In the past, the overvoltage peril was primarily to semiconductor substrates. Now, however, the metallization itself — the points to which leads connect — is subject to damage. Failures are the result of three primary overvoltage sources:

- External man-made. Overvoltages coupled into electronic hardware from utility company ac power feeds, or other ac or dc power sources.
- External natural. Overvoltages coupled into electronic hardware as a result of natural sources.
- Electrostatic discharge. Overvoltages coupled into electronic hardware as a result of static generation and subsequent discharge.

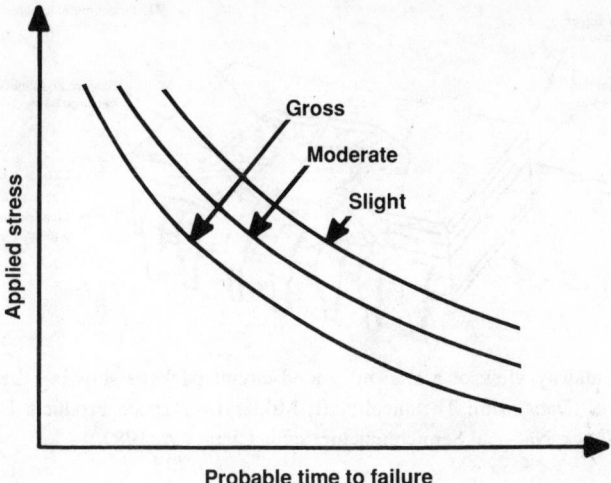

Figure 3.22 Illustration of the likelihood of component failure based on applied stress and degree of latent defects.

Most semiconductor failures are of a random nature. That is, different devices respond differently to a specific stress. Figure 3.22 illustrates how built-in (latent) defects in a given device affect the time-to-failure point of the component. Slight imperfections require greater stress than gross imperfections to reach a quantifiable failure mode.

Integrated circuits intended for microcomputer applications have been a driving force in the semiconductor industry. Figure 3.23 plots the dramatic increase in device counts that have occurred during the past two decades. The 80286 microprocessor

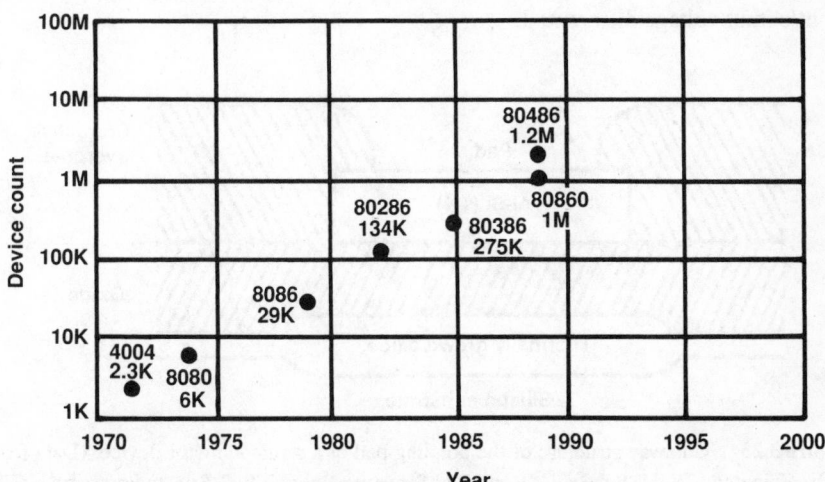

Figure 3.23 Microcomputer transistor count per chip as a function of time. (Data from: D. L. Crook, "Evolution of VLSI Reliability Engineering," *Proceedings* of the IEEE Reliability Physics Symposium, IEEE, New York, 1990.)

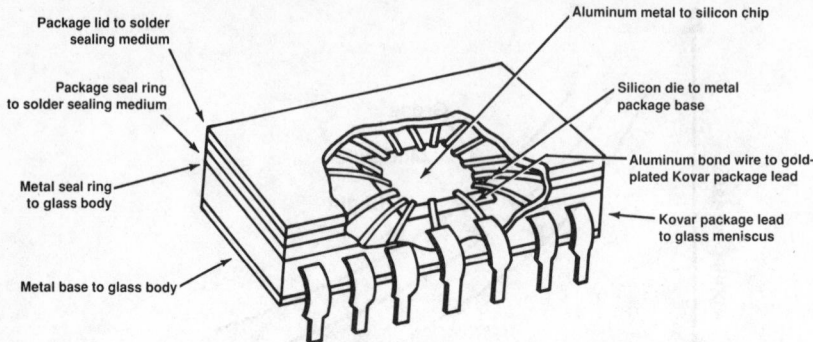

Figure 3.24   Cutaway view of a DIP integrated circuit package showing the internal-to-external interface. (Data from: Technical staff, Military/Aerospace Products Division, *The Reliability Handbook*, National Semiconductor, Santa Clara, CA, 1987.)

chip, for example, contains the equivalent of more than 1.2 million transistors. Figure 3.24 shows a simplified cutaway view of a DIP IC package. Connections between the die itself and the outside world are made with bonding wires. Figure 3.25 shows a cutaway view of a bonding pad.

Hybrid microcircuits also have become common in consumer and industrial equipment. A hybrid typically utilizes a number of components from more than one technology to perform a function that could not be achieved in monolithic form with the same performance, efficiency, and/or cost. A simple multichip hybrid is shown in Figure 3.26.

The effects of high-voltage breakdown in a hybrid semiconductor chip are illustrated graphically in Figure 3.27 (a–c). Failure analysis demonstrated that the *pass transistor* in this voltage regulator device was overstressed because of excessive input/output voltage differential.

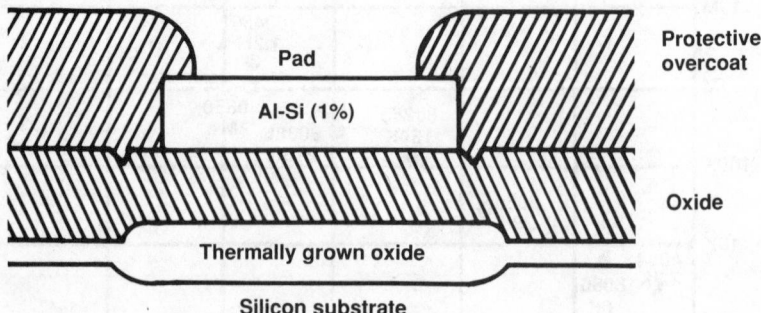

Figure 3.25   Cutaway structure of the bonding pad of a semiconductor device. (Data from: T. B. Ching and W. H. Schroen, "Bond Pad Structure Reliability," *Proceedings* of the IEEE Reliability Physics Symposium, IEEE, New York, 1988.)

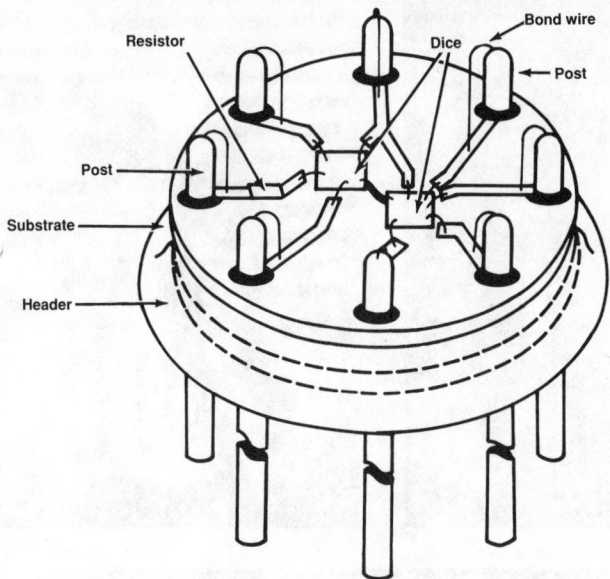

Figure 3.26   Basic construction of a multichip hybrid device. (Data from: Technical staff, Military/Aerospace Products Division, *The Reliability Handbook*, National Semiconductor, Santa Clara, CA, 1987.)

3.6.1 Chip Protection

With the push for faster and more complex ICs, it is unlikely that semiconductor manufacturers will return to thicker oxide layers or larger junctions. Overvoltage protection will come, instead, from circuitry built into individual chips to shunt transient energy to ground.

Most MOS circuits incorporate protective networks. These circuits can be made quite efficient, but there is a tradeoff between the amount of protection provided and device speed and packing density. Protective elements, usually diodes, must be physically large if they are to clamp adequately. Such elements take up a significant amount of chip space. The RC time constants of protective circuits also can place limits on switching speeds.

Protective networks for NMOS devices typically use MOS transistors as shunting elements, rather than diodes. Although diodes are more effective, fewer diffusions are available in the NMOS process, so not as many forward-biased diodes can be constructed. Off-chip protective measures, including electromagnetic shielding, filters, and discrete diode clamping, are seldom used because they are bulky and expensive.

Figure 3.28 shows the protective circuitry used in a 54HC high-speed CMOS (complementary metal-oxide silicon) device. Polysilicon resistors are placed in series

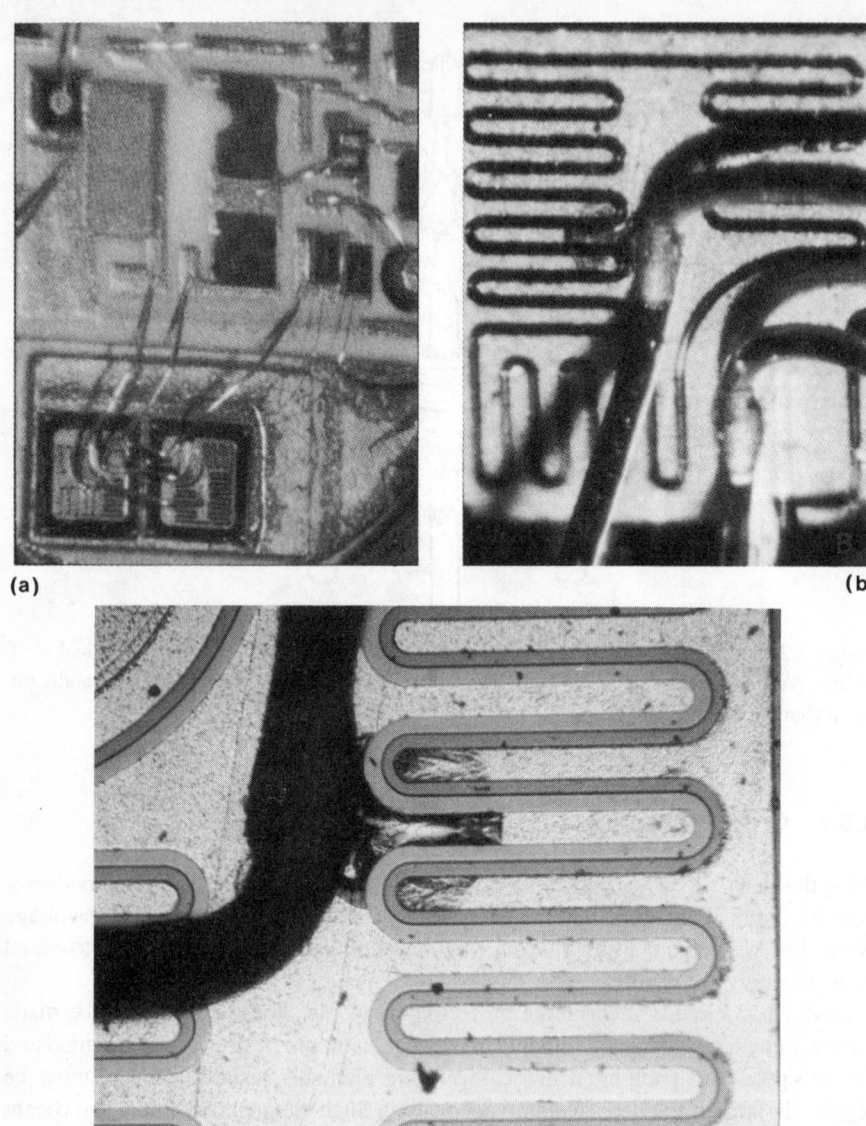

Figure 3.27   Three views of a hybrid voltage regulator that failed because of a damaged pass transistor: (a, *top left*) the overall circuit geometry; (b, *top right*) a closeup of the damaged pass transistor area; (c, *bottom*) an enlarged view of the damage point.

with each input pin, and relatively large-geometry diodes are added as clamps on the IC side of the resistors. Clamping diodes also are used at the output. The diodes restrict the magnitude of the voltages that can reach the internal circuitry. Protective features such as these have allowed CMOS devices to withstand ESD test voltages in excess of 2 kV.

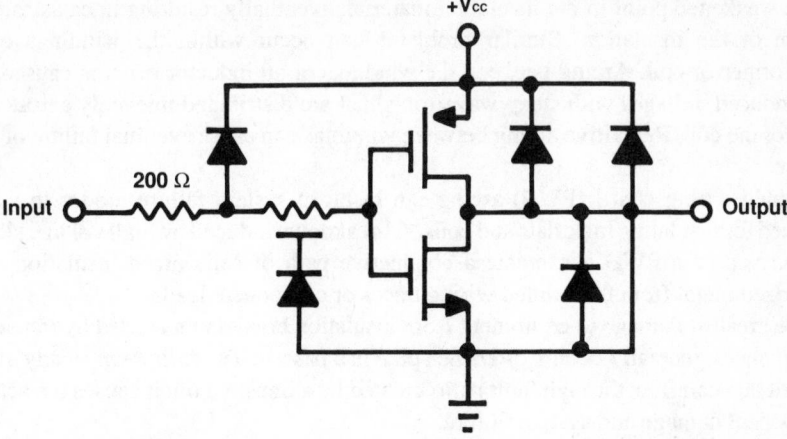

Figure 3.28   CMOS transistor with built-in ESD protection circuitry. (Data from: Technical staff, Military/Aerospace Products Division, *The Reliability Handbook*, National Semiconductor, Santa Clara, CA, 1987.)

3.7 SWITCH-ARCING EFFECTS

High voltages often are generated by breaking current to an inductor with a mechanical switch. They can, with time, cause pitting, corrosion, or material transfer of the switch contacts. In extreme cases, the contacts even can be welded together. The actual wear (or failure) of a mechanical switch is subject to many factors, including:

- Contact construction and the type of metal used
- Amount of contact bounce that typically occurs with the switching mechanism
- Atmosphere
- Temperature
- Steady-state and in-rush currents
- Whether ac or dc voltages are being switched by the mechanism

Effective transient suppression can significantly reduce the amount of energy dissipated during the operation of switch contacts. This reduction will result in a corresponding increase in switch life. In applications where relay contacts are acting as power-switching elements, the use of effective transient-suppression techniques will reduce the amount of maintenance (contact cleaning) required for the device.

3.7.1 Insulation Breakdown

The breakdown of a solid insulating material usually results in localized carbonization, which may be catastrophic, or may result in decreased dielectric strength at the arc-over point. The occurrence of additional transients often will cause a breakthrough at the weakened point in the insulating material, eventually resulting in catastrophic failure of the insulation. Similar problems can occur within the windings of a transformer or coil. Arcing between the windings of an inductor often is caused by self-induced voltages with steep wavefronts that are distributed unevenly across the turns of the coil. Repetitive arcing between windings can cause eventual failure of the device.

Printed wiring board (PWB) arcing can result in system failure modes in ways outlined for insulating materials and coils. A breakdown induced by high voltage along the surface of a PWB can create a conductive path of carbonized insulation and vaporized metal from the printed wiring traces or component leads.

The greatest damage to equipment from insulation breakdown caused by transient disturbances generally occurs *after* the spike has passed. The *follow-on* steady-state current that can flow through fault paths created by a transient often causes the actual component damage and system failure.

3.8 BIBLIOGRAPHY

Antinone, Robert J.: "How to Prevent Circuit Zapping," *IEEE Spectrum*, IEEE, New York, April 1987.

Benson, K. B., and J. Whitaker: *Television and Audio Handbook for Engineers and Technicians*, McGraw-Hill, New York, 1989.

Blackburn, David L.: "Turn-Off Failure of Power MOSFETs," *IEEE Transactions on Power Electronics*, vol. PE-2, no. 2, IEEE, New York, April 1987.

Boxleitner, Warren: "How to Defeat Electrostatic Discharge," *IEEE Spectrum*, IEEE, New York, August 1989.

Crook, D. L.: "Evolution of VLSI Reliability Engineering," *Proceedings* of the IEEE Reliability Physics Symposium, IEEE, New York, 1990.

Federal Information Processing Standards Publication no. 94, *Guideline on Electrical Power for ADP Installations*, U.S. Department of Commerce, National Bureau of Standards, Washington, DC, 1983.

Frank, Donald: "Please Keep Your EMC Out of My ESD," *Proceedings* of the IEEE Reliability and Maintainability Symposium, IEEE, New York, 1986.

Gloer, H. Niles: "Voltage Transients and the Semiconductor," *The Electronic Field Engineer*, vol. 2, 1979.

Jordan, Edward C.: *Reference Data for Engineers: Radio, Electronics, Computer, and Communications*, 7th ed., Howard W. Sams Company, Indianapolis, 1985.

Kanarek, Jess: "Protecting Against Static Electricity Damage," *Electronic Servicing & Technology* magazine, Intertec Publishing, Overland Park, KS, March 1986.

Koch, T., W. Richling, J. Witlock, and D. Hall: "A Bond Failure Mechanism," *Proceedings* of the IEEE Reliability Physics Conference, IEEE, New York, April 1986.

Meeldijk, Victor: "Why Do Components Fail?," *Electronic Servicing & Technology* magazine, Intertec Publishing, Overland Park, KS, November 1986.

Motorola TMOS Power MOSFET Data Handbook, Motorola Semiconductor, Phoenix.

Nenoff, Lucas: "Effect of EMP Hardening on System R&M Parameters," *Proceedings* of the IEEE Reliability and Maintainability Symposium, IEEE, New York, 1986.

SCR Applications Handbook, International Rectifier Corporation, El Segundo, CA, 1977.

SCR Manual, 5th ed., General Electric Company, Auburn, NY.

Sydnor, Alvin: "Voltage Breakdown in Transistors," *Electronic Servicing & Technology* magazine, Intertec Publishing, Overland Park, KS, July 1986.

Technical staff, *Bipolar Power Transistor Reliability Report*, Motorola Semiconductor, Phoenix, 1988.

Technical staff, Military/Aerospace Products Division, *The Reliability Handbook*, National Semiconductor, Santa Clara, CA, 1987.

Technical staff, *MOV Varistor Data and Applications Manual*, General Electric Company, Auburn, NY.

Voss, Mike: "The Basics of Static Control," *Electronic Servicing & Technology*, Intertec Publishing, Overland Park, KS, July 1988.

Whitaker, Jerry: *Radio Frequency Transmission Systems: Design and Operation*, McGraw-Hill, New York, 1990.

————: *Maintaining Electronic Systems*, CRC Press, Boca Raton, FL, 1991.

4

POWER-SYSTEM COMPONENTS

4.1 INTRODUCTION

The design of any piece of electronic equipment is a complicated process that involves a number of tradeoffs. Most manufacturers of professional and industrial equipment now are building transient protection into their products. This work is welcomed because effective transient suppression is a *systems problem* that extends from the utility company ac input to the circuit boards in each piece of hardware. Although progress has been made, more work needs to be accomplished on circuit-level transient suppression.

There is nothing magical about effective transient suppression. Disturbances on the ac line can be suppressed if the protection method used has been designed carefully and installed properly. Whether the protection method involves a systems approach or discrete devices at key points in the facility, the time and money spent incorporating protection will yield a good return on investment.

4.1.1 Power-Supply Components

The circuit elements most vulnerable to failure in any given piece of electronic hardware are those exposed to the outside world. In most systems, the greatest threat generally involves the ac-to-dc power supply. The power supply is subject to high-energy surges from lightning and other sources. Because power-supply systems often are exposed to extreme voltage and environmental stresses, *derating* of individual components is a key factor in improving supply reliability. The goal of derating is to reduce the electrical, mechanical, thermal, and other environmental stresses on a

Table 4.1 Statistical Distribution of Component Failures in an EMP Protection Circuit. Note that types of components tend to fail in predictable ways. (Data from: Lucas Nenoff, "Effect of EMP Hardening on System R&M Parameters," *Proceedings* of the 1986 Reliability and Maintainability Symposium, IEEE, New York, 1986)

COMPONENT	MODE OF FAILURE	DISTRIBUTION
CAPACITOR (ALL TYPES)	OPEN SHORT	.01 .99
COIL	OPEN SHORT	.75 .25
DIODE (ZENER)	OPEN SHORT	.01 .99
GE-MOV	OPEN SHORT	.01 .99
TRANSZORB	OPEN SHORT	.01 .99
CONNECTOR PIN	OPEN SHORT TO GND	.99 .01
SOLDER JOINT	OPEN	1.00
LUG CONNECTION	OPEN	1.00
SURGE PROTECTOR	OPEN SHORT	.99 .01

component to decrease the degradation rate and to prolong expected life. Through derating, the margin of safety between the operating stress level and the maximum permissible stress level for a given part is increased. This consideration provides added protection from system overstress, unforeseen during design.

Experience has demonstrated that types of components tend to fail in predictable ways. Table 4.1 shows the statistical distribution of failures for a transient-suppression (EMP) protection circuit. Although the data presented applies only to a specific product, some basic conclusions can be drawn:

- The typical failure mode for a capacitor is a short circuit.
- The typical failure mode for a zener diode is a short circuit.
- The typical failure mode for a connector pin is an open circuit.
- The typical failure mode for a solder joint is an open circuit.

These conclusions present no great surprises, but they point out the predictability of equipment failure modes. The first step in solving a problem is knowing what is likely to fail, and what the typical failure modes are.

4.2 RELIABILITY OF POWER RECTIFIERS

Virtually all power supplies use silicon rectifiers as the primary ac-to-dc converting device. Rectifier parameters generally are expressed in terms of reverse-voltage ratings and mean-forward-current ratings in a 1/2-wave rectifier circuit operating from a 60 Hz supply and feeding a purely resistive load. The three primary reverse-voltage ratings are:

- *Peak transient reverse voltage* (V_{RM}). The maximum value of any nonrecurrent surge voltage. This value must never be exceeded, even for a microsecond.
- *Maximum repetitive reverse voltage* [$V_{RM(rep)}$]. The maximum value of reverse voltage that may be applied recurrently (in every cycle of 60 Hz power). This includes oscillatory voltages that may appear on the sinusoidal supply.
- *Working peak reverse voltage* [$V_{RM(wkg)}$]. The crest value of the sinusoidal voltage of the ac supply at its maximum limit. Rectifier manufacturers generally recommend a value that has a significant safety margin, relative to the peak transient reverse voltage (V_{RM}), to allow for transient overvoltages on the supply lines.

There are three forward-current ratings of similar importance in the application of silicon rectifiers:

- *Nonrecurrent surge current* [$I_{FM(surge)}$]. The maximum device transient current that must not be exceeded at any time. $I_{FM(surge)}$ is sometimes given as a single value, but often is presented in the form of a graph of permissible surge-current values vs. time. Because silicon diodes have a relatively small thermal mass, the potential for short-term current overloads must be given careful consideration.
- *Repetitive peak forward current* [$I_{FM(rep)}$]. The maximum value of forward current reached in each cycle of the 60 Hz waveform. This value does not include random peaks caused by transient disturbances.
- *Average forward current* [$I_{FM(av)}$]. The upper limit for average load current through the device. This limit is always well below the repetitive peak forward current rating to ensure an adequate margin of safety.

Rectifier manufacturers generally supply curves of the instantaneous forward voltage vs. instantaneous forward current at one or more specific operating temperatures. These curves establish the forward-mode upper operating parameters of the device.

Figure 4.1 shows a typical rectifier application in a bridge rectifier circuit.

4.2.1 Operating Rectifiers in Series

High-voltage power supplies (5 kV and greater) often require rectifier voltage ratings well beyond those typically available from the semiconductor industry. To meet the

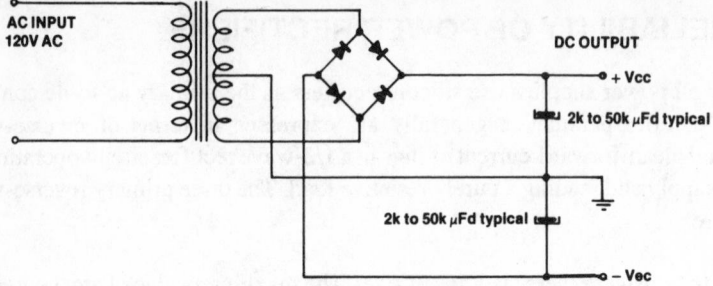

Figure 4.1   Conventional capacitor input filter full-wave bridge.

requirements of the application, manufacturers commonly use silicon diodes in a series configuration to give the required working peak reverse voltage. For such a configuration to work properly, the voltage across any one diode must not exceed the rated peak transient reverse voltage (V_{RM}) at any time. The dissimilarity commonly found between the reverse leakage current characteristics of different diodes of the same type number makes this objective difficult to achieve. The problem normally is overcome by connecting shunt resistors across each rectifier in the chain, as shown in Figure 4.2. The resistors are chosen so that the current through the shunt elements (when the diodes are reverse-biased) will be several times greater than the leakage current of the diodes themselves.

The *carrier storage* effect also must be considered in the use of a series-connected rectifier stack. If precautions are not taken, different diode recovery times (caused by the carrier storage phenomenon) will effectively force the full applied reverse voltage across a small number of diodes, or even a single diode. This problem can be prevented by connecting small-value capacitors across each diode in the rectifier stack. The capacitors equalize the transient reverse voltages during the carrier storage recovery periods of the individual diodes.

Figure 4.3 illustrates a common circuit configuration for a high-voltage three-phase rectifier bank. A photograph of a high-voltage series-connected three-phase rectifier assembly is shown in Figure 4.4.

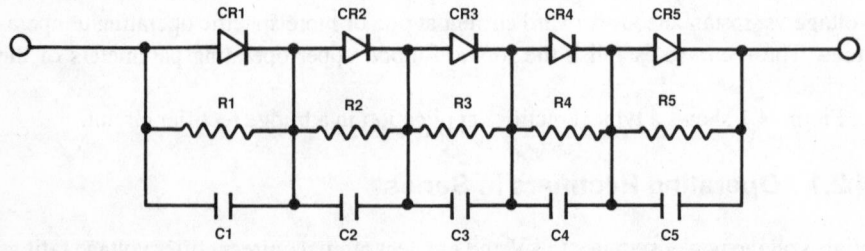

Figure 4.2   A portion of a high-voltage series-connected rectifier stack.

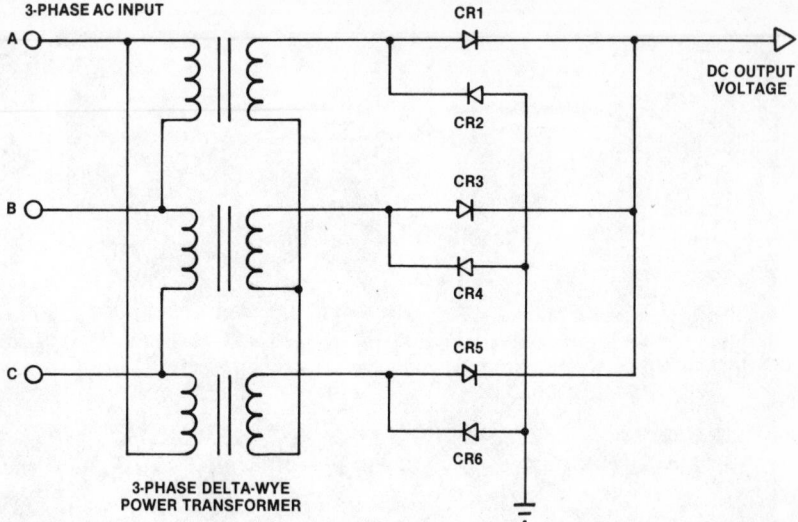

Figure 4.3   Three-phase delta-connected high-voltage rectifier.

4.2.2   Operating Rectifiers in Parallel

Silicon rectifiers are used in a parallel configuration when a large amount of current is required from the power supply. Parallel assemblies normally are found in low-voltage, high-current supplies. *Current sharing* is the major design problem with a parallel rectifier assembly because diodes of the same type number do not necessarily exhibit the same forward characteristics.

Semiconductor manufacturers often divide production runs of rectifiers into tolerance groups, matching forward characteristics of the various devices. When parallel diodes are used, devices from the same tolerance group must be selected to avoid unequal sharing of the load current. As a margin of safety, designers allow a substantial derating factor for devices in a parallel assembly to ensure that the maximum operating limits of any one component are not exceeded.

The problems inherent in a parallel rectifier assembly can be reduced through the use of a resistance or reactance in series with each component, as shown in Figure 4.5. The buildout resistances (R1 through R4) force the diodes to share the load current equally. Such assemblies can, however, be difficult to construct and may be more expensive than simply adding diodes or going to higher-rated components.

4.2.3   Silicon Avalanche Rectifiers

The silicon avalanche diode is a special type of rectifier that can withstand high reverse power dissipation. For example, an avalanche diode with a normal forward rating of

Figure 4.4   High-voltage rectifier assembly for a three-phase delta-connected circuit.

10 A can dissipate a reverse transient of 8 kW for 10 ms without damage. This characteristic of the device allows elimination of the surge-absorption capacitor and voltage-dividing resistor networks needed when conventional silicon diodes are used in a series rectifier assembly. Because fewer diodes are needed for a given applied reverse voltage, significant underrating of the device (to allow for reverse voltage transient peaks) is not required.

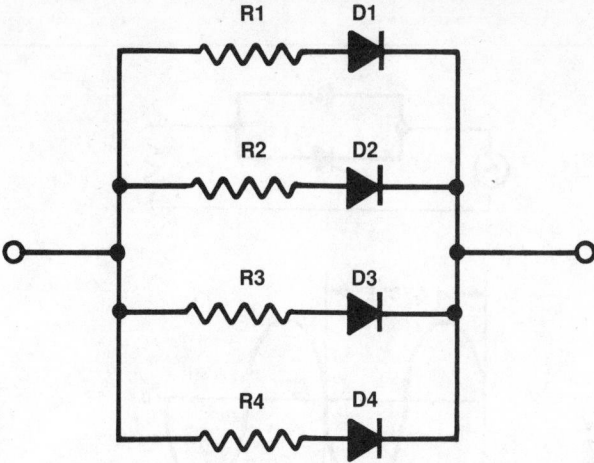

Figure 4.5   Using buildout resistances to force current sharing in a parallel rectifier assembly.

When an extra-high-voltage rectifier stack is used, it is still advisable to install shunt capacitors — but not resistors — in an avalanche diode assembly. The capacitors are designed to compensate for the effects of carrier storage and stray capacitance in a long series assembly.

4.3 THYRISTOR SERVO SYSTEMS

Thyristor control of ac power has become a popular method of regulating high-voltage and/or high-current power supplies. The type of servo system employed depends on the application. Figure 4.6 shows a basic single-phase ac control circuit using discrete thyristors. The rms load current (I_{rms}) at any specific phase delay angle ($\emptyset$) is given in terms of the normal full-load rms current at a phase delay of zero (I_{rms-0}):

$$I_{rms} = I_{rms-0} \, [1 - (\alpha/\pi) + (2\pi)^{-1} \sin 2\pi]1/2$$

The load rms voltage at any particular phase-delay angle bears the same relationship to the full-load rms voltage at zero phase delay as the previous equation illustrates for load current. An analysis of the mathematics shows that although the theoretical delay range for complete control of a resistive load is 0° to 180°, a practical span of 20° to 160° gives a power-control range of approximately 99 percent to 1 percent of maximum output to the load. Figure 4.7 illustrates typical phase-control waveforms.

The circuit shown in Figure 4.6 requires a source of gate trigger pulses that must be isolated from each other by at least the peak value of the applied ac voltage. The two gate pulse trains must also be phased 180° with respect to each other. Also, the gate pulse trains must shift together with respect to the ac supply voltage phase when power throughput is adjusted.

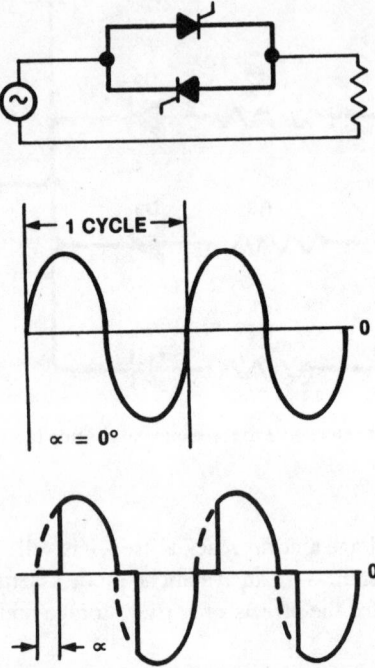

Figure 4.6   Inverse-parallel thyristor ac power control: (a, *top*) circuit diagram; (b, *bottom*) voltage and current waveforms for full and reduced thyristor control angles. The waveforms apply for a purely resistive load.

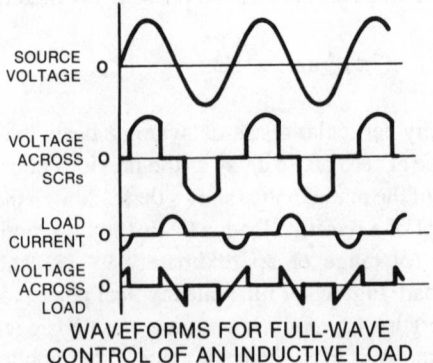

Figure 4.7   Waveforms in an ac circuit using thyristor power control.

Some power-control systems use two identical, but isolated, gate pulse trains operating at a frequency of twice the applied supply voltage (120 Hz for a 60 Hz system). Under such an arrangement, the forward-biased thyristor will fire when the gate pulses are applied to the SCR pair. The reverse-biased thyristor will not fire. Normally, it is considered unsafe to drive a thyristor gate positive while its anode-cathode is reverse-biased. In this case, however, it may be permissible because the thyristor that is fired immediately conducts and removes the reverse voltage from the other device. The gate of the reverse-biased device is then being triggered on a thyristor that essentially has no applied voltage.

4.3.1 Inductive Loads

The waveforms shown in Figure 4.8 illustrate effects of phase control on an inductive load. When inductive loads are driven at a reduced conduction angle, a sharp transient change of load voltage occurs at the end of each current pulse (or loop). The transients generally have no effect on the load, but they can be dangerous to proper operation of the thyristors. When the conducting thyristor turns off, thereby disconnecting the load from the ac line supply, the voltage at the load rapidly drops to zero. This rapid voltage change, in effect, applies a sharply rising positive anode voltage to the thyristor opposing the device that has been conducting. If the thyristor dv/dt rating is exceeded,

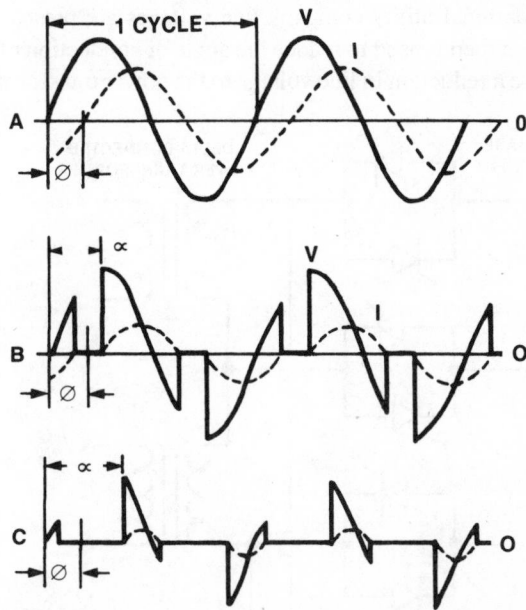

Figure 4.8 Voltage and current waveforms for inverse-parallel thyristor power control with an inductive load: (a) full conduction; (b) small-angle phase reduction; (c) large-angle phase reduction.

the opposing device will turn on and conduction will take place, independent of any gate drive pulse.

A common protective approach involves the addition of a resistor-capacitor (RC) snubber circuit to control the rate of voltage change seen across the terminals of the thyristor pair. Whenever a thyristor pair is used to drive an inductive load, such as a power transformer, it is critically important that each device fires at a point in the applied waveform exactly 180° relative to the other. If proper timing is not achieved, the positive and negative current loops will differ in magnitude, causing a dc current to flow through the primary side of the transformer. A common trigger control circuit should, therefore, be used to determine gate timing for thyristor pairs.

4.3.2 Applications

Several approaches are possible for thyristor power control in a three-phase ac system. The circuit shown in Figure 4.9 consists of essentially three independent, but inter-locked, single-phase thyristor controllers. This circuit is probably the most common configuration found in industrial equipment.

In a typical application, the thyristor pairs feed a power transformer with multitap primary windings, thereby giving the user an adjustment range to compensate for variations in utility company line voltages from one location to another. A common procedure specifies selection of transformer tap positions that yield a power output of 105 percent when nominal utility company line voltages are present. The thyristor power-control system then is used to reduce the angle of conduction of the SCR pairs as necessary to cause a reduction in line voltage to the power transformer to yield 100

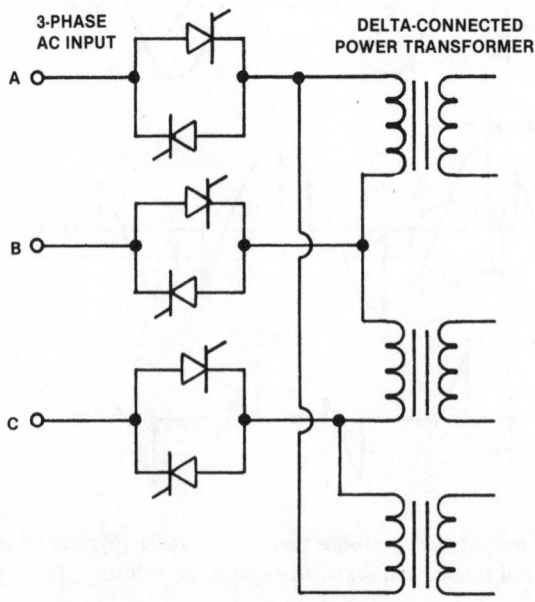

Figure 4.9   Modified full-thyristor three-phase ac control of an inductive delta load.

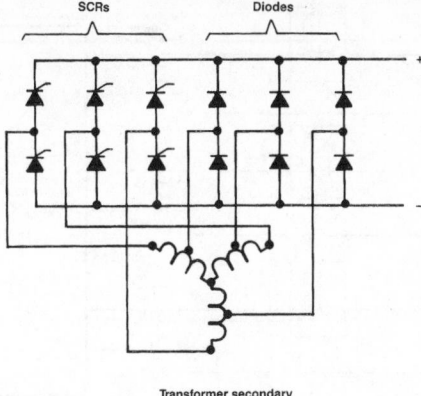

Transformer secondary

Figure 4.10   Six-phase boost rectifier circuit.

percent rated power output from the power supply. A servo loop from a sample point at the load may be used to automatically compensate for line-voltage variations. With such an arrangement, the thyristors are kept within a reasonable degree of retarded phase operation. Line voltages will be allowed to sag 5 percent or so without affecting the dc supply output. Utility supply voltage excursions above nominal value simply will result in delayed triggering of the SCR pairs.

Thyristor control of high-power loads (200 kW and above) typically uses special transformers that provide 6- or 12-phase outputs. Although they are more complicated and expensive, such designs allow additional operational control, and filtering requirements are reduced significantly. Figure 4.10 shows a six-phase *boost rectifier* circuit. The configuration consists basically of a full-wave three-phase SCR bridge connected to a wye-configured transformer secondary. A second bridge, consisting of six diodes, is connected to low-voltage taps on the same transformer. When the SCRs are fully on, their output is at a higher voltage than the diode bridge. As a result, the diodes are reverse-biased and turned off. When the SCRs are partially on, the diodes are free to conduct. The diodes improve the quality of the output waveform during low-voltage (reduced conduction angle) conditions. The minimum output level of the supply is determined by the transformer taps to which the diodes are connected.

A thyristor-driven three-phase power-control circuit is shown in Figure 4.11. A single-phase power-control circuit is shown in Figure 4.12.

4.3.3   Triggering Circuits

Accurate, synchronized triggering of the gate pulses is a critical element in thyristor control of a three-phase power supply. The gate signal must be synchronized properly with the phase of the ac line that it is controlling. The pulse also must properly match the phase angle delay of the gates of other thyristors in the power-control system. Lack of proper synchronization of gate pulse signals between thyristor pairs can result in improper current sharing ("current hogging") among individual legs of the three-phase supply.

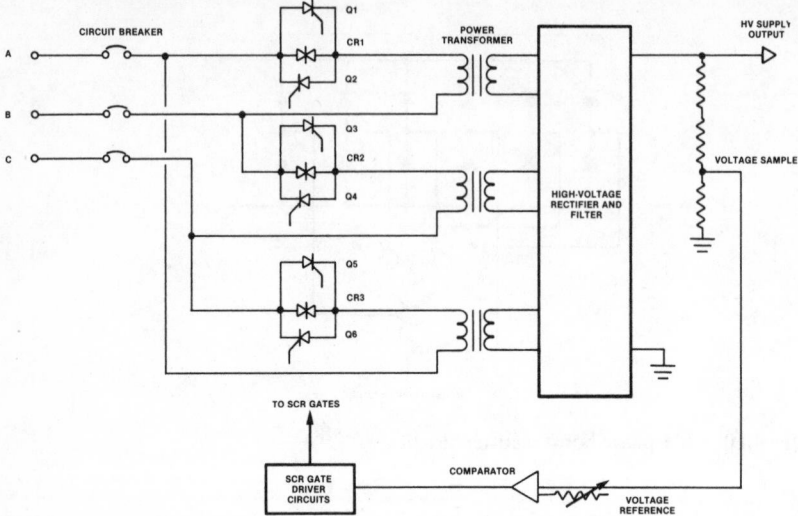

Figure 4.11 Thyristor-controlled high-voltage servo power supply.

The gate circuit must be protected against electrical disturbances that could make proper operation of the power-control system difficult or unreliable. Electrical isolation of the gate is a common approach. Standard practice calls for the use of gate pulse transformers in thyristor servo system gating cards. Pulse transformers are ferrite-cored devices with a single primary winding and (usually) multiple secondary windings that feed, or at least control, the individual gates of a back-to-back thyristor pair. This concept is illustrated in Figure 4.13. Newer thyristor designs may use optocouplers (primarily for low-power systems) to achieve the necessary electrical isolation between the trigger circuit and the gate.

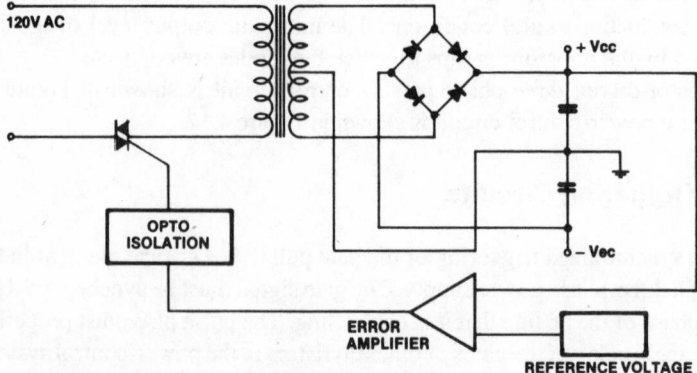

Figure 4.12 Phase-controlled power supply with primary regulation.

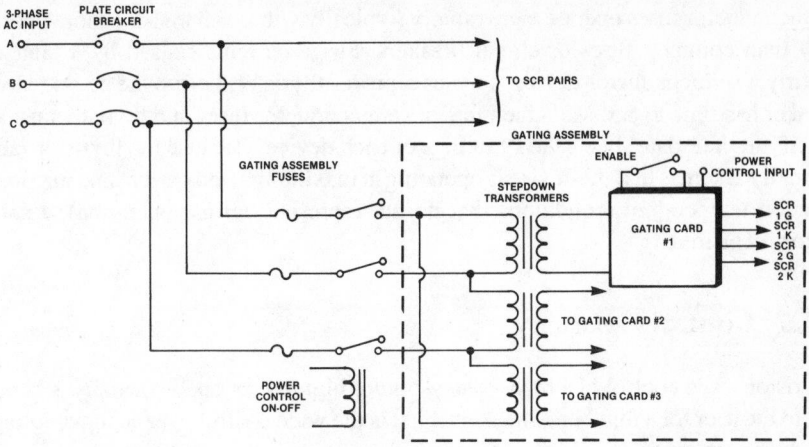

Figure 4.13 Simplified block diagram of the gating circuit for a phase-control system using back-to-back SCRs.

It is common practice to tightly twist together the leads from the gate and cathode of a thyristor to the gating card assembly. This practice provides a degree of immunity to high-energy pulses that might inadvertently trigger the thyristor gate. The gate circuit must be designed and configured carefully to reduce inductive and capacitive coupling that might occur between power and control circuits. Because of the high di/dt conditions commonly found in thyristor-controlled power circuits, power wiring and control (gate) wiring must be separated physically as much as possible. Shielding of gating cards in metal card cages is advisable.

Equipment manufacturers use various means to decrease gate sensitivity to transient sources, including placement of a series resistor in the gate circuit and/or a shunting capacitor between the gate and cathode. A series resistor has the effect of decreasing gate sensitivity, increasing the allowable dv/dt of the thyristor and reducing the turn-off time, which simultaneously increases the required holding and latching currents. The use of a shunt capacitor between the gate and cathode leads reduces high-frequency noise components that might be present on the gate lead and increases the dv/dt withstand capability of the thyristor. The application of these techniques is the exclusive domain of the design engineer. Users should not consider modifying a design without detailed consultation with the engineering department of the original equipment manufacturer.

4.3.4 Fusing

Current-limiting is a basic method of protection for any piece of equipment operated from the utility ac line. The device typically used for breaking fault currents is either a fuse or a circuit breaker. Some designs incorporate both components. *Semiconductor fuses* often are used in conjunction with a circuit breaker to provide added protection.

Semiconductor fuses operate more rapidly (typically within 8.3 ms) and more predictably than common fuses or circuit breakers. Surge currents caused by a fault can destroy a semiconductor device, such as a power thyristor, before the ac line circuit breaker has time to act. Manufacturers of semiconductor fuses and thyristors usually specify in their data sheets the I^2t ratings of each device. Because the thyristor rating normally assumes that the device is operating at maximum rated current and maximum junction temperature (conditions that do not represent normal operation), a safety factor is ensured.

4.3.5   Control Flexibility

Thyristor servo control of a high-voltage and/or high-current power supply is beneficial to the user for a number of reasons. First is the wide control over ac input voltages that such systems provide. A by-product of this feature is the capability to compensate automatically for line-voltage variations. Other benefits include the capability to soft-start the dc supply. Thyristor control circuits typically include a ramp generator that increases the ac line voltage to the power transformer from zero to full value within 2 to 5 s. This prevents high-surge currents through rectifier stacks and filter capacitors during system startup.

Although thyristor servo systems are preferred over other power-control approaches from an operational standpoint, they are not without their drawbacks. The control system is complex and can be damaged by transient activity on the ac power line. Conventional power contactors are simple and straightforward. They either make contact or they do not. For reliable operation of the thyristor servo system, attention must be given to transient suppression at the incoming power lines.

4.4   TRANSFORMER FAILURE MODES

The failure of a power transformer is almost always a catastrophic event that will cause the system to fail, and the result will be a messy cleanup job. The two primary enemies of power transformers are transient overvoltages and heat. Power input to a transformer is not all delivered to the secondary load. Some is expended as copper losses in the primary and secondary windings. These I^2R losses are practically independent of voltage; the controlling factor is current flow. To keep the losses as small as possible, the coils of a power transformer are wound with wire of the largest cross section that space will permit.

A practical transformer also will experience core-related losses, also known as *iron losses*. Repeated magnetizing and demagnetizing of the core (which occurs in an ac waveform) results in power loss because of the repeated realignment of the magnetic domains. This factor (hysteresis loss) is proportional to frequency and flux density. Silicon steel alloy is used for the magnetic circuit to minimize hysteresis loss. The changing magnetic flux also induces circulating currents (eddy currents) in the core material. Eddy current loss is proportional to the square of the frequency and the square

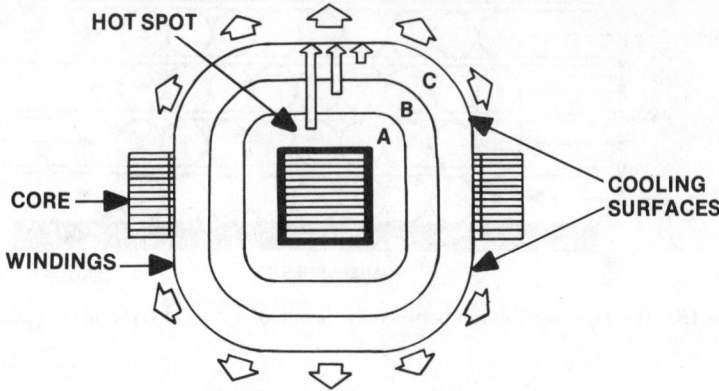

Figure 4.14   The dynamic forces of heat generation in a power transformer.

of the flux density. To minimize eddy currents, the core is constructed of laminations or layers of steel that are clamped or bonded together to form a single magnetic mass.

4.4.1 Thermal Considerations

Temperature rise inside a transformer is the result of power losses in the windings and the core. The insulation within and between the windings tends to blanket these heat sources and prevent efficient dissipation of the waste energy, as illustrated in Figure 4.14. Each successive layer of windings (shown as *A, B,* and *C* in the figure) acts to prevent heat transfer from the hot core to the local environment (air).

The hot spot shown in the figure can be dangerously high even though the outside transformer case and winding are relatively cool to the touch. Temperature rise is the primary limiting factor in determining the power-handling capability of a transformer. To ensure reliable operation, a large margin of safety must be designed into a transformer. Design criteria include winding wire size, insulation material, and core size.

4.4.2 Voltage Considerations

Transformer failures resulting from transient overvoltages typically occur between layers of windings within a transformer. See Figure 4.15. At the end of each layer, where the wire rises from one layer to the next, zero potential voltage exists. However, as the windings move toward the opposite end of the coil in a typical layer-wound device, a potential difference of up to twice the voltage across one complete layer exists. The greatest potential difference, therefore, is found at the far opposite end of the layers.

This voltage distribution applies to continuous 60 Hz signals. When the transformer is first switched on or when a transient overvoltage is impressed upon the device, the voltage distribution from one "hot layer" to the next can increase dramatically, raising

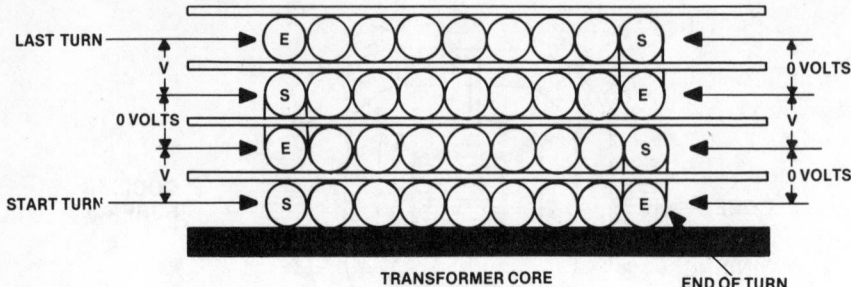

Figure 4.15   Voltage distribution between the layers of a typical layer-wound power transformer.

the possibility of arc-over. This effect is caused by the inductive nature of the transformer windings and the inherent distributed capacitance of the coil. Insulation breakdown may result from one or more of the following:

- Puncture through the insulating material of the device.
- Tracking across the surface of the windings.
- Flashing through the air.

Any of these modes may result in catastrophic failure. Figure 4.16 illustrates the mechanisms involved. A transformer winding can be modeled as a series of inductances and shunt capacitances. The interturn and turn-to-ground capacitances are shown by C_s and C_g, respectively. During normal operation, the applied voltage is distributed evenly across the full winding. However, if a steep front wave is impressed upon the device, the voltage distribution radically changes. For the voltage wave to start distributing itself along the winding, the line-to-ground capacitance (C_g) must be charged. This charging is dependent upon the device winding-to-ground capacitance and the impedance of the supply line.

4.4.3 Mechanical Considerations

Current flow through the windings of a transformer applies stress to the coils. The individual turns in any one coil tend to be crushed together when current flows through

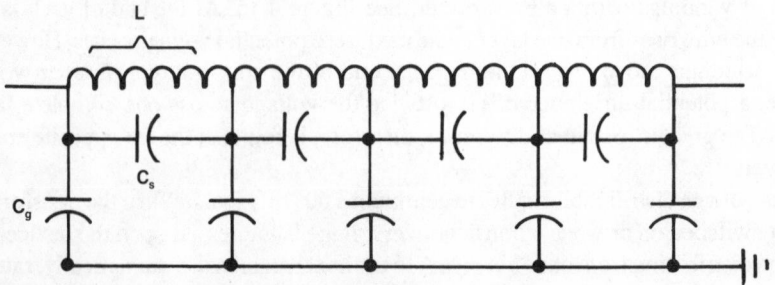

Figure 4.16   Equivalent circuit of a transformer winding.

them. There also may be large repulsion forces between the primary and secondary windings. These mechanical forces are proportional to the square of the instantaneous current; they are, therefore, vibratory in nature under normal operating conditions. These forces, if not controlled, can lead to failure of the transformer through insulation breakdown. Vibration over a sufficient period of time can wear the insulation off adjacent conductors and create a short circuit. To prevent this failure mode, power transformers routinely are coated or dipped into an insulating varnish to solidify the windings and the core into one element.

4.5 CAPACITOR FAILURE MODES

Experience has shown that capacitor failures are second only to semiconductors and vacuum tubes in components prone to malfunction in electronic equipment. Of all the various types of capacitors used today, it is estimated that electrolytics present the greatest potential for problems to equipment users.

4.5.1 Electrolytic Capacitors

Electrolytic capacitors are popular because they offer a large amount of capacitance in a small physical size. They are widely used as filters in low-voltage power supplies and as coupling devices in audio and RF stages. An aluminum electrolytic capacitor consists of two aluminum foil plates separated by a porous strip of paper (or other material) soaked with a conductive electrolyte solution. Construction of a typical device is illustrated in Figure 4.17. The separating material between the capacitor plates does not form the dielectric, but instead, serves as a spacer to prevent the plates from mechanically short-circuiting. The dielectric consists of a thin layer of aluminum oxide that is electrochemically formed on the positive foil plate. The electrolyte conducts the charge applied to the capacitor from the negative plate, through the paper spacer and into direct contact with the dielectric. This sandwich arrangement of foil-spacer-foil is then rolled up and encapsulated.

Problems with electrolytic capacitors fall into two basic categories: mechanical failure and failure of electrolyte.

4.5.2 Mechanical Failure

Mechanical failures relate to poor bonding of the leads to the outside world, contamination during manufacture, and shock-induced short-circuiting of the aluminum foil plates. Typical failure modes include short circuits caused by foil impurities, manufacturing defects (such as burrs on the foil edges or tab connections), breaks or tears in the foil, and breaks or tears in the separator paper.

Short circuits are the most frequent failure mode during the useful life period of an electrolytic capacitor. Such failures are the result of random breakdown of the dielectric oxide film under normal stress. Proper capacitor design and processing will minimize such failures. Short circuits also can be caused by excessive stress, where voltage, temperature, or ripple conditions exceed specified maximum levels.

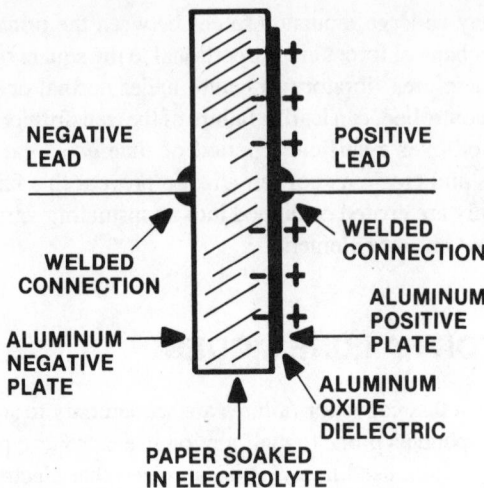

Figure 4.17   The basic design of an aluminum electrolytic capacitor. (Data from: Technical staff, *Sencore News*, Issue 122, Sencore Corporation, Sioux Falls, SD.)

Open circuits, although infrequent during normal life, can be caused by failure of the internal connections joining the capacitor terminals to the aluminum foil. Mechanical connections can develop an oxide film at the contact interface, increasing contact resistance and eventually producing an open circuit. Defective weld connections also can cause open circuits. Excessive mechanical stress will accelerate weld-related failures.

Temperature Cycling.   Like semiconductor components, capacitors are subject to failures induced by thermal cycling. Experience has shown that thermal stress is a major contributor to failure in aluminum electrolytic capacitors. Dimensional changes between plastic and metal materials may result in microscopic ruptures at termination joints, possible electrode oxidation, and unstable device termination (changing series resistance). The highest-quality capacitor will fail if its voltage and/or current ratings are exceeded. Appreciable heat rise (20°C during a 2-hour period of applied sinusoidal voltage) is considered abnormal and may be a sign of incorrect application of the component or impending failure of the device.

Figure 4.18 illustrates the effects of high ambient temperature on capacitor life. Note that operation at 33 percent duty cycle is rated at 10 years when the ambient temperature is 35°C, but the life expectancy drops to just 4 years when the same device is operated at 55°C. A common rule of thumb is this: In the range of + 75°C through the full-rated temperature, stress and failure rate double for each 10°C increase in operating temperature. Conversely, the failure rate is reduced by half for every 10°C decrease in operating temperature.

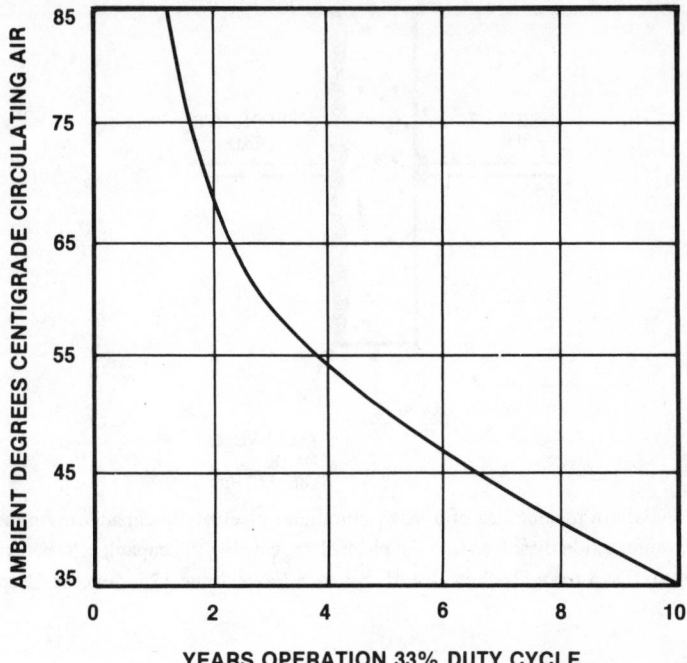

Figure 4.18   Life expectancy of an electrolytic capacitor as a function of operating temperature.

4.5.3 Electrolyte Failures

Failure of the electrolyte can be the result of application of a reverse bias to the component, or of a drying of the electrolyte itself. Electrolyte vapor transmission through the end seals occurs on a continuous basis throughout the useful life of the capacitor. This loss has no appreciable effect on reliability during the useful life period of the product cycle. When the electrolyte loss approaches 40 percent of the initial electrolyte content of the capacitor, however, the electrical parameters deteriorate and the capacitor is considered to be worn out.

As a capacitor dries out, three failure modes may be experienced: leakage, a downward change in value, or *dielectric absorption*. Any one of these may cause a system to operate out of tolerance or fail altogether.

The most severe failure mode for an electrolytic is increased leakage, illustrated in Figure 4.19. Leakage can cause loading of the power supply, or upset the dc bias of an amplifier. Loading of a supply line often causes additional current to flow through the capacitor, possibly resulting in dangerous overheating and catastrophic failure.

A change of device operating value has a less devastating effect on system performance. An aluminum electrolytic has a typical tolerance range of about ±20 percent. A capacitor suffering from drying of the electrolyte can experience a drastic

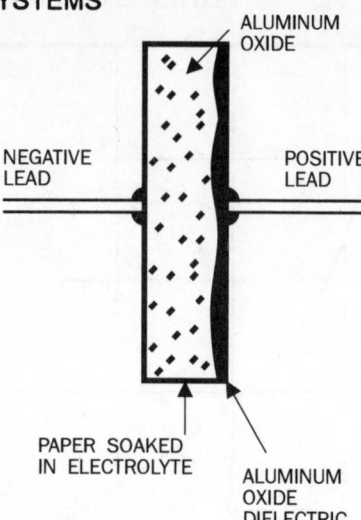

Figure 4.19   Failure mechanism of a leaky aluminum electrolytic capacitor. As the device ages, the aluminum oxide dissolves into the electrolyte, causing the capacitor to become leaky at high voltages. (Data from: Technical staff, *Sencore News*, Issue 122, Sencore Corporation, Sioux Falls, SD.)

drop in value (to just 50 percent of its rated value, or less). The reason for this phenomenon is that after the electrolyte has dried to an appreciable extent, the charge on the negative foil plate has no way of coming in contact with the aluminum oxide dielectric. This failure mode is illustrated in Figure 4.20. Remember, it is the aluminum oxide layer on the positive plate that gives the electrolytic capacitor its large rating. The dried-out paper spacer, in effect, becomes a second dielectric, which significantly reduces the capacitance of the device.

4.5.4 Capacitor Life Span

The life expectancy of a capacitor — operating in an ideal circuit and environment — will vary greatly, depending upon the grade of device selected. Typical operating life, according to capacitor manufacturer data sheets, ranges from a low of 3 to 5 years for inexpensive electrolytic devices, to a high of greater than 10 years for computer-grade products. Catastrophic failures aside, expected life is a function of the rate of electrolyte loss by means of vapor transmission through the end seals, and the operating or storage temperature. Properly matching the capacitor to the application is a key component in extending the life of an electrolytic capacitor. The primary operating parameters include:

- *Rated voltage.* The sum of the dc voltage and peak ac voltage that can be applied continuously to the capacitor. Derating of the applied voltage will decrease the failure rate of the device.

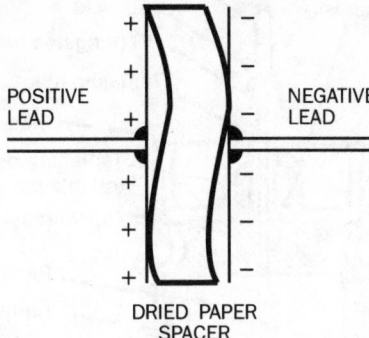

POSITIVE
LEAD

NEGATIVE
LEAD

DRIED PAPER
SPACER

Figure 4.20 Failure mechanism of an electrolytic capacitor exhibiting a loss of capacitance. After the electrolyte dries, electrons can no longer come in contact with the aluminum oxide. The result is a decrease in capacitor value. (Data from: Technical staff, *Sencore News*, Issue 122, Sencore Corporation, Sioux Falls, SD.)

- *Ripple current.* The rms value of the maximum allowable ac current, specified by product type at 120 Hz and + 85°C (unless otherwise noted). The ripple current may be increased when the component is operated at higher frequencies or lower ambient temperatures.
- *Reverse voltage.* The maximum voltage that may be applied to an electrolytic without damage. Electrolytic capacitors are polarized, and must be used accordingly.

4.5.5 Tantalum Capacitors

Tantalum electrolytic capacitors have become the preferred type of device where high reliability and long service life are primary considerations. The *tantalum pentoxide* compound possesses high dielectric strength and a high dielectric constant. As the components are being manufactured, a film of tantalum pentoxide is applied to the electrodes by means of an electrolytic process. The film is applied in various thicknesses. Figure 4.21 shows the internal construction of a typical tantalum capacitor. Because of the superior properties of tantalum pentoxide, tantalum capacitors tend to have as much as 3 times higher capacitance per volume efficiency as an aluminum electrolytic capacitor. This, coupled with the fact that extremely thin films can be deposited during the electrolytic process, makes tantalum capacitors efficient with respect to the number of microfarads per unit volume.

The capacitance of any device is determined by the surface area of the conducting plates, the distance between the plates, and the dielectric constant of the insulating material between the plates. In the tantalum capacitor, the distance between the plates is small; it is just the thickness of the tantalum pentoxide film. Tantalum capacitors contain either liquid or solid electrolytes.

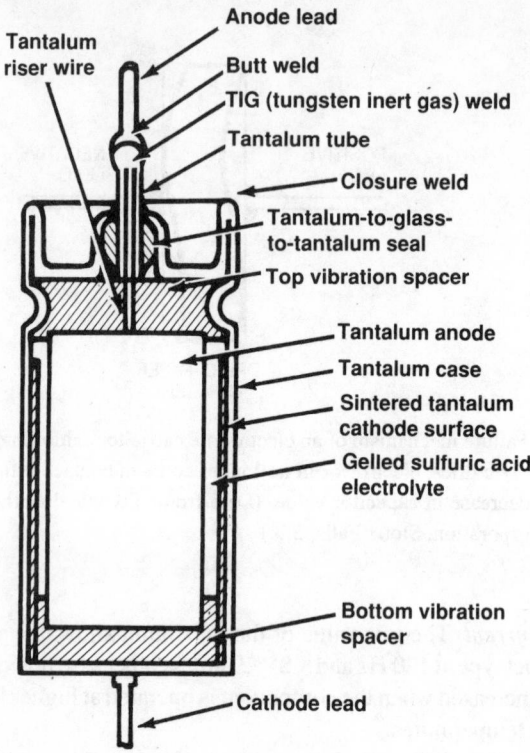

Figure 4.21   Basic construction of a tantalum capacitor.

4.6   FAULT PROTECTORS

Fuses and circuit breakers are the two most common methods used in electronic equipment to prevent system damage in the event of a component failure. Although it is hardly new technology, there are still a lot of misconceptions about fuse and circuit-breaker ratings and operation.

4.6.1   Fuses

Fuses are rated according to the current they can pass safely. This may give the wrong idea — that excessive current will cause a fuse to blow. Actually, there is no amount of current that can cause a fuse to blow. Rather, the cause is power dissipation in the form of heat. Put in more familiar terms, it is the I^2R loss across the fuse element that causes the linkage to melt. The current rating of a given device, however, is not the brick wall protection value that many operators think it is. Consider the graph shown in Figure 4.22, which illustrates the relationship of rated current across a fuse to the blowing time of the device.

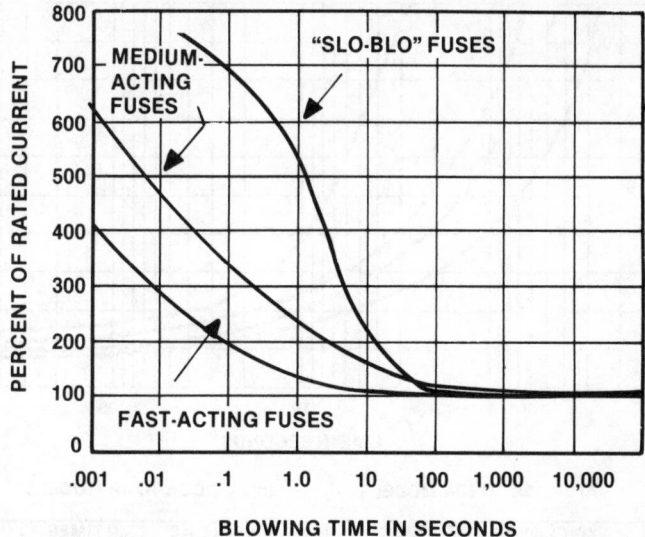

Figure 4.22 The relationship between the rated current of a fuse and its blowing time. Curves are given for three types of devices: fast-acting, medium-acting, and slow-blow.

Fuse characteristics can be divided into three general categories: fast-acting, medium-acting, and slow-blow. Circuit protection for each type of device is a function of both current and time. For example, a slow-blow fuse will allow 6 times the rated current through a circuit for a full second before opening. Such delay characteristics have the benefit of offering protection against nuisance blowing due to high inrush currents during system startup. This feature, however, comes with the price of possible exposure to system damage in the event of a component failure.

4.6.2 Circuit Breakers

Circuit breakers are subject to similar current let-through constraints. Figure 4.23 illustrates device load current as a percentage of breaker rating vs. time. The "A" and "B" curves refer to breaker load capacity product divisions. Note the variations possible in trip time for the two classifications. The minimum clearing time for the "A" group (the higher classification devices) is 1 s for a 400 percent overload. Similar to fuses, these delays are designed to prevent nuisance tripping caused by normally occurring current surges from (primarily) inductive loads. Most circuit breakers are designed to carry 100 percent of their rated load continuously without tripping. They normally are specified to trip at between 101 and 135 percent of rated load after a period of time determined by the manufacturer. In this example, the must-trip point at 135 percent is 1 hour.

Circuit breakers are available in both thermal and magnetic designs. Magnetic protectors offer the benefit of relative immunity to changes in ambient temperature. Typically, a magnetic breaker will operate over a temperature range of − 40°C to

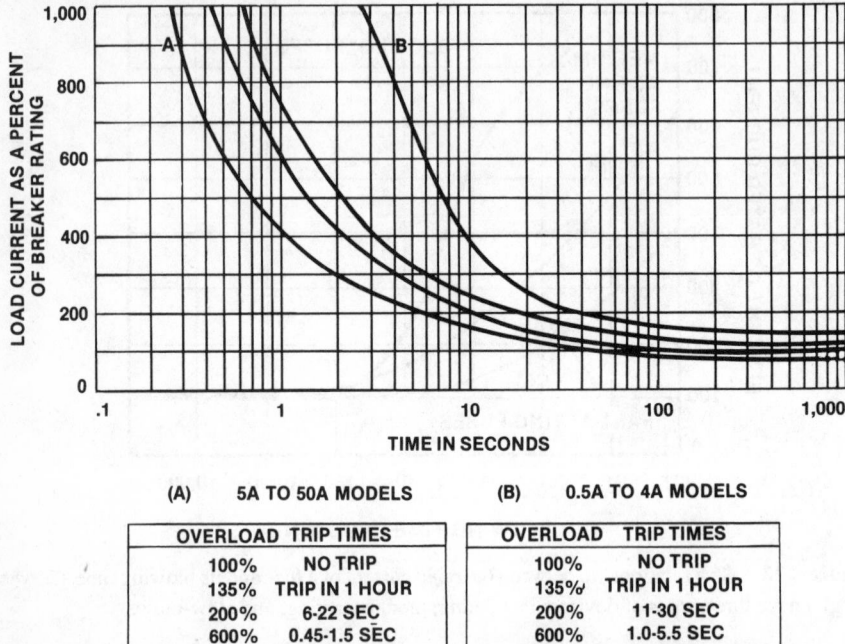

(A) 5A TO 50A MODELS	(B) 0.5A TO 4A MODELS
OVERLOAD TRIP TIMES	**OVERLOAD TRIP TIMES**
100% NO TRIP	100% NO TRIP
135% TRIP IN 1 HOUR	135% TRIP IN 1 HOUR
200% 6-22 SEC	200% 11-30 SEC
600% 0.45-1.5 SEC	600% 1.0-5.5 SEC
1,000% 0.25-0.6 SEC	1,000% 0.4-2.5 SEC

Figure 4.23   The relationship between the rated current of a circuit breaker and its blowing time. Curves (a) and (b) represent different product current ranges, as shown.

+ 85°C without significant variation of the trip point. Time delays usually are provided for magnetic breakers to prevent nuisance tripping caused by startup currents from inductive loads. Trip-time delay ratings range from instantaneous (under 100 ms) to slow (10 to 100 s).

4.6.3   Semiconductor Fuses

The need for a greater level of protection for semiconductor-based systems has led to the development of semiconductor fuses. Figure 4.24 shows the clearing characteristics of a typical fuse of this type. The total clearing time of the device, designed to be less than 8.3 ms, consists of two equal time segments: the *melting time* and the *arcing time*. The rate of current decrease during the arcing time must be low enough that high induced voltages, which could destroy some semiconductor components, are not generated.

4.6.4   Application Considerations

Although fuses and circuit breakers are a key link in preventing equipment damage during the occurrence of a system fault, they are not without some built-in disadvantages. Lead alloy fuses work well and are the most common protection device found,

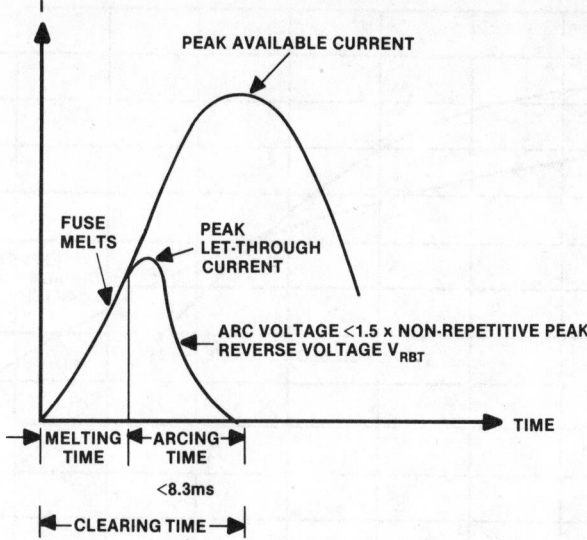

Figure 4.24 The current let-through characteristics of a semiconductor fuse. Note that the clearing time of the device is less than 8.3 ms.

but because the current-interrupting mechanism is dependent on the melting of a metal link, their exact blow point is not constant. The interrupting current may vary, depending on the type and size of fuse clip or holder, conductor size, physical condition of the fuse element, extent of vibration present, and ambient temperature. Figure 4.25 illustrates the effects of ambient temperature on blowing time and current-carrying capacity.

Transient Currents.   Trip delays for fuses and circuit breakers are necessary because of the high startup or inrush currents that occur when inductive loads or tungsten filament lamps are energized. The resistance of a tungsten lamp is high when it is hot, but low when it is cold. A current surge of as much as 15 times the rated steady-state value may occur when the lamp is energized (pulse duration approximately 4 ms).

Transformer inrush currents may measure as high as 30 times the normal rated current for newer types of transformers that have grain-oriented high silicon steel cores. These transformer designs are becoming popular because of their favorable size and weight characteristics. Older transformers of more conventional design typically exhibit an inrush current approximately 18 times greater than the steady-state value. This transient current surge reaches its peak in the first half-cycle of applied ac power and decreases with each successive half-cycle. Such transients are relatively insensitive to the load placed on the secondary. The transient may, in fact, be smaller when the transformer is loaded than unloaded. A worst-case turn-on current surge will not occur every time the transformer is energized; instead, it will occur in a random fashion (perhaps once in every 5 or 10 turn-on events). Among the determining factors involved is the magnitude of the applied voltage at the instant the transformer is

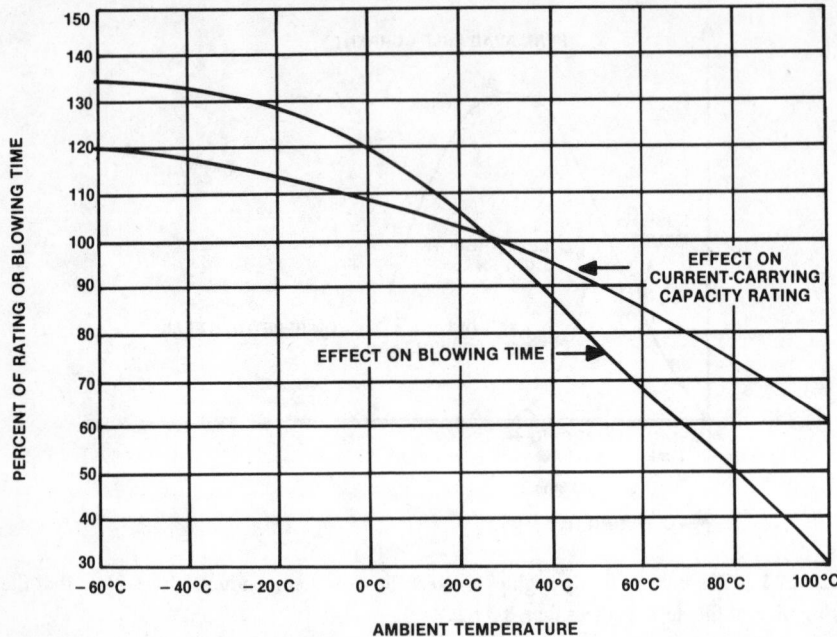

Figure 4.25   The effects of ambient temperature on fuse-blowing time and current-carrying capacity.

connected. Minimum transient energy occurs when the transformer is switched on at the zero-crossing point of the sine wave.

The magnitude of the turn-on current surge of an ac-to-dc power supply is determined mainly by the transformer transient and the capacitive load placed after the rectifier elements. Large filter capacitors commonly found in low-voltage, high-current supplies can place severe stress on the rectifiers and transformer. A fully discharged capacitor appears as a virtual short circuit when power is first applied. Some form of surge-limiting circuit often is provided for power supplies containing more than 10,000 µF total capacitance.

The surge current of an ac motor is spread over tenths of a second, rather than milliseconds, and varies considerably with the load connected to the unit. Table 4.2 lists typical motor surge currents for various types of devices. Note that the single-phase induction motor has the highest surge rating (7 times the running value for a period of 750 ms). Three-phase motors exhibit a relatively low-surge current during startup (350 percent for 167 ms).

Delay-Trip Considerations.   The occurrence of turn-on surges for inductive loads, ac-to-dc power supplies, and tungsten filament lamps calls for the installation of protective devices that exhibit delayed-trip characteristics that match the given load. The problem, however, is that high-surge currents of brief duration — not related to turn-on activities — can occur without tripping the circuit breaker or opening the fuse element. The result may be damage to other circuit devices, such as semiconductors.

Table 4.2   The Starting Surge Current Characteristics for Various Types of AC Motors Selected from a Random Test Group

Motor type	Start current peak ampl. rms	Duration of start surge in sec	Load-sec %I·t sec
Shaded pole	150%	2.0	0.3
Split phase no. 1	600%	0.116	0.7
Split phase no. 2	425%	0.500	2.0
Capacitor (loaded) no. 1	400%	0.600	2.4
Capacitor (no load)	300%	0.100	0.3
Capacitor (loaded) no. 2	420%	0.500	2.1
Induction	700%	0.750	5.0
3 phase	350%	0.167	0.6
Cap. start split-phase run	290%	0.083	0.24

To provide full protection to an electronic system, the overload withstand characteristics of all components should match. This is not always an easy goal to accomplish.

For example, consider a simple SCR-controlled ac-to-dc power supply. The transformer will set the upper limit on surge current presented to the protective device and the SCR (assuming light capacitive loading). If that surge is 18 times the normal steady-state current for a period of 8 ms, then a protective device must be selected that will allow the surge to pass without tripping. An SCR must be selected for the circuit, therefore, that can withstand at least 18 times the normal rated current for 8 ms. If not, the SCR will become the weak link in the system, not the protective device.

4.7 BIBLIOGRAPHY

Anderson, Leonard R.: *Electric Machines and Transformers*, Reston Publishing Company, Reston, VA.

Davis, William: "Selecting Circuit Protective Devices," *Plant Electrical Systems*, Intertec Publishing, Overland Park, KS, March 1980.

Fink, D., and D. Christiansen: *Electronics Engineers' Handbook*, 3rd ed., McGraw-Hill, New York, 1989.

Jordan, Edward C.: *Reference Data for Engineers: Radio, Electronics, Computer, and Communications*, 7th ed., Howard W. Sams Company, Indianapolis, 1985.

Lawrie, Robert: *Electrical Systems for Computer Installations*, McGraw-Hill, New York, 1988.

Lowdon, Eric: *Practical Transformer Design Handbook*, Howard W. Sams Company, Indianapolis, 1980.

Meeldijk, Victor: "Why Do Components Fail?," *Electronic Servicing & Technology* magazine, Intertec Publishing, Overland Park, KS, November 1986.

Nenoff, Lucas: "Effect of EMP Hardening on System R&M Parameters," *Proceedings* of the 1986 Reliability and Maintainability Symposium, IEEE, New York, 1986.

Pearman, Richard: *Power Electronics*, Reston Publishing Company, Reston, VA, 1980.

SCR Applications Handbook, International Rectifier Corporation, El Segundo, CA, 1977.

SCR Manual, 5th ed., General Electric Company, Auburn, NY.

Stephens, Dan: "Surge Protection for Electric Motors," *Technical Briefs* newsletter, Burns & McDonnell, Kansas City, MO, February 1991.

Technical staff, "Aluminum Electrolytic Capacitors: Reliability, Expected Life, and Shelf Capability," Sprague Applications Guide, Sprague Electric Company, Lansing, NC, 1989.

Technical staff, "Introduction to Tantalum Capacitors," Sprague Applications Guide, Sprague Electric Company, Lansing, NC, 1989.

Technical staff, *Sencore News*, Issue 122, Sencore Corp., Sioux Falls, SD.

Whitaker, Jerry: *Radio Frequency Transmission Systems: Design and Operation*, McGraw-Hill, New York, 1990.

———: *Maintaining Electronic Systems*, CRC Press, Boca Raton, FL, 1991.

Wilson, Sam: "What Do You Know About Electronics?," *Electronic Servicing & Technology* magazine, Intertec Publishing, Overland Park, KS, September 1985.

5

POWER-SYSTEM PROTECTION ALTERNATIVES

5.1 INTRODUCTION

Utility companies make a good-faith attempt to deliver clean, well-regulated power to their customers. Most disturbances on the ac line are beyond the control of the utility company. Large load changes imposed by customers on a random basis, PF correction switching, lightning, and accident-related system faults all combine to produce an environment in which tight control over ac power quality is difficult to maintain. Therefore, the responsibility for ensuring ac power quality must rest with the users of sensitive equipment.

The selection of a protection method for a given facility is as much an economic question as it is a technical one. A wide range of power-line conditioning and isolation equipment is available. A logical decision about how to proceed can be made only with accurate, documented data on the types of disturbances typically found on the ac power service to the facility. The protection equipment chosen must be matched to the problems that exist on the line. Using inexpensive basic protectors may not be much better than operating directly from the ac line. Conversely, the use of a sophisticated protector designed to shield the plant from every conceivable power disturbance may not be economically justifiable.

Purchasing transient-suppression equipment is only one element in the selection equation. Consider the costs associated with site preparation, installation, and maintenance. Also consider the operating efficiency of the system. Protection units that are placed in series with the load consume a certain amount of power and, therefore, generate heat. These considerations may not be significant, but they should be taken into account. Prepare a complete life-cycle cost analysis of the protection methods proposed. The study may reveal that the long-term operating expense of one system outweighs the lower purchase price of another.

The amount of money a facility manager is willing to spend on protection from utility company disturbances generally depends on the engineering budget and how much the plant has to lose. Spending $125,000 on systemwide protection for a highly computerized manufacturing center is easily justified. At smaller operations, justification may not be so easy.

5.1.1   The Key Tolerance Envelope

The susceptibility of electronic equipment to failure because of disturbances on the ac power line has been studied by many organizations. The benchmark study was conducted by the Naval Facilities Engineering Command (Washington, DC). The far-reaching program, directed from 1968 to 1978 by Lt. Thomas Key, identified three distinct categories of recurring disturbances on utility company power systems. As shown in Table 5.1, it is not the magnitude of the voltage, but the duration of the disturbance, that determines the classification.

In the study, Key found that most data processing (DP) equipment failure caused by ac line disturbances occurred during bad weather. According to a report on the findings, the incidence of thunderstorms in an area may be used to predict the number of failures. The type of power-transmission system used by the utility company also

Table 5.1   Types of Voltage Disturbances Identified in the Key Report

Parameter	Type 1	Type 2	Type 3
Definition	Transient and oscillatory overvoltage	Momentary undervoltage or overvoltage	Power outage
Causes	Lightning, power network switching, operation of other loads	Power system faults, large load changes, utility company equipment malfunctions	Power system faults, unacceptable load changes, utility equipment malfunctions
Threshold (Note 1)	200–400% of rated rms voltage or higher (peak instantaneous above or below rated rms)	Below 80–85% and above 110% of rated rms voltage	Below 80–85% of rated rms voltage
Duration	Transients 0.5–200 μs wide and oscillatory up to 16.7 ms at frequencies of 200 Hz to 5 kHz and higher	From 4–6 cycles, depending on the type of power system distribution equipment	From 2–60 sec if correction is automatic; from 15 min to 4 hrs if manual

Note 1: The approximate limits beyond which the disturbance is considered to be harmful to the load equipment.

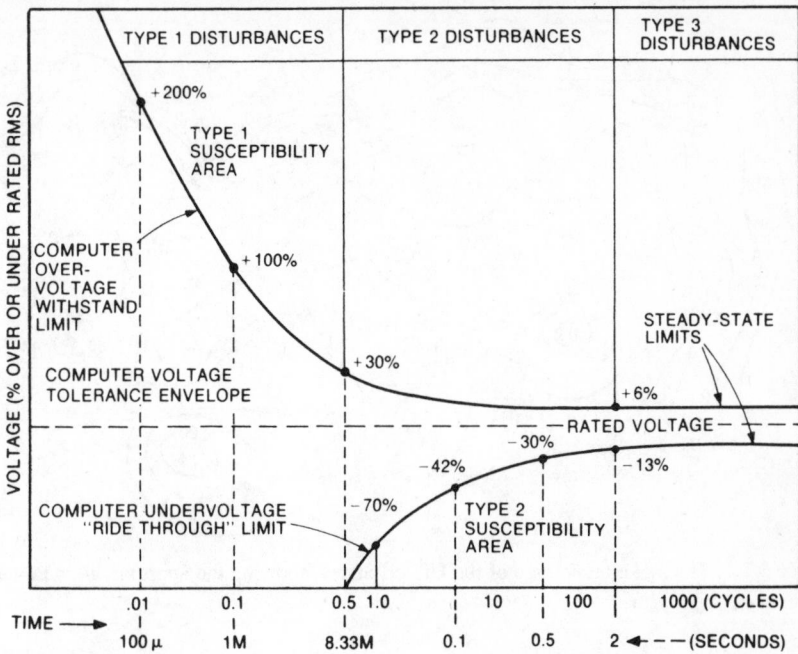

Figure 5.1 The recommended voltage tolerance envelope for computer equipment. This chart is based on pioneering work done by the Naval Facilities Engineering Command. The study identified how the magnitude *and* duration of a transient pulse must be considered in determining the damaging potential of a spike. The design goals illustrated in the chart are recommendations to computer manufacturers for implementation in new equipment. (Adapted from: Lt. Thomas Key, "The Effects of Power Disturbances on Computer Operation," IEEE Industrial and Commercial Power Systems Conference, Cincinnati, June 7, 1978.)

was found to affect the number of disturbances observed on power company lines. For example, an analysis of utility system problems in Washington, DC, Norfolk, VA, and Charleston, SC, demonstrated that underground power-distribution systems experienced one-third fewer failures than overhead lines in the same areas. Based on his research, Key developed the "recommended voltage tolerance envelope" shown in Figure 5.1. The design goals illustrated are recommendations to computer manufacturers for implementation in new equipment.

5.1.2 Assessing the Lightning Hazard

As identified by Key in his Naval Facilities study, the extent of lightning activity in an area significantly affects the probability of equipment failure caused by transient activity. The threat of a lightning flash to a facility is determined, in large part, by the type of installation and its geographic location. The type and character of the lightning flash are also important factors.

The *Keraunic number* of a geographic location describes the likelihood of lightning activity in that area. Figure 5.2 shows the *Isokeraunic map* of the United States, which

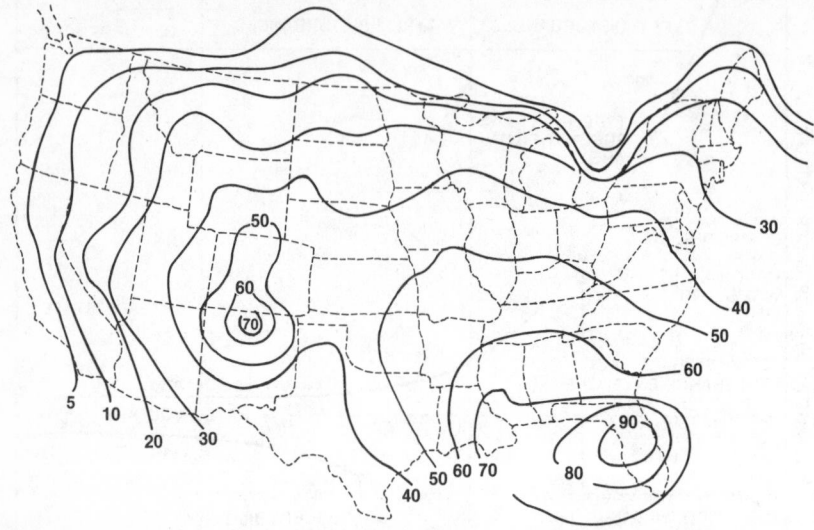

Figure 5.2   The Isokeraunic map of the United States, showing the approximate number of lightning days per year.

estimates the number of lightning days per year across the country. On average, 30 storm days occur per year across the continental United States. This number does not fully describe the lightning threat because many individual lightning flashes occur during a single storm.

The structure of a facility has a significant effect on the exposure of equipment to potential lightning damage. Higher structures tend to collect and even trigger localized lightning flashes. Because storm clouds tend to travel at specific heights above the earth, conductive structures in mountainous areas more readily attract lightning activity. The *plant exposure factor* is a function of the size of the facility and the Isokeraunic rating of the area. The larger the physical size of an installation, the more likely it is to be hit by lightning during a storm. The longer a transmission line (ac or RF), the more lightning flashes it is likely to receive.

The relative frequency of power problems is seasonal in nature. As shown in Figure 5.3, most problems are noted during June, July, and August. These high-problem rates can be traced primarily to increased thunderstorm activity.

5.1.3   Transient Protection Alternatives

A facility can be protected from transient disturbances in two basic ways: the *systems* approach or the *discrete device* approach. Table 5.2 outlines the major alternatives available:

- UPS (uninterruptible power system) and standby generator
- UPS stand-alone system

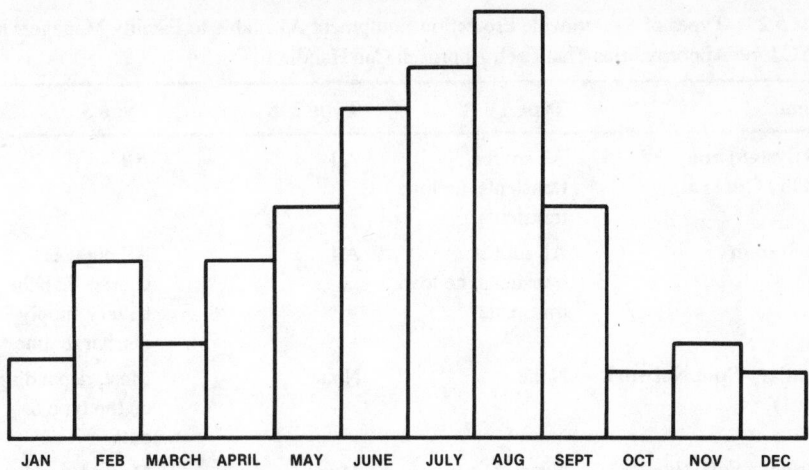

Figure 5.3 The relative frequency of power problems in the United States, classified by month.

- Secondary ac spot network
- Secondary selective ac network
- Motor-generator set
- Shielded isolation transformer
- Suppressors, filters, and lightning arresters
- Solid-state line-voltage regulator

Table 5.3 lists the relative benefits of each protection method. Because each installation is unique, conduct a thorough investigation of facility needs before purchasing equipment. The systems approach offers the advantages of protection engineered to a particular application and need, and (usually) high-level factory support during equipment design and installation. The systems approach also means higher costs for the end-user.

Specifying System-Protection Hardware. Developing specifications for system-wide power-conditioning/backup hardware requires careful analysis of various factors before a particular technology or a specific vendor is selected. Key factors in this process relate to the load hardware and load application. The electrical power required by a sensitive load may vary widely, depending on the configuration of the system. The principal factors that apply to system specification include the following:

- Power requirements, including: voltage, current, power factor, harmonic content, and transformer configuration.
- Voltage-regulation requirements of the load.

Table 5.2   Types of Systemwide Protection Equipment Available to Facility Managers and the AC Line Abnormalities That Each Approach Can Handle

System	Type 1	Type 2	Type 3
UPS System and Standby Generator	All source transients; no load transients	All	All
UPS System	All source transients; no load transients	All	All outages shorter than the battery supply discharge time
Secondary Spot Network (Note 1)	None	None	Most, depending on the type of outage
Secondary Selective Network (Note 2)	None	Most	Most, depending on the type of outage
Motor-Generator Set	All source transients; no load transients	Most	Only brown-out conditions
Shielded Isolation Transformer	Most source transients; no load transients	None	None
Suppressors, Filters, Lightning Arresters	Most transients	None	None
Solid-State Line-Voltage Regulator	Most source transients; no load transients	Some, depending on the response time of the system	Only brown-out conditions

Note 1: Dual-power feeder network.
Note 2: Dual-power feeder network using a static (solid-state) transfer switch.

- Frequency stability required by the load, and the maximum permissible *slew rate* (the rate of change of frequency per second).
- Effects of unbalanced loading.
- Overload and inrush current capacity.
- Bypass capability.
- Primary/standby path transfer time.
- Maximum standby power reserve time.
- System reliability and maintainability.
- Operating efficiency.

An accurate definition of *critical applications* will aid in the specification process for a given site. The potential for future expansion also must be considered in all plans.

Table 5.3   Relative Benefits of Systemwide Protection Equipment Installation and Operation

System	Strong Points	Weak Points	Technical Profile
UPS System and Standby Generator	Full protection from power outage failures and transient disturbances; ideal for critical DP and life-safety loads	Hardware is expensive and may require special construction; electrically and mechanically complex; noise may be a problem; high annual maintenance costs	Efficiency 80–90%; typical high impedance presented to the load may be a consideration; frequency stability good; harmonic distortion determined by UPS system design
UPS System	Completely eliminates transient disturbances; eliminates surge and sag conditions; provides power outage protection up to the limits of the battery supply; ideal for critical load applications	Hardware is expensive; depending on battery supply requirements, special construction may be required; noise may be a problem; periodic maintenance required	Efficiency 80–90%; typical high impedance presented to the load may be a consideration; frequency stability good; harmonic content determined by inverter type
Secondary Spot Network (Note 1)	Simple; inexpensive when available in a given area; protects against local power interruptions; no maintenance required by user	Not available in all locations; provides no protection from area-wide utility failures; provides no protection against transient disturbances or surge/sag conditions	Virtually no loss, 100% efficient; presents low impedance to the load; no effect on frequency or harmonic content
Secondary Selective Network (Note 2)	Same as above; provides faster transfer from one utility line to the other	Same as above	Same as above

(continued)

Table 5.3   Continued

System	Strong Points	Weak Points	Technical Profile
Motor-Generator Set	Electrically simple; reliable power source; provides up to 0.5 sec power-fail ride-through; completely eliminates transient and surge/sag conditions	Mechanical system requires regular maintenance; noise may be a consideration; hardware is expensive; depending on m-g set design, power-fail ride-through may be less than typically quoted by manufacturer	Efficiency 80–90%; typical high impedance presented to the load may be a consideration; frequency stability may be a consideration, especially during momentary power-fail conditions; low harmonic content
Shielded Isolation Transformer	Electrically simple; provides protection against most types of transients and noise; moderate hardware cost; no maintenance required	Provides no protection from brown-out or outage conditions	No significant loss, essentially 100% efficient; presents low impedance to the load; no effect on frequency stability; usually low harmonic content
Suppressors, Filters, Lightning Arresters	Components inexpensive; units can be staged to provide transient protection exactly where needed in a plant; no periodic maintenance required	No protection from Type 2 or 3 disturbances; transient protection only as good as the installation job	No loss, 100% efficient; some units subject to power-follow conditions; no effect on impedance presented to the load; no effect on frequency or harmonic content
Solid-State Line-Voltage Regulator	Moderate hardware cost; uses a combination of technologies to provide transient suppression and voltage regulation; no periodic maintenance required	No protection against power outage conditions; slow response time may be experienced with some designs	Efficiency 92–98%; most units present low impedance to the load; usually no effect on frequency; harmonic distortion content may be a consideration

Note 1: Dual-power feeder network.
Note 2: Dual-power feeder network using a static (solid-state) transfer switch.

Power requirements can be determined either by measuring the actual installed hardware or by checking the nameplate ratings. Most nameplate ratings include significant safety margins. Moreover, the load normally will include a *diversity factor*; all individual elements of the load will not necessarily be operating at the same time.

Every load has a limited tolerance to noise and harmonic distortion. *Total harmonic distortion* (THD) is a measure of the quality of the waveform applied to the load. It is calculated by taking the geometric sum of the harmonic voltages present in the waveform, and expressing that value as a percentage of the fundamental voltage. Critical DP loads typically can withstand 5 percent THD, where no single harmonic exceeds 3 percent. The power-conditioning system must provide this high-quality output waveform to the load, regardless of the level of noise and/or distortion present at the ac input terminals.

If a power-conditioning/standby system does not operate with high reliability, the results often can be disastrous. In addition to threats to health and safety, there is a danger of lost revenue or inventory, and hardware damage. Reliability must be considered from three different viewpoints:

- Reliability of utility ac power in the area.
- Impact of line-voltage disturbances on DP loads.
- Ability of the protection system to maintain reliable operation when subjected to expected and unexpected external disturbances.

The environment in which the power-conditioning system operates will have a significant effect on reliability. Extremes of temperature, altitude, humidity, and vibration can be encountered in various applications. Extreme conditions can precipitate premature component failure and unexpected system shutdown. Most power-protection equipment is rated for operation from 0°C to 40°C. During a commercial power failure, however, the ambient temperature of the equipment room can easily exceed either value, depending on the exterior temperature. Operating temperature derating typically is required for altitudes in excess of 1000 ft.

Table 5.4 lists key power-quality attributes that should be considered when assessing the need for power-conditioning hardware.

5.2 MOTOR-GENERATOR SET

As the name implies, a motor-generator (m-g) set consists of a motor powered by the ac utility supply that is mechanically tied to a generator, which feeds the load. See Figure 5.4. Transients on the utility line will have no effect on the load when this arrangement is used. Adding a flywheel to the motor-to-generator shaft will protect against brief power dips (up to 1/2 s on many models). Figure 5.5 shows the construction of a typical m-g set. The attributes of an m-g include the following:

- An independently generated source of voltage can be regulated without interaction with line-voltage changes on the power source. Utility line changes of ± 20 percent commonly can be held to within ± 1 percent at the load.

Table 5.4   Power-Quality Attributes for Data Processing Hardware *(Adapted from: Federal Information Processing Standards Publication no. 94*, Guideline on Electrical Power for ADP Installations, *U.S. Department of Commerce, National Bureau of Standards, Washington, DC, 1983)*

Environmental Attribute	Typical Environment	Acceptable Limits for DP Systems	
		Normal	Critical
Line frequency	± 0.1 to ± 3%	± 1%	± 0.3%
Rate of frequency change	0.5 to 20 Hz/s	1.5 Hz/s	0.3 Hz/s
Over and undervoltage	± 5% to + 6, – 13.3%	+ 5% to – 10%	± 3%
Phase imbalance	2 to 10%	5% max	3% max
Tolerance to low power factor	0.85 to 0.6 lagging	0.8 lagging	Less than 0.6 lagging or 0.9 leading
Tolerance to high steady-state peak current	1.3 to 1.6 peak/rms	1.0 to 2.5 peak/rms	Greater than 2.5 peak/rms
Harmonic voltages	0 to 20% total rms	10 to 20% total 5 to 10% largest	5% max total 3% largest
Voltage deviation from sine wave	5 to 50%	5 to 10%	3 to 5%
Voltage modulation	Negligible to 10%	3% max	1% max
Surge/sag conditions	+ 10%, - 15%	+ 20%, - 30%	+ 5%, - 5%
Transient impulses	2 to 3 times nominal peak value (0 to 130% V_s)	Varies; 1.0 to 1.5 kV typical	Varies; 200 to 500 V typical
RFI/EMI normal and common modes	10 V up to 20 kHz, less at high freq.	Varies widely; 3 V typical	Varies widely; 0.3 V typical
Ground currents	0 to 10 A plus impulse noise current	0.001 to 0.5 A or more	0.0035 A or less

- The rotational speed and inertial momentum of the rotating mass represents a substantial amount of stored rotational energy, preventing sudden changes in voltage output when the input is momentarily interrupted.
- The input and output windings are separated electrically, preventing transient disturbances from propagating from the utility company ac line to the load.

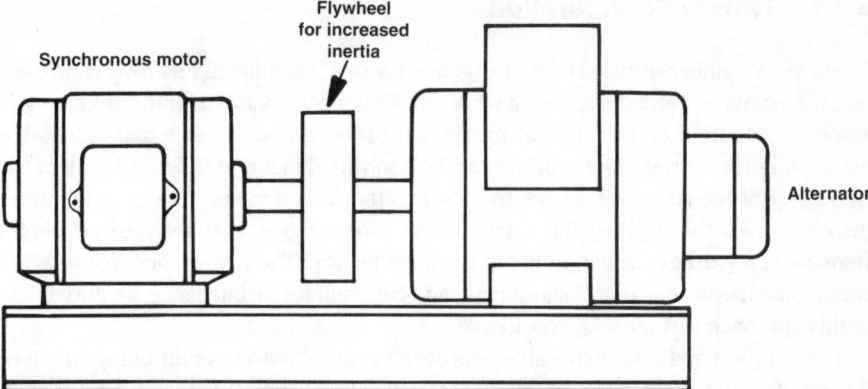

Figure 5.4 Two-machine motor-generator set with an optional flywheel to increase inertia and carry-through capability. (Adapted from documentation provided by Dranetz Technologies, Edison, NJ.)

- Stable electrical characteristics for the load: (1) output voltage and frequency regulation, (2) ideal sine wave output, and (3) true 120° phase shift for three-phase models.
- Reduced problems relating to the power factor presented to the utility company power source.

The efficiency of a typical m-g ranges from 65 to 89 percent, depending on the size of the unit and the load. Motor-generator sets have been used widely to supply 415 Hz power to mainframe computers that require this frequency.

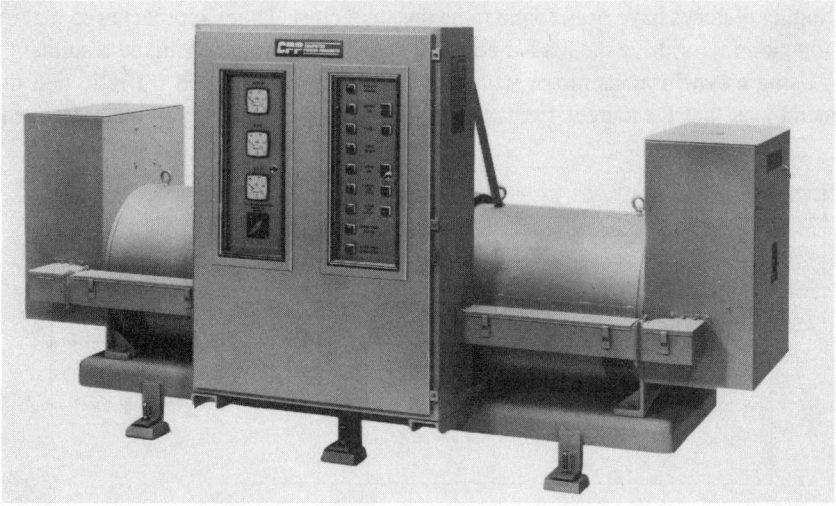

Figure 5.5 Construction of a typical m-g set. (Courtesy of Computer Power Protection)

5.2.1 System Configuration

There are a number of types of motor-generator sets, each having its own characteristics, advantages, and disadvantages. A simplified schematic diagram of an m-g is
shown in Figure 5.6. The type of motor that drives the set is an important design
element. Direct-current motor drives can be controlled in speed independently of the
frequency of the ac power source from which the dc is derived. Use of a dc motor,
thereby, gives the m-g set the capability to produce power at the desired output
frequency, regardless of variations in input frequency. The requirement for rectifier
conversion hardware, control equipment, and commutator maintenance are drawbacks
to this approach that must be considered.

The simplest and least expensive approach to rotary power conditioning involves
the use of an induction motor as the mechanical source. Unfortunately, the rotor of an
induction motor turns slower than the rotating field produced by the power source.
This results in the generator being unable to produce 60 Hz output power if the motor
is operated at 60 Hz and the machines are directly coupled end-to-end at their shafts,
or are belted in a 1:1 ratio. Furthermore, the shaft speed and output frequency of the
generator decreases as the load on the generator is increased. This potential for varying
output frequency may be acceptable where the m-g set is used solely as the input to a
power supply in which the ac is rectified and converted to dc. However, certain loads,
such as some disk memory drives, cannot tolerate frequency changes greater than 1
Hz/s and frequency deviations of more than 0.5 Hz from the nominal 60 Hz value.

Low-slip induction motor-driven generators are available that can produce 59.7 Hz
at full load, assuming 60 Hz input. During power interruptions, the output frequency
will drop further depending upon the length of the interruption. The capability of the
induction motor to restart after a momentary power interruption is valuable. Various
systems of variable-speed belts have been tried successfully. Magnetically controlled
slipping clutches have been found to be unsatisfactory. Other approaches to make the
induction motor drive the load at constant speed have produced mixed results.

Using a synchronous motor with direct coupling or a cogged 1:1 ratio belt drive
guarantees that the output frequency will be equal to the motor input frequency.

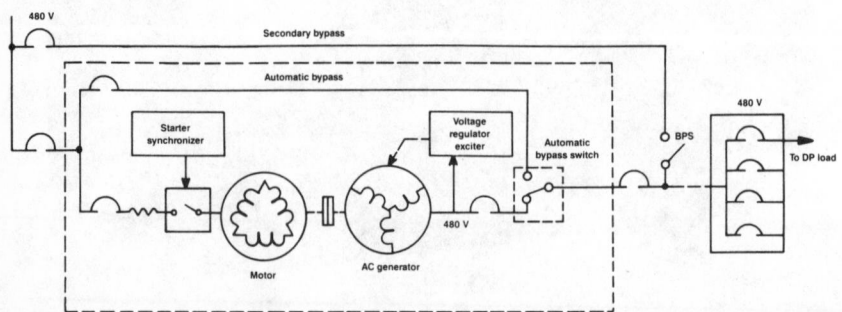

Figure 5.6   Simplified schematic diagram of an m-g set with automatic and secondary bypass
capability. (Adapted from: Federal Information Processing Standards Publication no. 94,
Guideline on Electrical Power for ADP Installations, U.S. Department of Commerce, National
Bureau of Standards, Washington, DC, 1983.)

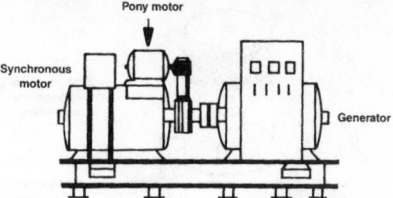

Figure 5.7   Use of a pony motor for an m-g set to aid in system starting and restarting. (Adapted from: Federal Information Processing Standards Publication no. 94, *Guideline on Electrical Power for ADP Installations,* U.S. Department of Commerce, National Bureau of Standards, Washington, DC, 1983.)

Although the synchronous motor is more expensive, it is more efficient and can be adjusted to provide a unity PF load to the ac source. The starting characteristics and the mechanical disturbance following a short line-voltage interruption depends, to a large extent, on motor design. Many synchronous motors that are not required to start under load have weak starting torque, and may use a *pony motor* to aid in starting. This approach is shown in Figure 5.7. Those motors designed to start with a load have starting pole face windings that provide starting torque comparable to that of an induction motor. Such motors can be brought into synchronism while under load with proper selection of the motor and automatic starter system. Typical utility company ac interruptions are a minimum of six cycles (0.1 s). Depending upon the design and size of the flywheel used, the ride-through period can be as much as 0.5 s or more. The generator will continue to produce output power for a longer duration, but the frequency and rate of frequency change will most likely fall outside of the acceptable range of most DP loads after 0.5 s.

If input power is interrupted and does not return before the output voltage and frequency begin to fall outside acceptable limits, the generator output controller can be programmed to disconnect the load. Before this event, a warning signal is sent to the DP control circuitry to warn of impending shutdown and to initiate an orderly interruption of active computer programs. This facilitates easy restart of the computer after the power interruption has passed.

It is important for users to accurately estimate the length of time that the m-g set will continue to deliver acceptable power without input to the motor from the utility company. This data facilitates accurate power-fail shutdown routines. It is also important to ensure that the m-g system can handle the return of power without operating overcurrent protection as a result of high inrush currents that may be required to accelerate and synchronize the motor with the line frequency. Protection against the latter problem requires proper programming of the synchronous motor controller to correctly disconnect and then reconnect the field current supply. It may be worthwhile to delay an impending shutdown for 100 ms or so. This would give the computer time to prepare for the event through an orderly interruption. It also would be useful if the computer were able to resume operation without shutdown, in case utility power returns within the ride-through period. Control signals from the m-g controller should be configured to identify these conditions and events to the DP system.

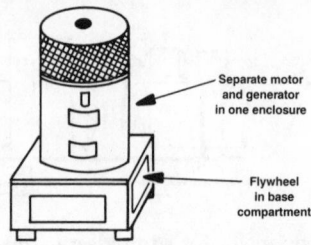

Separate motor
and generator
in one enclosure

Flywheel
in base
compartment

Figure 5.8   Vertical m-g set with an enclosed flywheel. (Adapted from: Federal Information Processing Standards Publication no. 94, *Guideline on Electrical Power for ADP Installations*, U.S. Department of Commerce, National Bureau of Standards, Washington, DC, 1983.)

Generators typically used in m-g sets have substantially higher internal impedance than equivalent kVA-rated transformers. Because of this situation, m-g sets sometimes are supplied with an oversize generator that will be lightly loaded, coupled with a smaller motor that is adequate to drive the actual load. This approach reduces the initial cost of the system, decreases losses in the motor, and provides a lower operating impedance for the load.

Motor-generator sets may be configured for either horizontal installation, as shown previously, or for vertical installation, as illustrated in Figure 5.8.

The most common utility supply voltage used to drive the input of an m-g set is 480 V. The generator output for systems rated at about 75 kVA or less is typically 208 Y/120 V. For larger DP systems, the most economical generator output is typically 480 V. A 480-208 Y/120 V three-phase isolating transformer usually is included to provide 208 Y/120 V power to the computer equipment.

5.2.2   Motor-Design Considerations

Both synchronous and induction motors have been used successfully to drive m-g sets. Each has advantages and disadvantages. The major advantage of the synchronous motor is that while running normally, it is synchronized with the supply frequency. An 1800 rpm motor rotates at exactly 1800 rpm for a supply frequency of exactly 60 Hz. The generator output, therefore, will be exactly 60 Hz. Utility frequencies *average* 60 Hz; utilities vary the frequency slowly to maintain this average value under changing load conditions. Research has shown that utility operating frequencies typically vary from 58.7 to 60.7 Hz. Although frequency tolerances permitted by most computer manufacturers are usually given as ± 0.5 Hz on a nominal 60 Hz system, these utility variations are spread over a 24-hour period or longer, and generally do not result in problems for the load.

The major disadvantage of a synchronous motor is that the device is difficult to start. A synchronous motor must be started and brought up to *pull-in* speed by an auxiliary winding on the armature, known as the *armortisseur* winding. The pull-in speed is the minimum speed (close to synchronous speed) at which the motor will pull into synchronization if excitation is applied to the field. The armortisseur winding is usually a squirrel-cage design, although it may be of the wound-rotor type in some cases. This winding allows the synchronous motor to start and come up to speed as an induction motor. When pull-in speed is achieved, automatic sensing equipment applies field excitation, and the motor locks in and runs as a synchronous machine. As discussed previously, some large synchronous motors are brought up to speed by an auxiliary pony motor.

The armortisseur winding can produce only limited torque, so synchronous motors usually are brought up to speed without a load. This requirement presents no problem for DP systems upon initial startup. However, in the event of a momentary power outage, problems can develop. When the utility ac fails, the synchronous motor must be disconnected from the input immediately, or it will act as a generator and feed power back into the line, thus rapidly depleting its stored (kinetic) rotational energy. During a power failure, the speed of the motor rapidly drops below the pull-in speed, and when the ac supply returns, the armortisseur winding must reaccelerate the motor under load until the field can be applied again. This requires a large winding and a sophisticated control system. While the speed of the m-g set is below synchronous operation, the generator output frequency may be too low for proper computer operation.

The induction motor has no startup problems, but it does have *slip*. To produce torque, the rotor must rotate at slightly lower speed than the stator field. For a nominal 1800 rpm motor, the actual speed will be about 1750 rpm, varying slightly with the load and the applied input voltage. This represents a slip of about 2.8 percent. The generator, if driven directly or on a common shaft, will have an output frequency of about 58.3 Hz. This is below the minimum permissible operating frequency for most computer hardware. Special precision-built low-slip induction motors are available with a slip of approximately 0.5 percent at a nominal motor voltage of 480 V. With 0.5 percent slip, speed at full load will be about 1791 rpm, and the directly driven or common-shaft generator will have an output frequency of 59.7 Hz. This frequency is within tolerance, but close to the minimum permissible frequency.

A belt-and-pulley system adjustable-speed drive is a common solution to this problem. By making the pulley on the motor slightly larger in diameter than the pulley on the generator (with the actual diameters adjustable) the generator can be driven at synchronous speed.

Voltage sags have no effect on the output frequency of a synchronous motor-driven m-g set until the voltage gets so low that the torque is reduced to a point at which the machine pulls out of synchronization. Resynchronization then becomes a problem. On an induction motor, when the voltage sags, slip increases and the machine slows down. The result is a drop in generator output frequency. The adjustable-speed drive between an induction motor and the generator solves the problem for separate machines. If severe voltage sags are anticipated at a site, the system can be set so that nominal input voltage from the utility company produces a frequency of 60.5 Hz, 0.5 Hz on the high side of nominal frequency. Figure 5.9 charts frequency vs. motor voltage for three operating conditions:

- Slip compensation set high (curve *A*)
- Slip compensation set for 60 Hz (curve *B*)
- No slip compensation (curve *C*)

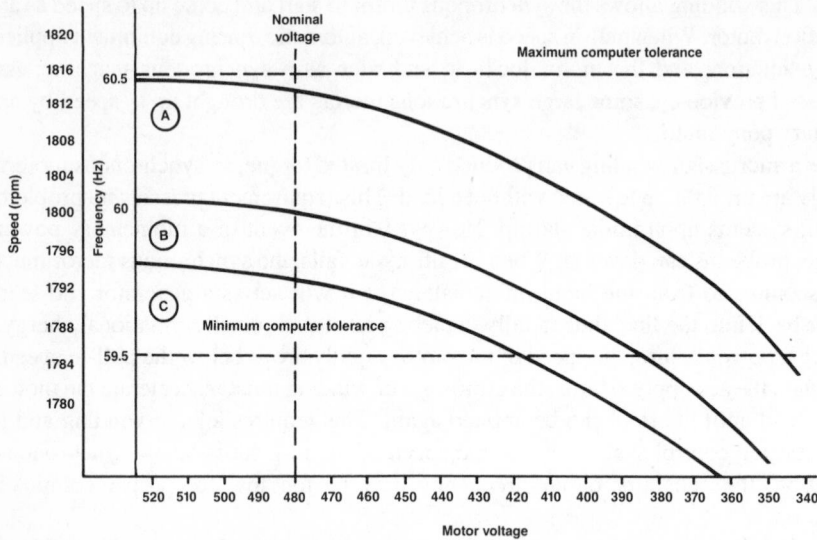

Figure 5.9   Generator output frequency vs. motor input voltage for an induction-motor-based m-g set.

Through proper adjustment of slip compensation, considerable input-voltage margins can be achieved.

Single-Shaft Systems.   There are two basic m-g set machine designs used for DP applications: (1) separate motor-generator systems, and (2) single-shaft, single-housing units. Both designs can use either a synchronous or induction motor. In each case, there are advantages and disadvantages. The separate machine design (discussed previously) uses a motor driving a physically separate generator by means of a coupling shaft or pulley. In an effort to improve efficiency and reduce costs, manufacturers also have produced various types of single-shaft systems.

The basic concept of a single-shaft system is to combine the motor and generator elements into a single unit. A common stator eliminates a number of individual components, making the machine less expensive to produce and mechanically more efficient. The common-stator set substantially reduces mechanical energy losses associated with traditional m-g designs, and it improves system reliability as well. In one design, the stator is constructed so that alternate slots are wound with input and output windings. When it is fed with a three-phase supply, a rotating magnetic field is created, causing the dc-excited rotor to spin at a synchronous speed. By controlling the electrical characteristics of the rotor, control of the output at the secondary stator windings is accomplished.

Common-stator machines offer lower working impedance for the load than a comparable two-machine system. For example, a typical 400 kVA machine has approximately an 800 kVA frame size. The larger frame size yields a relatively low-impedance power source capable of clearing subcircuit fuses under fault conditions. The output of the unit typically can supply up to seven times the full-load current under fault conditions. Despite the increase in frame size, the set is smaller and lighter than comparable systems because of the reduced number of mechanical parts.

5.2.3 Maintenance Considerations

Because m-g sets require some maintenance that necessitates shutdown, most systems provide bypass capability so the maintenance work can be performed without having to take the computer out of service. If the automatic bypass contactor, solid-state switch, and control hardware are in the same cabinet as other devices that also need to be deenergized for maintenance, a secondary bypass is recommended. After the automatic bypass path has been established, transfer switching to the secondary bypass can be enabled, taking the m-g set and its automatic bypass system out of the circuit completely. Some automatic bypass control arrangements are designed to transfer the load of the generator to the bypass route with minimum disturbance. This requires the generator output to be synchronized with the bypass power before closing the switch and opening the generator output breaker. However, with the load taken off the generator, bypass power no longer will be synchronized with it. Consequently, retransfer of the load back to the generator may occur with some disturbance. Adjustment for minimum disturbance in either direction requires a compromise in phase settings, or a means to shift the phase before and after the transfer.

The use of rotating field exciters has eliminated the need for slip rings in most m-g designs. Brush inspection and replacement, therefore, are no longer needed. However, as with any rotating machinery, bearings must be inspected and periodically replaced.

5.2.4 Motor-Generator UPS

Critical DP applications that cannot tolerate even brief ac power interruptions can use the m-g set as the basis for an uninterruptible source of power through the addition of a battery-backed dc motor to the line-driven ac motor shaft. This concept is illustrated in Figure 5.10. The ac motor normally supplies power to drive the system from the utility company line. The shafts of the three devices all are interconnected, as shown in the figure. When ac power is present, the dc motor serves as a generator to charge the battery bank. When line voltage is interrupted, the dc motor is powered by the batteries. Figure 5.11 shows a modified version of this basic m-g UPS using only a dc motor as the mechanical power source. This configuration eliminates the inefficiency involved in having two motors in the system. Power from the utility source is rectified to provide energy for the dc motor, plus power for charging the batteries. A complex control system to switch the ac motor off and the dc motor on in the event of a utility power failure is not needed in this design.

The m-g UPS also can be built around a synchronous ac motor, as illustrated in Figure 5.12. Utility ac energy is rectified and used to drive an inverter, which provides a regulated frequency source to power the synchronous motor. The output from the dc-to-ac inverter need not be a well-formed sine wave, nor a well-regulated source.

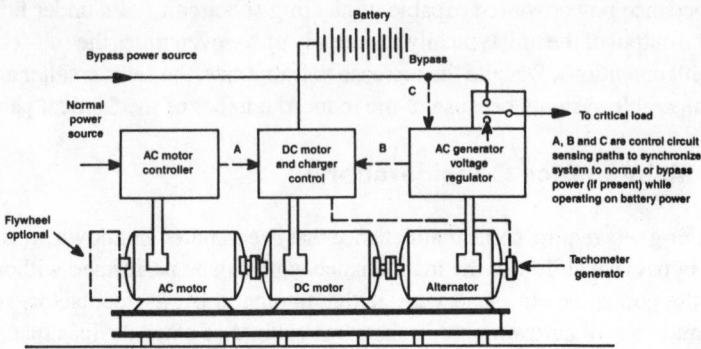

Figure 5.10   Uninterruptible m-g set with ac and dc motor drives. (Adapted from: Federal Information Processing Standards Publication no. 94, *Guideline on Electrical Power for ADP Installations*, U.S. Department of Commerce, National Bureau of Standards, Washington, DC, 1983.)

The output from the generator will provide a well-regulated sine wave for the load. The m-g set also can be operated in a bypass mode that eliminates the rectifier, batteries, and inverter from the current path, operating the synchronous motor directly from the ac line.

An m-g UPS set using a common stator machine is illustrated in Figure 5.13. A feedback control circuit adjusts the firing angle of the inverter to compensate for changes in input power. This concept is taken a step further in the system shown in Figure 5.14. A solid-state inverter bypass switch is added to improve efficiency. During normal operation, the bypass route is enabled, eliminating losses across the rectifier diodes. When the control circuitry senses a drop in utility voltage, the inverter

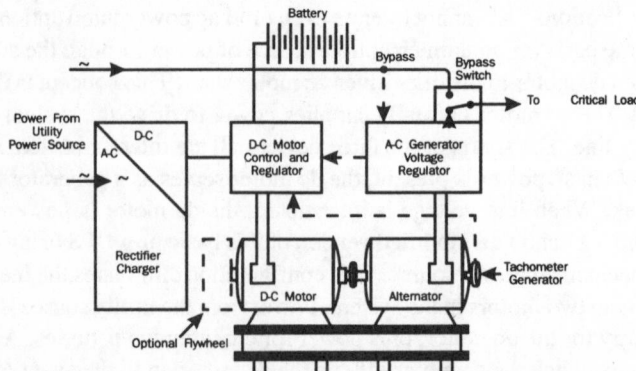

Figure 5.11   Uninterruptible m-g set using a single dc drive motor. (Adapted from: Federal Information Processing Standards Publication no. 94, *Guideline on Electrical Power for ADP Installations*, U.S. Department of Commerce, National Bureau of Standards, Washington, DC, 1983.)

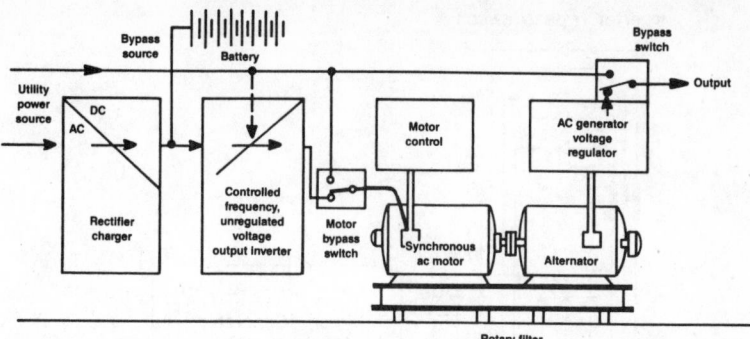

Figure 5.12   Uninterruptible m-g set using a synchronous ac motor. (Adapted from: Federal Information Processing Standards Publication no. 94, *Guideline on Electrical Power for ADP Installations*, U.S. Department of Commerce, National Bureau of Standards, Washington, DC, 1983.)

is switched on, and the bypass switch is deactivated. A simplified installation diagram of the inverter/bypass system is shown in Figure 5.15. Magnetic circuit breakers and chokes are included as shown. An isolation transformer is inserted between the utility input and the rectifier bank. The static inverter is an inherently simple design; commutation is achieved via the windings. Six thyristors are used. Under normal operating conditions, 95 percent of the ac power passes through the static switch; 5 percent passes through the inverter. This arrangement affords maximum efficiency, while keeping the battery bank charged and the rectifiers and inverter thyristors preheated. Preheating extends the life of the components by reducing the extent of thermal cycling that occurs when the load suddenly switches to the battery backup supply. The static switch allows for fast disconnect of the input ac when the utility power fails.

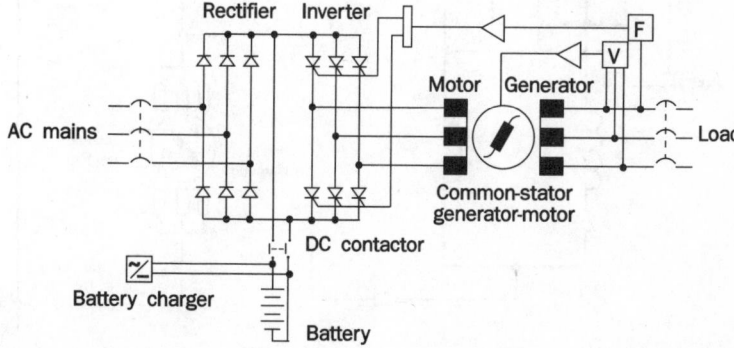

Figure 5.13   Common-stator UPS m-g set. The firing angle of the SCR inverters is determined by a feedback voltage from the generator output.

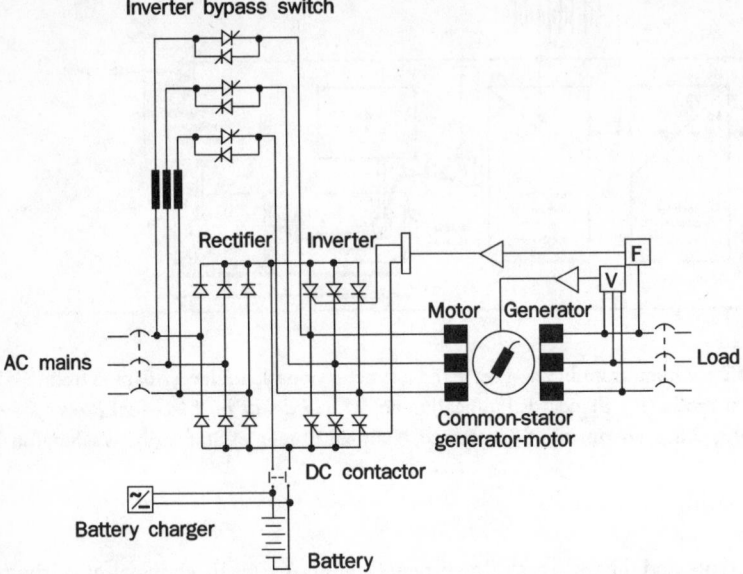

Figure 5.14   Common-stator UPS m-g set with a solid-state inverter bypass switch.

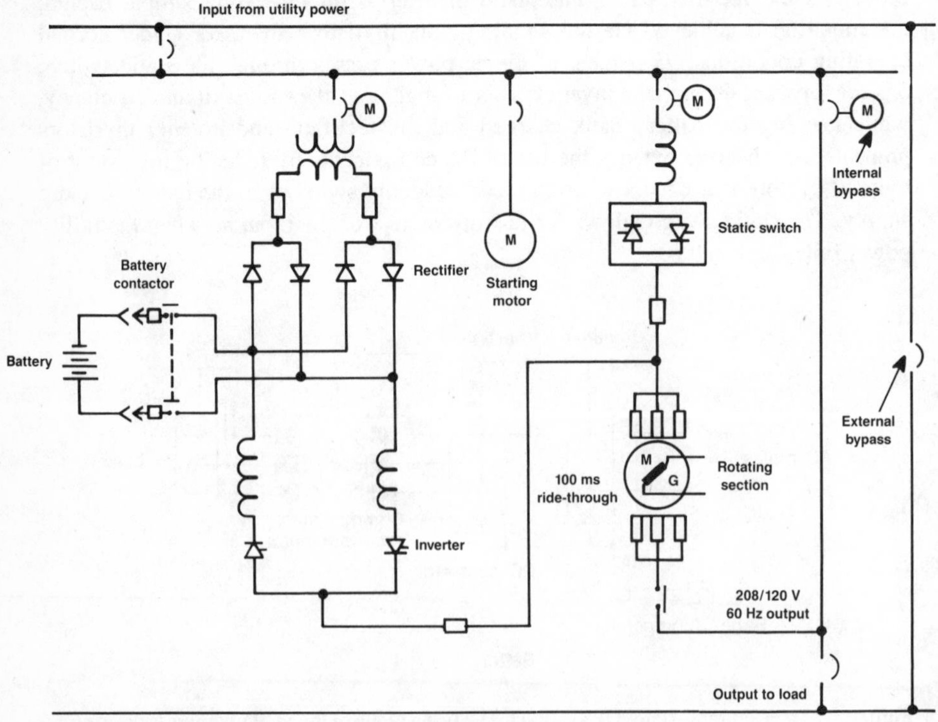

Figure 5.15   Power flow diagram of a full-featured UPS m-g set.

5.3 UNINTERRUPTIBLE POWER SYSTEMS

Uninterruptible power systems have become a virtual necessity for powering large or small computer systems where the application serves a critical need. Computers and data communications systems are no more reliable than the power from which they operate. The difference between UPS and emergency standby power is that the UPS is always in operation. It reverts to an alternative power source, such as the utility company, only if the UPS fails or needs to be deactivated for maintenance. Even then, the transfer of power occurs so quickly (within milliseconds) that it does not interrupt proper operation of the load.

Emergency standby power is normally off and does not start (manually or automatically) until the utility ac feed fails. A diesel generator can be started within 10 to 30 s if the system has been maintained properly. Such an interruption, however, is far too long for DP hardware. Most DP systems cannot ride through more than 8 to 22 ms of power interruption. Systems that can successfully ride through short-duration power breaks, as far as energy continuity is concerned, still may enter a fault condition because of electrical noise created by the disturbance.

UPS hardware is available in a number of different configurations. All systems, however, are variations of two basic designs:

- *Forward-transfer mode.* The load normally is powered by the utility power line, and the inverter is idle. If a commercial power failure occurs, the inverter is started and the load is switched. This configuration is illustrated in Figure 5.16. The primary drawback of this approach is the lack of load protection from power-line disturbances during normal (utility-powered) operation.
- *Reverse-transfer mode.* The load normally is powered by the inverter. In the event of an inverter failure, the load is switched directly to the utility line. This configuration is illustrated in Figure 5.17. The reverse-transfer mode is, by far, the most popular type of UPS system in use.

The type of load-transfer switch used in the UPS system is a critical design parameter. The continuity of ac power service to the load is determined by the type of switching circuit used. An electromechanical transfer switch, shown in Figure 5.18, is limited to switch times of 20 to 50 ms. This time delay may cause sensitive load equipment to malfunction, and perhaps shut down. A control circuit actuates the relay when the

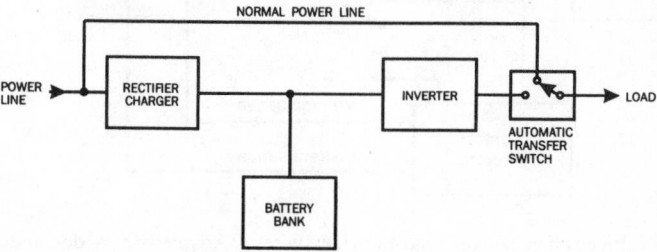

Figure 5.16   Forward-transfer UPS system.

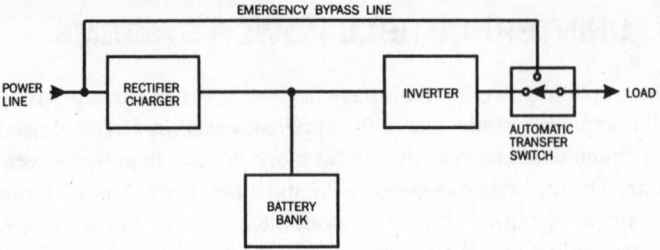

Figure 5.17   Reverse-transfer UPS system.

sensed output voltage falls below a preset value, such as 94 percent of nominal. A static transfer switch, shown in Figure 5.19, can sense a failure and switch the load in about 4 ms. Most loads will ride through this short delay without any malfunction. To accomplish a smooth transition, the inverter output must be synchronized with the power line.

5.3.1   UPS Configuration

The basic uninterruptible power system is built around a battery-driven inverter, with the batteries recharged by the utility ac line. As shown in Figure 5.20, ac from the utility feed is rectified and applied to recharge or *float* a bank of batteries. This dc power drives a single- or multiphase closed-loop inverter, which regulates output voltage and frequency. The output of the inverter is generally a sine wave, or pseudo

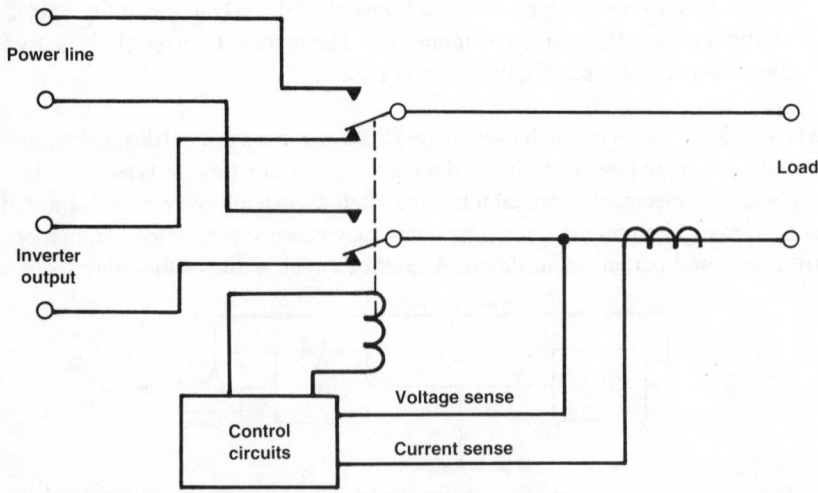

Figure 5.18   Electromechanical load-transfer switch. (Adapted from documentation provided by Dranetz Technologies.)

POWER-SYSTEM PROTECTION ALTERNATIVES 155

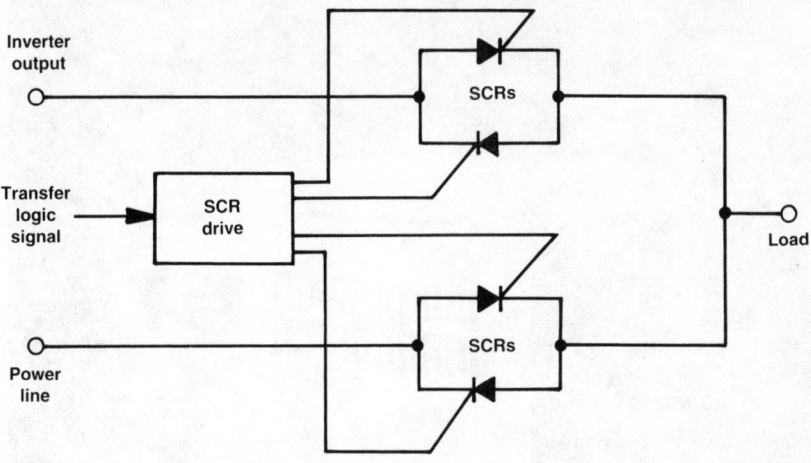

Figure 5.19 Static load-transfer switch. (Adapted from documentation provided by Dranetz Technologies.)

sine wave (a stepped square wave). If the utility voltage drops or disappears, current is drawn from the batteries. When ac power is restored, the batteries are recharged. Many UPS systems incorporate a standby diesel generator that starts as soon as the utility company feed is interrupted. With this arrangement, the batteries are called upon to supply operating current for only 30 s or so, until the generator gets up to speed. A UPS system designed for small loads is shown in Figure 5.21. A larger system, intended to power a computer center, is illustrated in Figure 5.22.

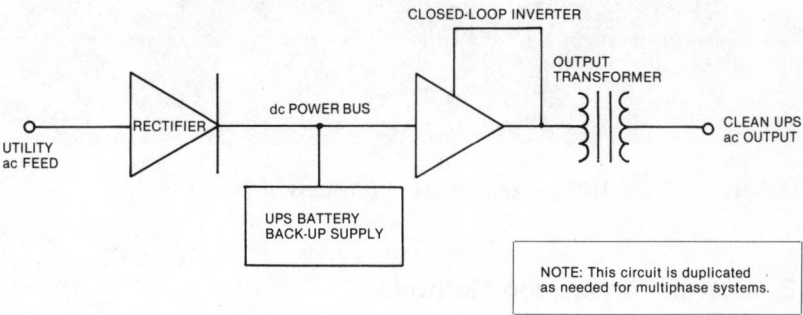

Figure 5.20 Block diagram of an uninterruptible power system using ac rectification to float the battery supply. A closed-loop inverter draws on this supply and delivers clean ac power to the protected load.

Figure 5.21   A 3 kVA UPS power conditioner. (Courtesy of Topaz)

5.3.2 Power-Conversion Methods

Solid-state UPS systems that do not employ rotating machinery utilize one of several basic concepts. The design of an inverter is determined primarily by the operating power level. The most common circuit configurations include:

- Ferroresonant inverter
- Delta magnetic inverter

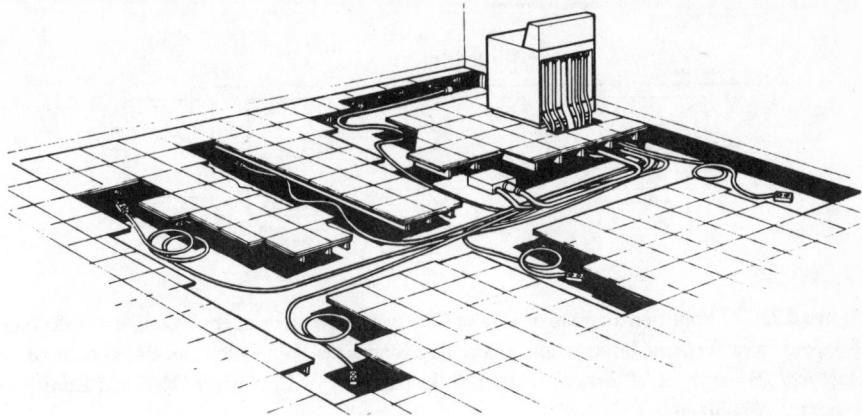

Figure 5.22 Installation details for a computer-room UPS power conditioner.

- Inverter-fed L/C tank
- Quasi-square wave inverter
- Step wave inverter
- Pulse-width modulation inverter
- Phase modulation inverter

Ferroresonant Inverter. Illustrated in Figure 5.23, the ferroresonant inverter is popular for low- to medium-power applications. A ferroresonant transformer can be driven with a distorted, unregulated input voltage and deliver a regulated, sinusoidal output when filtered properly. The ferroresonant transformer core is designed so that the secondary section is magnetically saturated at the desired output voltage. As a result, the output level remains relatively constant over a wide range of input voltages and loads. Capacitors, connected across the secondary, help drive the core into saturation and, in conjunction with inductive coupling, provide harmonic filtering. Regulated, sinusoidal three-phase output voltages are derived from two inverters operating into ferroresonant transformers in a *Scott-T* configuration. The basic ferroresonant inverter circuit, shown in Figure 5.24, consists of an oscillator that controls SCR switches, which feed a ferroresonant transformer and harmonic filter. The saturated operating mode produces a regulated output voltage and inherent current limiting. Efficiency varies from 50 to 83 percent, depending on the load. Response time of the ferroresonant inverter is about 20 ms.

Although ferroresonant inverters are rugged, simple, and reliable, they do have disadvantages. First, such systems tend to be larger and heavier than electronically controlled inverters. Second, there is a phase shift between the inverter square wave and the output sine wave. This phase shift varies with load magnitude and power factor. When an unbalanced load is applied to a three-phase ferroresonant inverter, the output phases can shift from their normal 120° relationships. This results in a change in

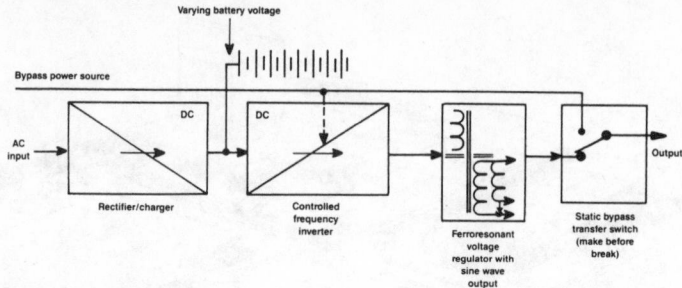

Figure 5.23   Simplified diagram of a static UPS system based on a ferroresonant transformer. (Adapted from: Federal Information Processing Standards Publication no. 94, *Guideline on Electrical Power for ADP Installations*, U.S. Department of Commerce, National Bureau of Standards, Washington, DC, 1983.)

line-to-line voltages, even if individual line-to-neutral voltages are regulated perfectly. This voltage imbalance cannot be tolerated by some loads.

Delta Magnetic Inverter.   While most multiphase inverters are single-phase systems adapted to three-phase operation, the delta magnetic inverter is inherently a three-phase system. A simplified circuit diagram of a delta magnetic inverter is shown in Figure 5.25. Inverter modules A1, B1, and C1 produce square wave outputs that are phase-shifted from each other by 120°. The waveforms are coupled to the primaries of transformer T1 through linear inductors. T1 is a conventional three-phase isolation transformer. The primaries of the device are connected in a delta configuration, reducing the third harmonic and all other harmonics that are odd-order multiples of the third. The secondaries of T1 are connected in a wye configuration to provide a four-wire three-phase output. Inductors L4-L9 form a network connected in a delta configuration to high-voltage taps on the secondary windings of T1. Inductors L4-L6 are single-winding saturating reactors, and L7-L9 are double-winding saturating reactors. Current drawn by this saturating reactor network is nearly sinusoidal and

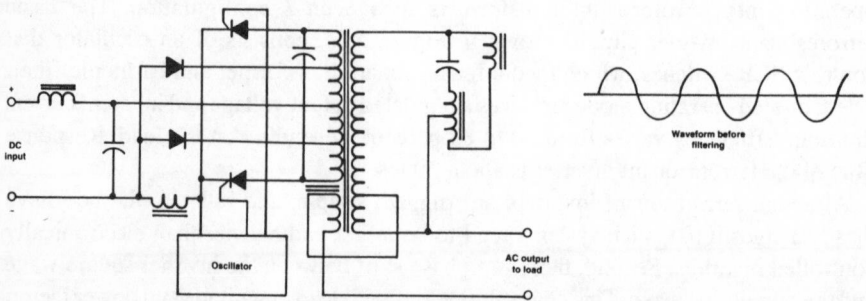

Figure 5.24   Simplified schematic diagram of a ferroresonant inverter. (Adapted from documentation provided by Dranetz Technologies.)

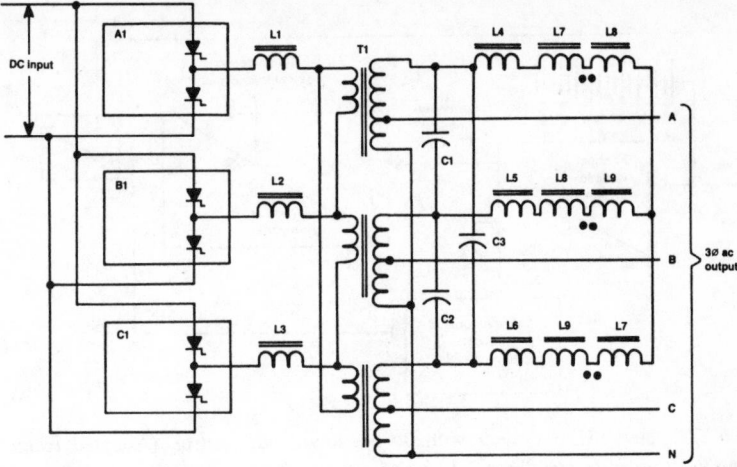

Figure 5.25   Simplified schematic diagram of the delta magnetic regulation system.

varies in magnitude, in a nonlinear manner, with voltage variations. For example, if an increase in load tended to pull the inverter output voltages down, the reduced voltage applied to the reactor network would result in a relatively large decrease in current drawn by the network. This, in turn, would decrease the voltage drop across inductors L1-L3 to keep the inverter output voltage at the proper level. The delta magnetic regulation technique is essentially a three-phase shunt regulator. Capacitors C1-C3 help drive the reactor network into saturation, as well as provide harmonic filtering in conjunction with the primary inductors.

Inverter-Fed L/C Tank.   An inductor/capacitor tank is driven by a dc-to-ac inverter, as illustrated in Figure 5.26. The tank circuit acts to reconstruct the sine wave at the output of the system. Regulation is accomplished by varying the capacitance or

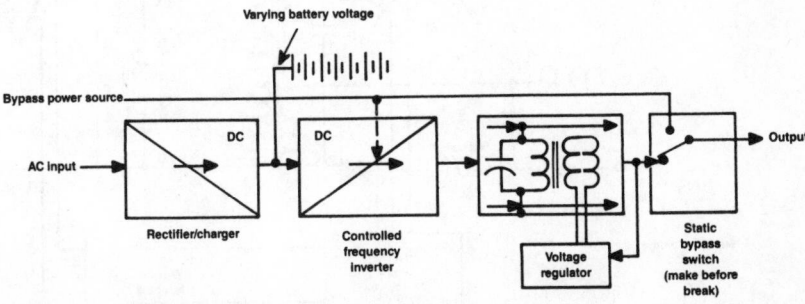

Figure 5.26   Static UPS system using saturable reactor voltage control. (Adapted from: Federal Information Processing Standards Publication no. 94, *Guideline on Electrical Power for ADP Installations*, U.S. Department of Commerce, National Bureau of Standards, Washington, DC, 1983.)

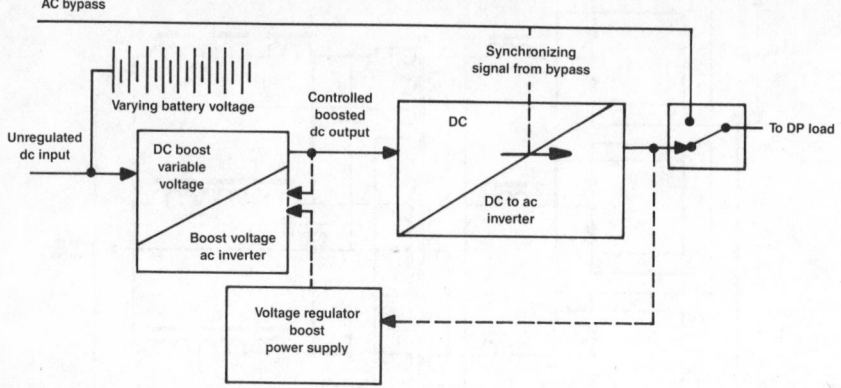

Figure 5.27   Static UPS system with dc boost voltage control. (Adapted from: Federal Information Processing Standards Publication no. 94, *Guideline on Electrical Power for ADP Installations*, U.S. Department of Commerce, National Bureau of Standards, Washington, DC, 1983.)

the inductance to control partial resonance and/or power factor. Some systems of this type use a saturable reactor in which electronic voltage-regulator circuits control the dc current in the reactor. Other systems, shown in Figure 5.27, use a dc-to-variable-dc inverter/converter to control the UPS output through adjustment of the boost voltage. This feature permits compensation for changes in battery level.

Quasi-Square Wave Inverter.   Shown in Figure 5.28, the quasi-square wave inverter produces a variable-duty waveshape that must be filtered by tuned series and parallel inductive-capacitive networks to reduce harmonics and form a sinusoidal output. Because of the filter networks present in this design, the inverter responds slowly to load changes; response time in the range of 150-200 ms is common. Efficiency is about 80 percent. This type of inverter requires voltage-regulating and

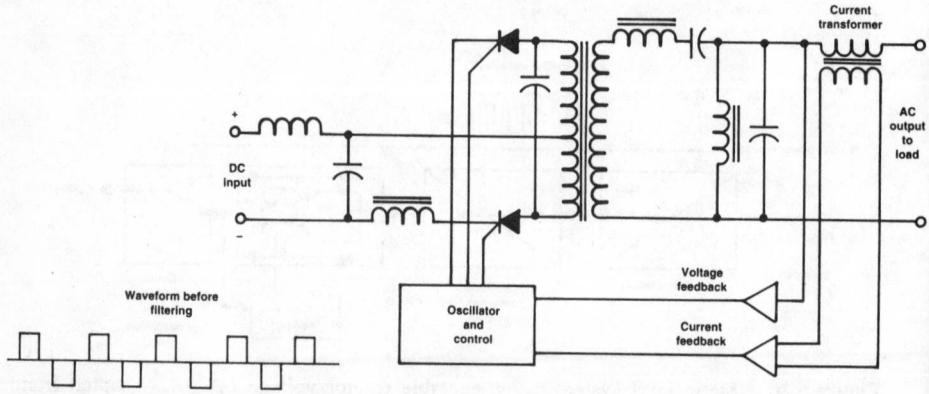

Figure 5.28   Simplified schematic diagram of a quasi-square wave inverter. (Adapted from documentation provided by Dranetz Technologies.)

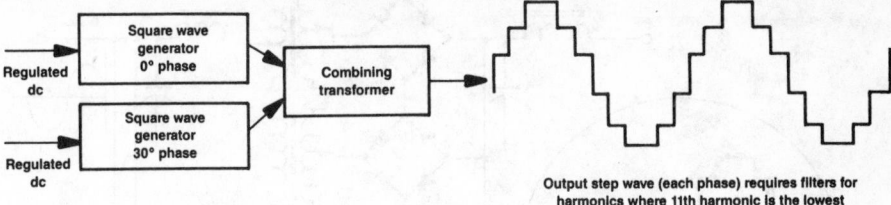

Output step wave (each phase) requires filters for
harmonics where 11th harmonic is the lowest

Figure 5.29   Static UPS system using a stepped wave output. (Adapted from: Federal Information Processing Standards Publication no. 94, *Guideline on Electrical Power for ADP Installations*, U.S. Department of Commerce, National Bureau of Standards, Washington, DC, 1983.)

current-limiting networks, which increase circuit complexity and make it relatively expensive.

Step Wave Inverter.   A multistep inverter drives a combining transformer, which feeds the load. The general concept is illustrated in Figure 5.29. The purity of the output sine wave is a function of the number of discrete steps produced by the inverter. Voltage regulation is achieved by a *boost* dc-to-dc power supply in series with the battery. Figures 5.30 and 5.31 show two different implementations of single-phase units. Each system uses a number of individual inverter circuits, typically three or multiples of three. The inverters are controlled by a master oscillator; their outputs are added in a manner that reduces harmonics, producing a near-sinusoidal output. These types of systems require either a separate voltage regulator on the dc bus or a phase shifter. Response time is about 20 ms. Little waveform filtering is required, and efficiency can be as high as 85 percent. The step wave inverter is complex and expensive. It usually is found only in large three-phase UPS systems.

Pulse-Width Modulation Inverter.   Illustrated in Figure 5.32, the pulse-width modulation (PWM) circuit incorporates two inverters, which regulate the output

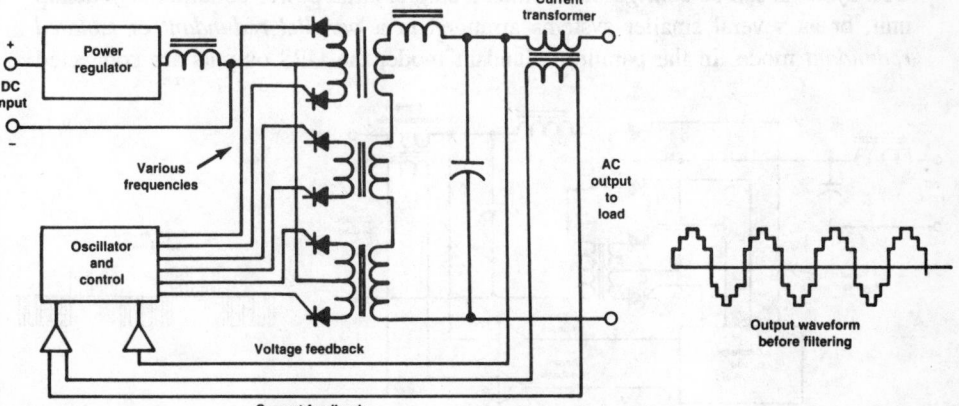

Figure 5.30   Step wave inverter schematic diagram. (Adapted from documentation provided by Dranetz Technologies.)

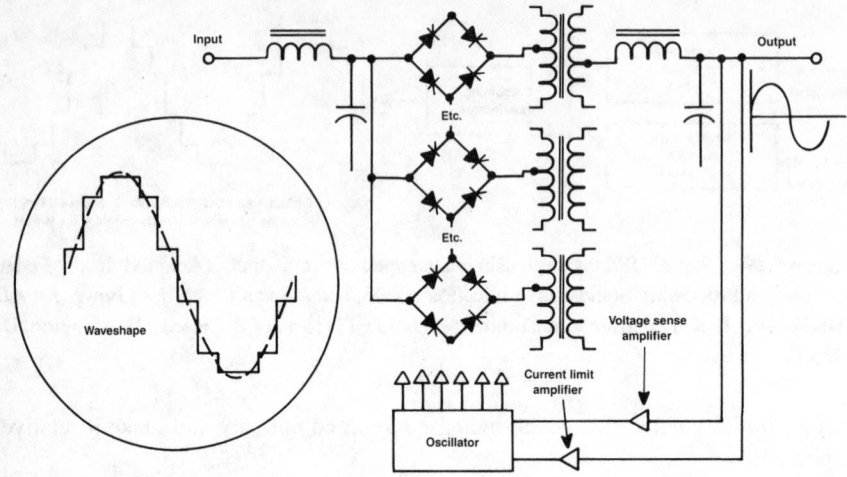

Figure 5.31   Step wave inverter and output waveform.

voltage by varying the pulse width. The output closely resembles a sine wave. Reduced filtering requirements result in good voltage-regulation characteristics. Response times close to 100 ms are typical. The extra inverter and feedback networks make the PWM inverter complex and expensive. Such systems usually are found at power levels greater than 50 kVA.

Phase Modulation Inverter.   Illustrated in Figure 5.33, this system uses dc-to-ac conversion through phase modulation of two square wave high-frequency signals to create an output waveform. The waveform then is filtered to remove the carrier signal and feed the load.

5.3.3 Redundant Operation

UPS systems can be configured as either a single, large power-conditioning/backup unit, or as several smaller systems arranged in a *parallel redundant* or *isolated redundant* mode. In the parallel redundant mode, the UPS outputs are connected

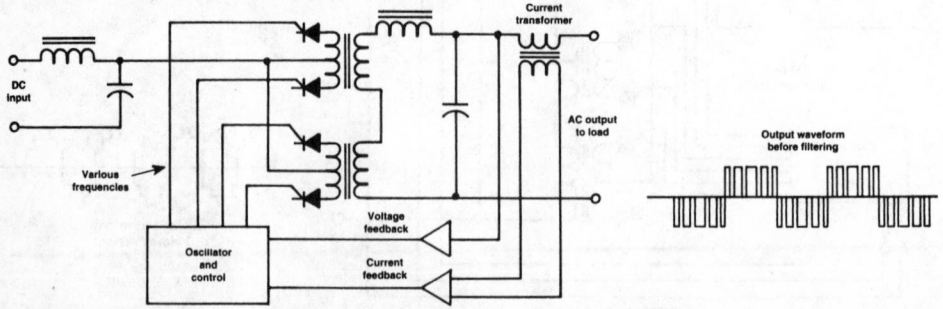

Figure 5.32   Simplified diagram of a pulse-width modulation inverter. (Adapted from documentation provided by Dranetz Technologies.)

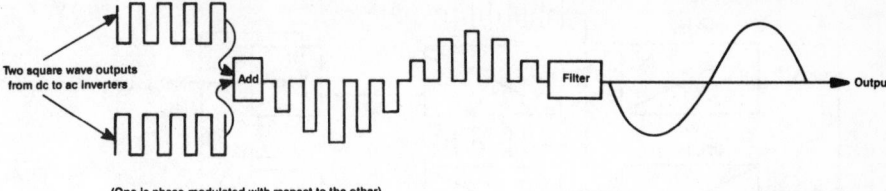

Two square wave outputs from dc to ac inverters

Add

Filter

Output

(One is phase-modulated with respect to the other)

Figure 5.33   Static UPS using a phase-demodulated carrier. (Adapted from: Federal Information Processing Standards Publication no. 94, *Guideline on Electrical Power for ADP Installations*, U.S. Department of Commerce, National Bureau of Standards, Washington, DC, 1983.)

together and share the total load. See Figure 5.34. The power-output ratings of the individual UPS systems are selected to provide for operation of the entire load with any one UPS unit out of commission. In the event of expansion of the DP facility, additional UPS units can be added to carry the load. The parallel system provides the ability to cope with the failure of any single unit.

An isolated redundant system, illustrated in Figure 5.35, divides the load among several UPS units. If one of the active systems fails, a static bypass switch will connect the affected load to a standby UPS system dedicated to that purpose. The isolated redundant system does not permit the unused capacity of one UPS unit to be utilized

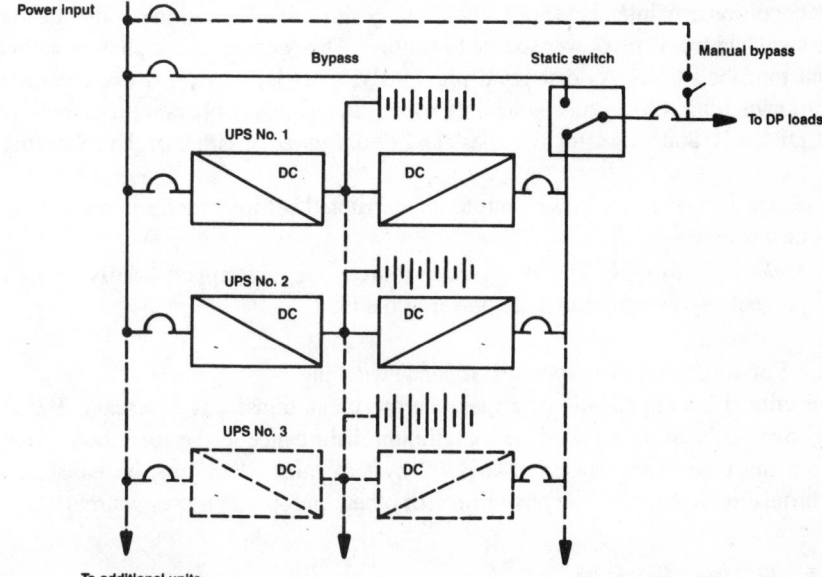

Power Input

Bypass

Static switch

Manual bypass

To DP loads

UPS No. 1

DC

DC

UPS No. 2

DC

DC

UPS No. 3

DC

DC

To additional units

Figure 5.34   Configuration of a parallel redundant UPS system. (Adapted from: Federal Information Processing Standards Publication no. 94, *Guideline on Electrical Power for ADP Installations*, U.S. Department of Commerce, National Bureau of Standards, Washington, DC, 1983.)

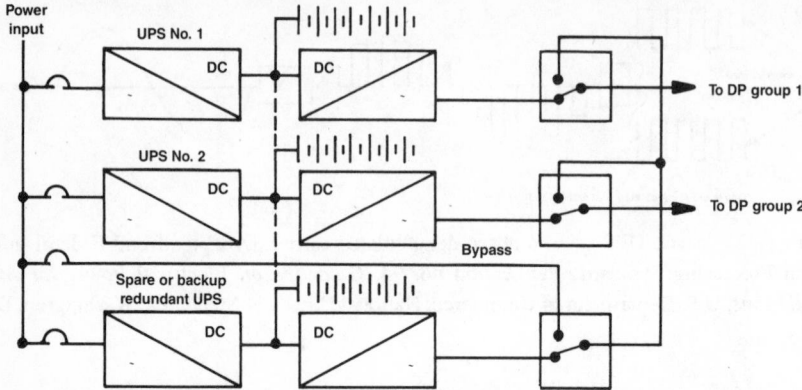

Figure 5.35   Configuration of an isolated redundant UPS system. (Adapted from: Federal Information Processing Standards Publication no. 94, *Guideline on Electrical Power for ADP Installations*, U.S. Department of Commerce, National Bureau of Standards, Washington, DC, 1983.)

on a DP system that is loading another UPS to full capacity. The benefit of the isolated configuration is its immunity to systemwide failures.

5.3.4   Output Transfer Switch

Fault conditions, maintenance operations, and system reconfiguration require the load to be switched from one power source to another. This work is accomplished with an output transfer switch. As discussed previously, most UPS systems use electronic (static) switching. Electromechanical or motor-driven relays operate too slowly for most DP loads. Static transfer switches can be configured as either of the following:

- *Break-before-make*. Power output is interrupted before transfer is made to the new source.
- *Make-before-break*. The two power sources are overlapped briefly so as to prevent any interruption in ac power to the load.

Figure 5.36 illustrates each approach to load switching.

For critical-load applications, a make-before-break transfer is necessary. For the switchover to be accomplished with minimum disturbance to the load, both power sources must be synchronized. The UPS system must, therefore, be capable of synchronizing to the utility ac power line (or other appropriate power source).

5.3.5   Battery Supply

UPS systems typically are supplied with sufficient battery capacity to carry a DP load for periods ranging from 5 min to 1 hour. Long backup time periods usually are

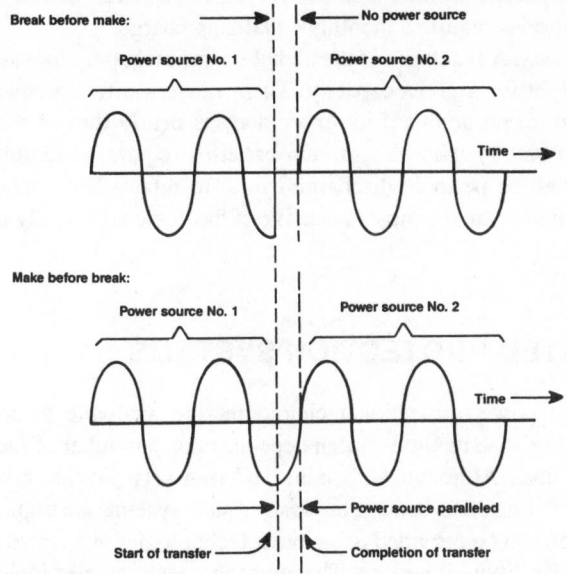

Figure 5.36   Static transfer switch modes.

handled by a standby diesel generator. Batteries require special precautions. For large installations, they almost always are placed in a room dedicated to that purpose. Proper temperature control is important for long life and maximum discharge capacity.

Most rectifier/charger circuits operate in a *constant-current* mode during the initial charge period, and automatically switch to a *constant-voltage* mode near the end of the charge cycle. This provides maximum battery life consistent with rapid recharge. It also prevents excessive battery outgassing and water consumption. The charger provides a *float voltage level* for maintaining the normal battery charge, and sometimes a higher voltage to equalize certain devices.

Four battery types typically are found in UPS systems:

- *Semisealed lead calcium.* A gel-type electrolyte is used that does not require the addition of water. There is no outgassing or corrosion. This type of battery is used when the devices are integral to small UPS units, or when the batteries must be placed in occupied areas. The lifespan of a semisealed lead calcium battery, under ideal conditions, is about 5 years.
- *Conventional lead calcium.* The most common battery type for UPS installations, these units require watering and terminal cleaning about every 6 months. Expected lifetime ranges up to 20 years. Conventional lead-calcium batteries outgas hydrogen under charge conditions and must be located in a secure, ventilated area.
- *Lead-antimony.* The traditional lead-acid batteries, these devices are equivalent in performance to lead-calcium batteries. Maintenance is required every

3 months. Expected lifetime is about 10 years. To retain their capacity, lead-antimony batteries require a monthly equalizing charge.
- *Nickel-cadmium.* Advantages of the nickel-cadmium battery includes small size and low weight for a given capacity. These devices offer excellent high- and low-temperature properties. Life expectancy is nearly that of a conventional lead-calcium battery. Nickel-cadmium batteries require a monthly equalizing charge, as well as periodic discharge cycles to retain their capacity. Nickel-cadmium batteries are the most expensive of the devices typically used for UPS applications.

5.4   DEDICATED PROTECTION SYSTEMS

A wide variety of power-protection technologies are available to solve specific problems at a facility. The method chosen depends upon a number of factors, not the least of which is cost. Although UPS units and m-g sets provide a high level of protection against ac line disturbances, the costs of such systems are high. Many times, adequate protection can be provided using other technologies at a fraction of the cost of a full-featured, facilitywide system. The applicable technologies include:

- Ferroresonant transformer
- Isolation transformer
- Tap-changing regulator
- Line conditioner

5.4.1   Ferroresonant Transformer

Ferroresonant transformers exhibit unique voltage-regulation characteristics that have proved valuable in a wide variety of applications. Voltage output is fixed by the size of the core, which saturates each half cycle, and by the turns ratio of the windings. Voltage output is determined at the time of manufacture and cannot be adjusted. The secondary circuit resonance depends upon capacitors, which work with the saturating inductance of the device to keep the resonance active. A single-phase ferroresonant transformer is shown in Figure 5.37.

Load currents tend to demagnetize the core, so output current is automatically limited. A ferroresonant transformer typically cannot deliver more than 125 to 150 percent of its full-load rated output without going into a current-limiting mode. Such transformers cannot support the normal starting loads of motors without a significant dip in output voltage.

Three-phase versions consisting of single-phase units connected in delta-delta, delta-wye, or wye-wye can be unstable when working into unbalanced loads. Increased stability for three-phase operation can be achieved with zigzag and other special winding configurations.

Shortcomings of the basic ferroresonant transformer have been overcome in advanced designs intended for low-power (2 kVA and below) voltage-regulator appli-

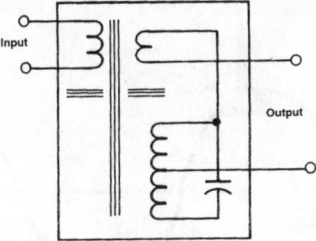

Figure 5.37   Basic design of a ferroresonant transformer. (Adapted from: Federal Information Processing Standards Publication no. 94, *Guideline on Electrical Power for ADP Installations*, U.S. Department of Commerce, National Bureau of Standards, Washington, DC, 1983.)

cations for DP equipment. Figure 5.38 illustrates one of the more common regulator designs. Two additional windings are included:

- *Compensating winding* W_c, which corrects for minor flux changes that occur after saturation has been reached.
- *Neutralizing winding* (W_n), which cancels out most of the harmonic content of the output voltage. Without some form of harmonic reduction, the basic ferroresonant transformer would be unsuitable for sensitive DP loads.

A unique characteristic of the ferroresonant regulator is its ability to reduce normal-mode impulses. Because the regulating capability of the device is based on driving the secondary winding into saturation, transients and noise bursts are clipped, as illustrated in Figure 5.39.

The ferroresonant regulator has a response time of about 25 ms. Because of the tuned circuit at the output, the ferroresonant regulator is sensitive to frequency changes. A 1 percent frequency change will result (typically) in a 1.5 percent change

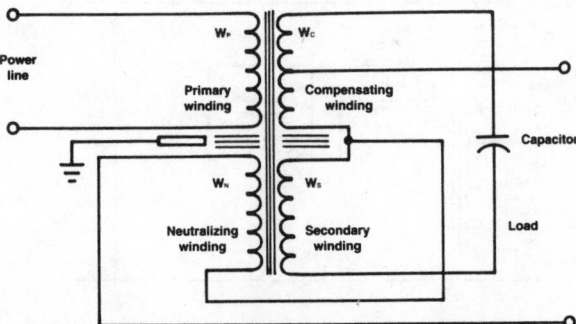

Figure 5.38   Improved ferroresonant transformer design incorporating a compensating winding and a neutralizing winding. (Adapted from documentation provided by Dranetz Technologies.)

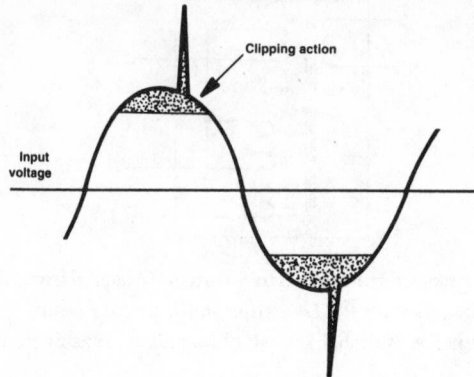

Figure 5.39   Clipping characteristics of a ferroresonant transformer. (Adapted from documentation provided by Dranetz Technologies.)

in output voltage. Efficiency of the device is about 90 percent at full rated load; efficiency declines as the load is reduced. The ferroresonant regulator is sensitive to leading and lagging power factors, and exhibits a relatively high output impedance.

Magnetic-Coupling-Controlled Voltage Regulator.   An electronic magnetic regulator uses dc to adjust the output voltage through magnetic saturation of the core. The direct-current fed winding changes the saturation point of the transformer. This action, in turn, controls the ac flux paths through boost or buck coils to raise or lower the output voltage in response to an electronic circuit that monitors the output of the device. A block diagram of the magnetic-coupling-controlled voltage regulator is shown in Figure 5.40. Changes in output voltage are smooth, although the typical response time of 5 to 10 cycles is too slow to prevent surge and sag conditions from reaching the load.

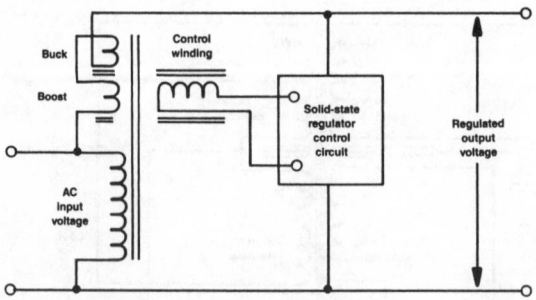

Figure 5.40   Block diagram of an electronic magnetic regulator. (Adapted from: Federal Information Processing Standards Publication no. 94, *Guideline on Electrical Power for ADP Installations*, U.S. Department of Commerce, National Bureau of Standards, Washington, DC, 1983.)

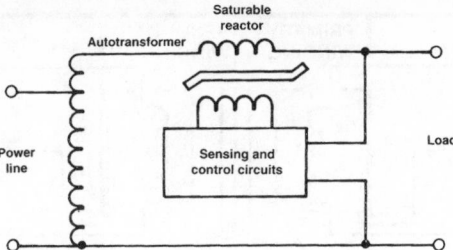

Figure 5.41   Autotransformer/saturable reactor voltage regulator.

Figure 5.41 illustrates an electromechanical version of a saturable regulating transformer. Output voltage is controlled by varying the impedance of the saturable reactor winding in series with a step-up autotransformer. Response to sag and surge conditions is 5 to 10 cycles. Such devices exhibit high output impedance and are sensitive to lagging load power factor.

5.4.2 Isolation Transformer

Transients, as well as noise (RF and low-level spikes) can pass through transformers, not only by way of the magnetic lines of flux between the primary and the secondary, but through resistive and capacitive paths between the windings as well. There are two basic types of noise signals with which transformer designers must cope:

- *Common-mode* noise. Unwanted signals in the form of voltages appearing between the local ground reference and each of the power conductors, including neutral and the equipment ground.
- *Normal-mode* noise. Unwanted signals in the form of voltages appearing in line-to-line and line-to-neutral signals.

Increasing the physical separation of the primary and secondary windings will reduce the resistive and capacitive coupling. However, it also will reduce the inductive coupling and decrease power transfer.

A better solution involves shielding the primary and secondary windings from each other to divert most of the primary noise current to ground. This leaves the inductive coupling basically unchanged. The concept can be carried a step further by placing the primary winding in a shielding box that shunts noise currents to ground and reduces the capacitive coupling between the windings.

One application of this technology is shown in Figure 5.42, in which transformer noise decoupling is taken a step further by placing the primary and secondary windings in their own wrapped foil box shields. The windings are separated physically as much as possible for the particular power rating and are placed between Faraday shields. This gives the transformer high noise attenuation from the primary to the secondary, and from secondary to the primary. Figure 5.43 illustrates the mechanisms involved. Capacitances between the windings, and between the windings and the frame, are

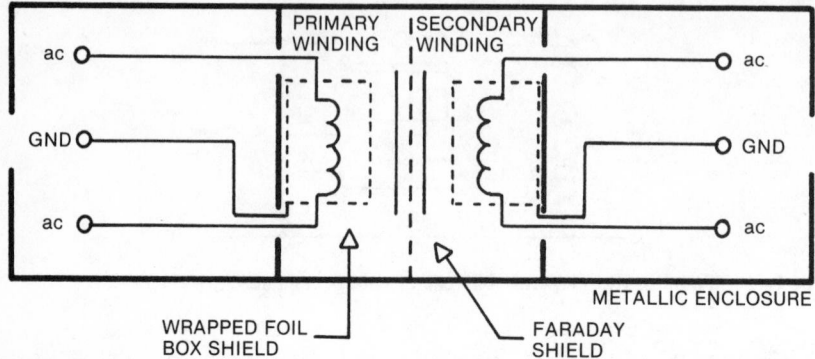

Figure 5.42   The shielding arrangement used in a high-performance isolation transformer. The design goal of this mechanical design is high common-mode and normal-mode noise attenuation.

broken into smaller capacitances and shunted to ground, thus minimizing the overall coupling. The interwinding capacitance of a typical transformer using this technique ranges from 0.001 pF to 0.0005 pF. Common-mode noise attenuation is generally in excess of 100 dB. This high level of attenuation prevents common-mode impulses on the power line from reaching the load. Figure 5.44 illustrates how an isolation transformer combines with the ac-to-dc power supply to prevent normal-mode noise impulses from affecting the load.

A 7.5 kVA noise-suppression isolation transformer is shown in Figure 5.45. High-quality isolation transformers are available in sizes ranging from 125 VA single-phase to 125 kVA (or more) three-phase. Usually, the input winding is tapped at 2.5 percent intervals to provide the rated output voltage, despite high or low average input voltages. The total tap adjustment range is typically from 5 percent above nominal to 10 percent below nominal. For three-phase devices, typical input-voltage nominal ratings are 600, 480, 240, and 208 V line-to-line for 15 kVA and larger.

5.4.3   Tap-Changing Regulator

The concept behind a tap-changing regulator is simple: Adjust the transformer input voltage to compensate for ac line-voltage variations. A tap-changing regulator is shown in Figure 5.46. Although simple in concept, the actual implementation of the system can become complex because of the timing waveforms and pulses necessary to control the SCR banks. Because of the rapid response time of SCRs, voltage adjustment can be made on a cycle-by-cycle basis in response to changes in both the utility input and the load. Tap steps are typically 2 to 3 percent. Such systems create no objectionable switching transients in unity power factor loads. For low-PF loads, however, small but observable transients may be generated in the output voltage at

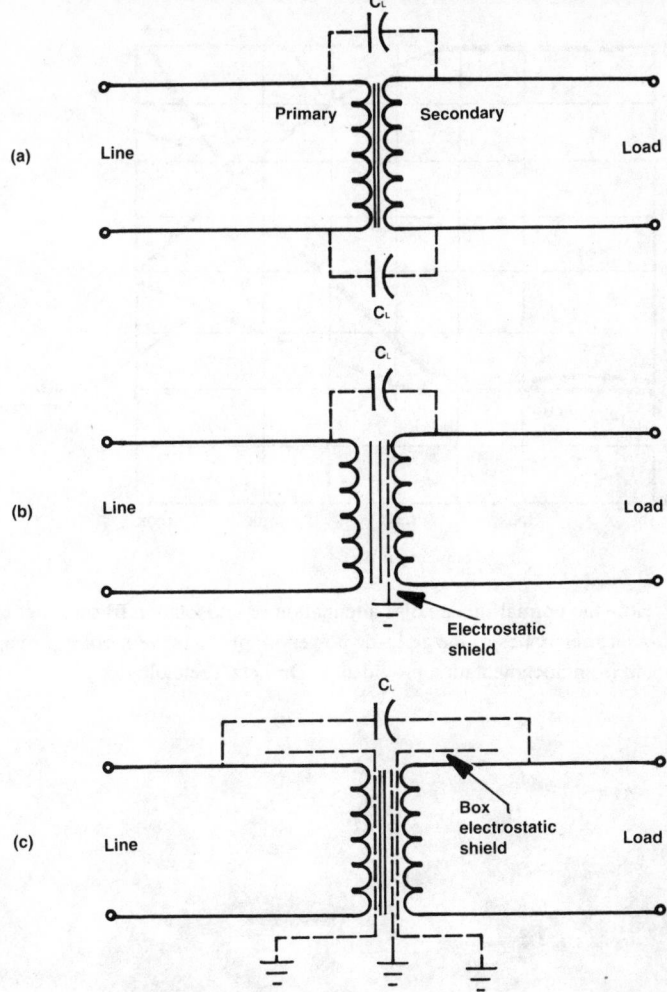

Figure 5.43 The elements involved in a noise-suppression isolation transformer: (a) conventional transformer, with capacitive coupling as shown; (b) the addition of an electrostatic shield between the primary and the secondary; (c) transformer with electrostatic box shields surrounding the primary and secondary windings. (Adapted from documentation provided by Dranetz Technologies.)

the moment the current is switched. This noise usually has no effect on DP loads. Other operating characteristics include:

- Low-internal impedance (similar to an equivalent transformer)
- High efficiency from full load to 25 percent or less

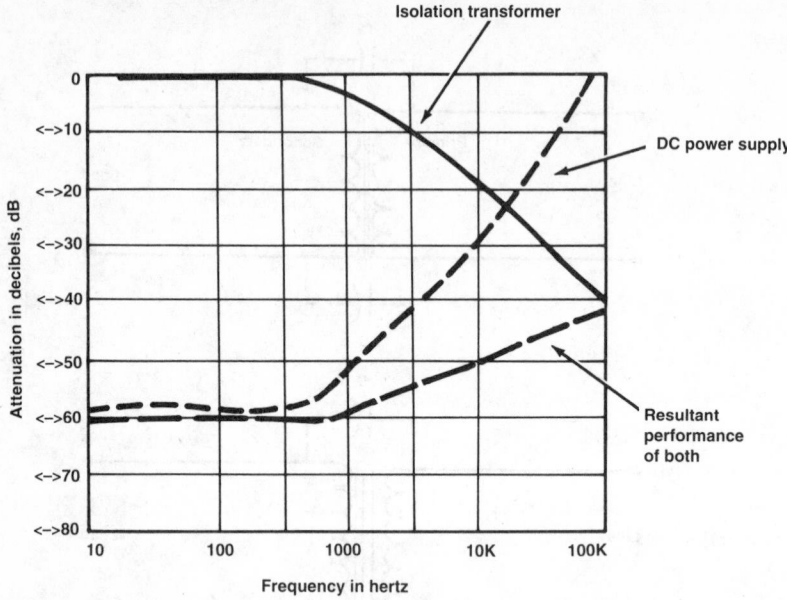

Figure 5.44   How the normal-mode noise attenuation of an isolation transformer combines with the filtering characteristics of the ac-to-dc power supply to prevent noise propagation to the load. (Adapted from documentation provided by Dranetz Technologies.)

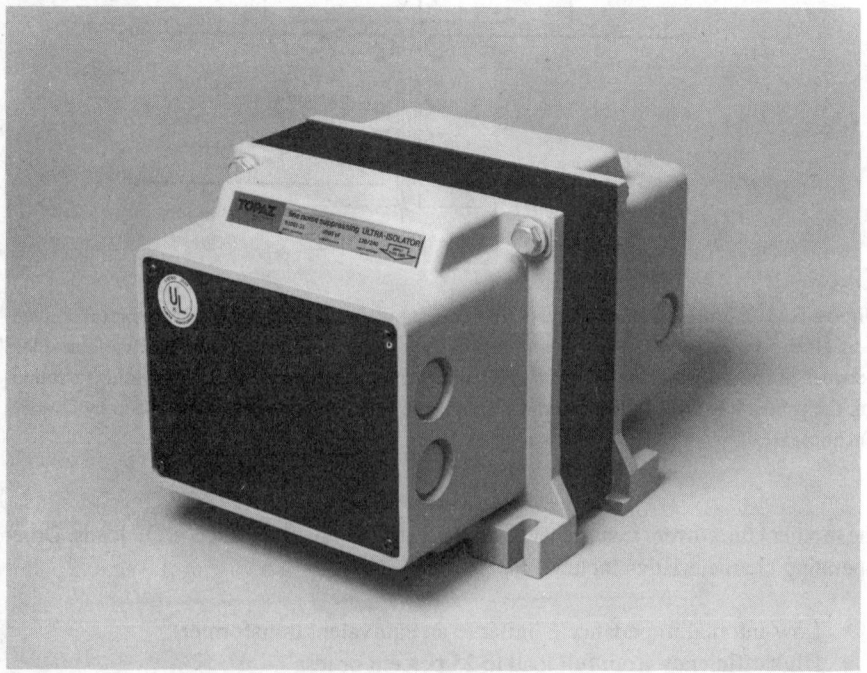

Figure 5.45   A 7.5 kVA noise-suppression isolation transformer. (Courtesy of Topaz)

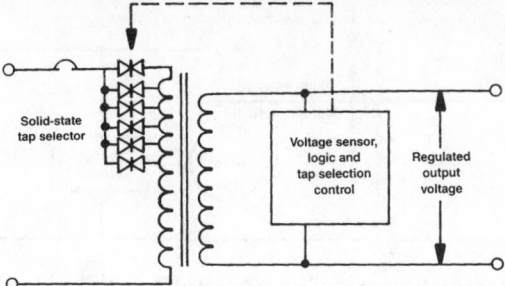

Figure 5.46   Simplified schematic diagram of a tap-changing voltage regulator. (Adapted from: Federal Information Processing Standards Publication no. 94. *Guideline on Electrical Power for ADP Installations*, U.S. Department of Commerce, National Bureau of Standards, Washington, DC, 1983.)

- Rapid response time (typically one to three cycles) to changes in the input ac voltage or load current
- Low acoustic noise level

An autotransformer version of the tap-changing regulator is shown in Figure 5.47.

Variable Ratio Regulator.   Functionally, the variable ratio regulator is a modified version of the tap-changer. Rather than adjusting output voltage in steps, the motor-driven regulator provides a continuously variable range of voltages. The basic concept is shown in Figure 5.48. The system is slow, but generally reliable. It is excellent for keeping DP hardware input voltages within the optimum operating range. Motor-driven regulators usually are able to follow the steady rise and fall of line voltages that typically are experienced on utility company lines. Efficiency is normally good, approaching that of a good transformer. The internal impedance is low, making it

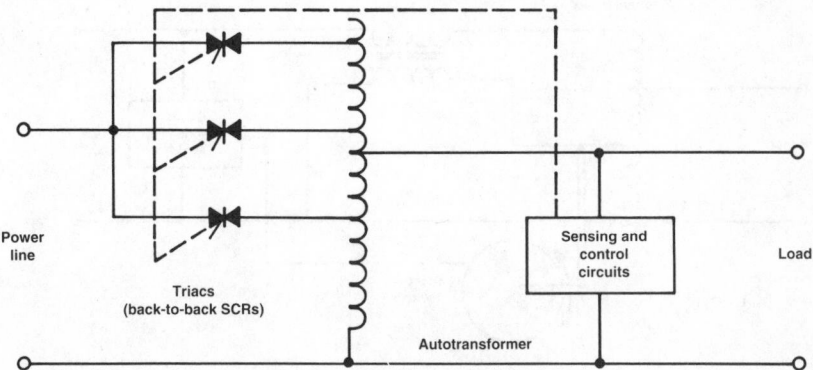

Figure 5.47   Tap-changing voltage regulator using an autotransformer as the power-control element. (Adapted from documentation provided by Dranetz Technologies.)

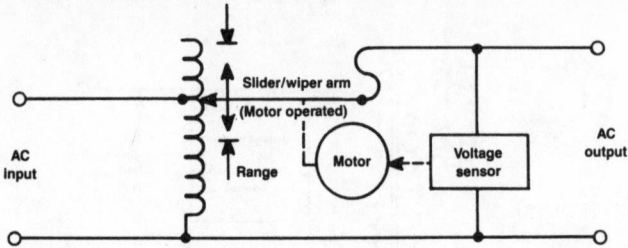

Figure 5.48   Variable ratio voltage regulator. (Adapted from: Federal Information Processing Standards Publication no. 94, *Guideline on Electrical Power for ADP Installations*, U.S. Department of Commerce, National Bureau of Standards, Washington, DC, 1983.)

possible to handle a sudden increase or decrease in load current without excessive undervoltages and/or overvoltages. Primary disadvantages of the variable ratio regulator are limited current ratings, determined by the moving brush assembly, and the need for periodic maintenance.

Variations on the basic design have been marketed, including the system illustrated in Figure 5.49. A motor-driven brush moves across the exposed windings of an autotransformer, causing the series transformer to buck or boost voltage to the load. The correction motor is controlled by a voltage-sensing circuit at the output of the device.

The induction regulator, shown in Figure 5.50, is still another variation on the variable ratio transformer. Rotation of the rotor in one direction or the other varies the magnetic coupling and raises or lowers the output voltage. Like the variable ratio transformer, the induction regulator (or *rototrol*) is slow, but it has no brushes and requires little maintenance. The induction regulator has higher inductive reactance and is slightly less efficient than the variable ratio transformer.

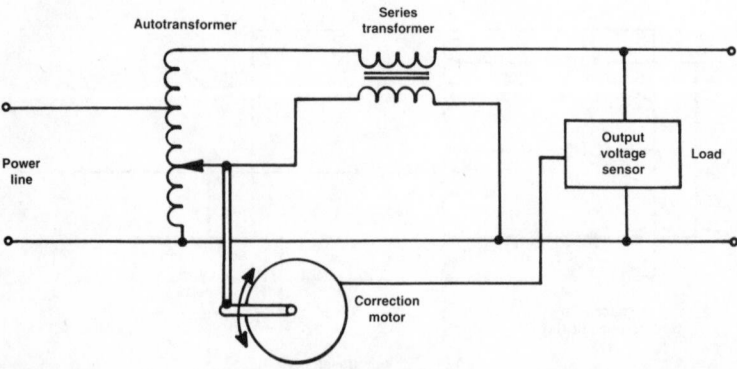

Figure 5.49   Motor-driven line-voltage regulator using an autotransformer with a buck/boost series transformer. (Adapted from documentation provided by Dranetz Technologies.)

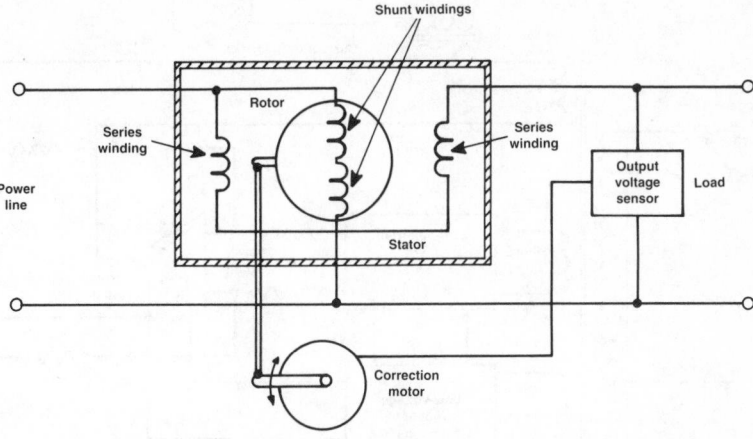

Figure 5.50   Rotary induction voltage regulator. (Adapted from documentation provided by Dranetz Technologies.)

5.4.4   Line Conditioner

A line conditioner combines the functions of an isolation transformer and a voltage regulator in one unit. The three basic types of line conditioners for DP applications are:

- *Linear amplifier correction system.* As illustrated in Figure 5.51, correction circuitry compares the ac power output to a reference source, derived from a 60 Hz sine wave generator. A correction voltage is developed and applied to the secondary power winding to cancel noise and voltage fluctuations. A box shield around the primary winding provides common-mode impulse rejection (80 to 100 dB typical). The linear amplifier correction system is effective and fast, but the overall regulating range is limited.

- *Hybrid ferroresonant transformer.* As shown in Figure 5.52, this system consists of a ferroresonant transformer constructed using the isolation techniques discussed in Section 5.4.2. The box and Faraday shields around the primary and compensating windings give the transformer essentially the characteristics of a noise-attenuating device, while preserving the voltage-regulating characteristics of a ferroresonant transformer.

- *Electronic tap-changing high-isolation transformer.* This system is built around a high-attenuation isolation transformer with a number of primary winding taps. SCR pairs control voltage input to each tap, as in a normal tap-changing regulator. Tap changing also can be applied to the secondary, as shown in Figure 5.53. The electronic tap-changing high-isolation transformer is an efficient design that effectively regulates voltage output and prevents noise propagation to the DP load.

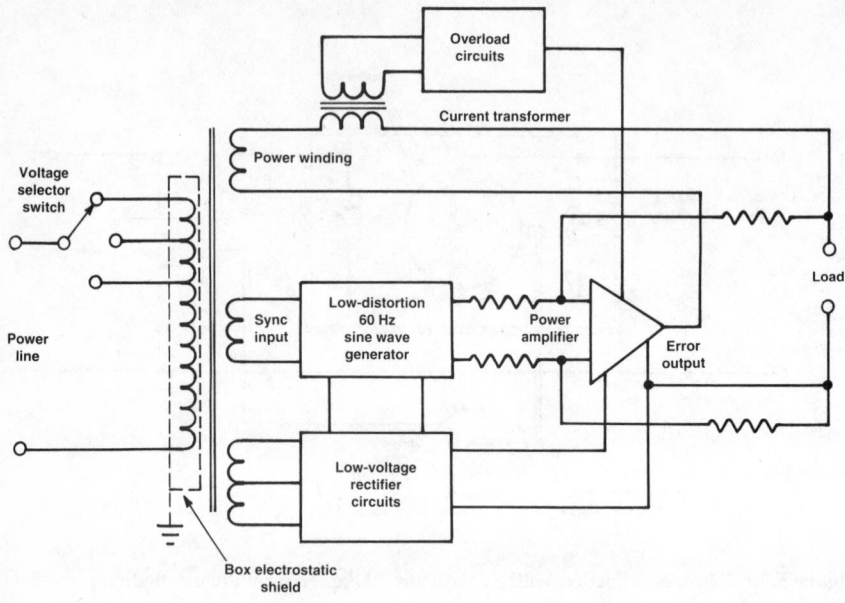

Figure 5.51   Power-line conditioner using linear amplifier correction. (Adapted from documentation provided by Dranetz Technologies.)

Hybrid Transient Suppressor.   A wide variety of ac power-conditioning systems are available, based on a combination of solid-state technologies. Most incorporate a combination of series and parallel elements to shunt transient energy. One such system is pictured in Figure 5.54.

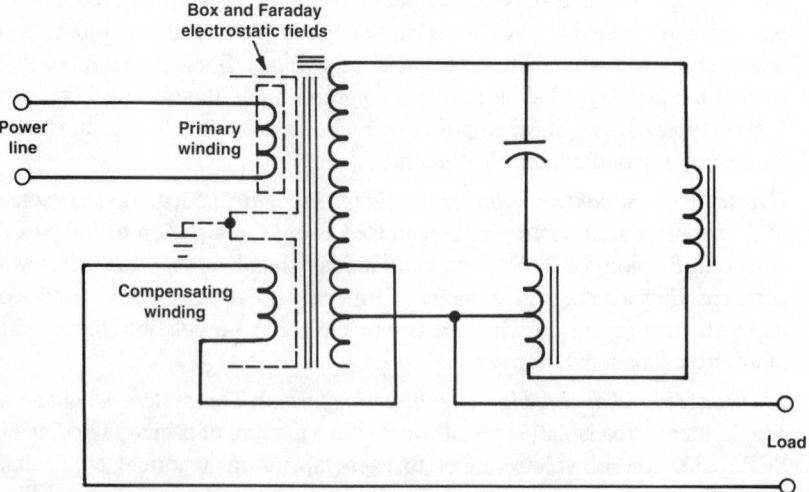

Figure 5.52   Line conditioner built around a shielded-winding ferroresonant transformer. (Adapted from documentation provided by Dranetz Technologies.)

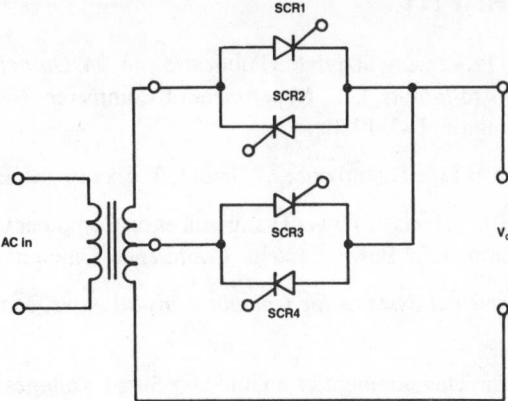

Figure 5.53   Secondary-side synchronous tap-changing transformer.

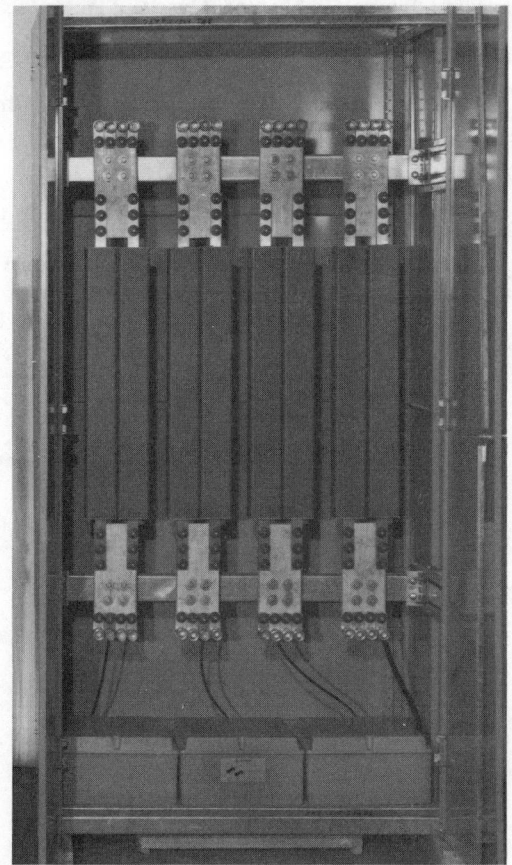

Figure 5.54   Interior view of a high-capacity hybrid power conditioner. (Courtesy of Control Concepts)

5.5   BIBLIOGRAPHY

Federal Information Processing Standards Publication no. 94, *Guideline on Electrical Power for ADP Installations*, U.S. Department of Commerce, National Bureau of Standards, Washington, DC, 1983.

"How to Correct Power Line Disturbances," Dranetz Technologies, Edison, NJ, 1985.

Key, Lt. Thomas: "The Effects of Power Disturbances on Computer Operation," IEEE Industrial and Commercial Power Systems Conference, Cincinnati, June 7, 1978.

Lawrie, Robert: *Electrical Systems for Computer Installations*, McGraw-Hill, New York, 1988.

Martzloff, F. D.: "The Development of a Guide on Surge Voltages in Low-Voltage AC Power Circuits," 14th Electrical/Electronics Insulation Conference, IEEE, Boston, October 1979.

Newman, Paul: "UPS Monitoring: Key to an Orderly Shutdown," *Microservice Management* magazine, Intertec Publishing, Overland Park, KS, March 1990.

Noise Suppression Reference Manual, Topaz Electronics, San Diego.

Nowak, Stewart: "Selecting a UPS," *Broadcast Engineering* magazine, Intertec Publishing, Overland Park, KS, April 1990.

Pettinger, Wesley: "The Procedure of Power Conditioning," *Microservice Management* magazine, Intertec Publishing, Overland Park, KS, March 1990.

Smeltzer, Dennis: "Getting Organized About Power," *Microservice Management* magazine, Intertec Publishing, Overland Park, KS, March 1990.

6

FACILITY-PROTECTION METHODS

6.1 INTRODUCTION

In the business of transient protection, you get what you pay for. Discrete devices are less expensive and usually provide less protection compared with a sophisticated systems approach. It is unrealistic for a user to expect a group of discrete transient suppressors to do the job of a much more expensive systems design. However, properly applied discrete devices can prevent equipment damage from all but the most serious transient disturbances. The key to achieving this level of performance lies in understanding and properly applying discrete protection components.

The performance of discrete transient-suppression components available to engineers has greatly improved within the past 10 years. The variety of reasonably priced devices now available makes it possible to exercise tight control over unwanted voltage excursions and allows the complicated electronic equipment being manufactured today to work as intended. Much of the credit for transient-suppression work goes to the computer industry, which has been dealing with this problem for more than 2 decades.

Discrete transient-suppression hardware can be divided into three basic categories:

- Filters
- Crowbar devices
- Voltage-clamping components

6.1.1 Filter Devices

Filters are designed to pass power-line frequency energy and reject unwanted harmonics and noise. The simplest type of ac power-line filter is a capacitor placed across the voltage source. The impedance of the capacitor forms a voltage divider with the impedance of the source, resulting in attenuation of high-frequency transients. This simple approach has definite limitations in transient-suppression capability and may introduce unwanted resonances with inductive components in the power-distribution system. The addition of a series resistance will reduce the undesirable resonant effects, but also will reduce the effectiveness of the capacitor in attenuating transient disturbances.

The conducting enclosure of an electronic device is required to be grounded for reasons of safety. However, this conductor often creates a path for noise currents. Any noise current flowing in a ground loop can create noise by electromagnetic coupling or direct connection through stray capacitances. Such noise voltages and currents are associated with the potential differences that appear throughout a large interconnected system of grounding conductors. It usually is impossible to eliminate all common-mode noise voltages in a practical ac power-distribution system. It is possible, however, to exert control over where ground loops will occur and the noise current paths that may result.

Filters intended for ac power applications are usually low-pass devices that remove high-frequency common-mode electrical noise. Figure 6.1 illustrates a common filter, known as a *balun*. The balun passes the flow of normal-mode current without significant voltage drop because the power conductors, wound on a common core, are wound in opposition. Load current does not saturate the core. High-frequency noise or impulse signals appearing between the equipment ground conductor and the DP load are attenuated by the device because of the high impedance presented to the noise signals. The ground voltage offset appears across the device, where it will do no harm. Almost identical voltages also appear across the neutral and line conductor coils, thereby offsetting their voltage-to-equipment ground by the same amount as the frame-to-ground offset.

Data processing hardware usually contains line filters at the ac input point. Figure 6.2 shows several levels of complexity in radio frequency interference (RFI) filter design. RFI filters can be regarded as mismatching networks at high frequencies. To operate properly without interaction, filters typically are designed to operate most effectively when the input and output impedances are matched to the respective

o Signifies winding polarity

Figure 6.1   Balun common-mode filter. (Adapted from: Federal Information Processing Standards Publication no. 94, *Guideline on Electrical Power for ADP Installations*, U.S. Department of Commerce, National Bureau of Standards, Washington, DC, 1983.)

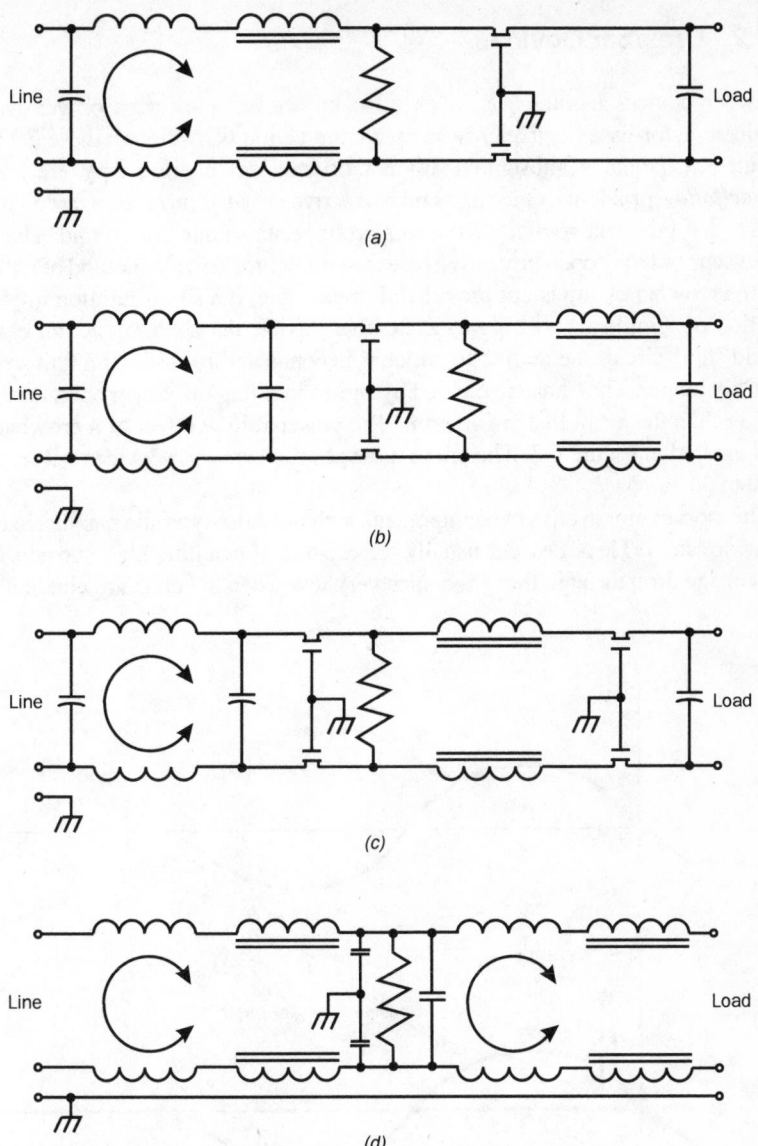

Figure 6.2   Power-line filter circuits of increasing complexity.

impedances of the power source and the load. These parameters are not always constant, however. Practical filters often must be designed for compromise conditions. The result may be unwanted ringing or line-voltage distortions at some frequency other than the disturbance frequency, which the filter is intended to correct.

6.1.2   Crowbar Devices

Crowbar devices include gas tubes (also known as spark-gaps or *gas-gaps*) and semiconductor-based *active crowbar* protection circuits. Although these devices and circuits can shunt a substantial amount of transient energy, they are subject to *power-follow* problems. Once a gas tube or active crowbar protection circuit has fired, the normal line voltage *and* the transient voltage are shunted to ground. This power-follow current may open protective fuses or circuit breakers if a method of extinguishing the crowbar clamp is not provided. For example, if a short-duration impulse of a fraction of a millisecond triggers a crowbar device, the shunting action essentially would short-circuit the ac line to which it is connected for at least a half cycle, and possibly longer. The transient created by short-circuiting the power conductor may be greater than the event that triggered it. The power-follow effect of a crowbar device is illustrated in Figure 6.3. The arc in most gas tube crowbar devices will extinguish at 20 to 30 V.

The most common crowbar components include pellet-type and gas-discharge-type surge arrestors. These devices usually are capable of handling high currents because the voltage drop through them becomes very low when the units are conducting.

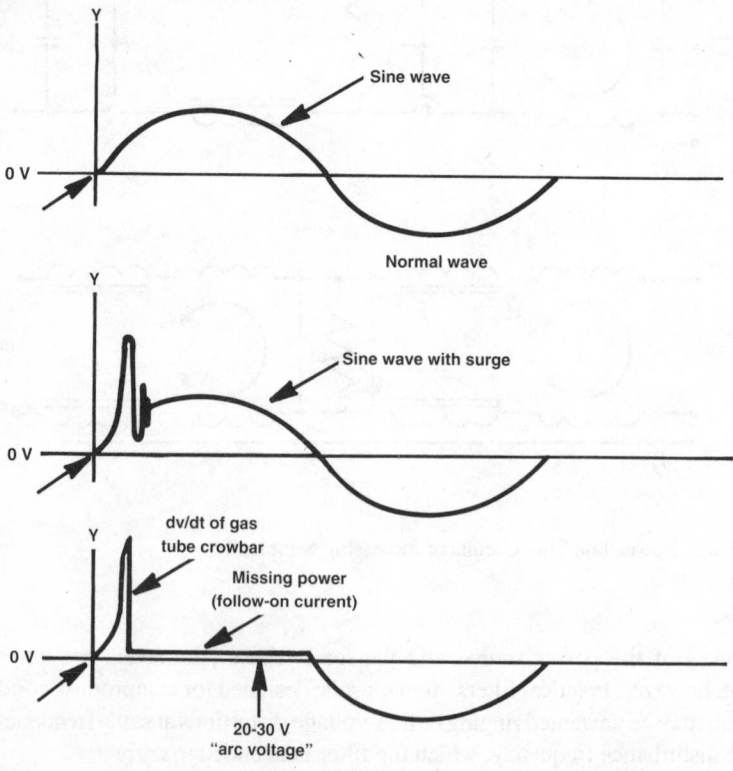

Figure 6.3   The power-follow effect of a crowbar device.

The selection of a crowbar device, the location(s) where it can be most effective, and applicable safety precautions require a solid understanding of surge voltages and the nature of traveling pulse waves. In general, crowbar devices are most appropriate where current in an inductive circuit element can be interrupted. They also may be warranted where long cables interconnect widely separated ac systems.

Characteristics of Arcs.   Defined very broadly, an arc can be described as a discharge of positive ions and electrons between electrodes in air, vapor, or both, which has a potential drop at the cathode in the order of magnitude of the minimum ionizing potential of the air, vapor, or both. A unique characteristic of an arc is the *negative resistance* effect; increasing current reduces the arc resistance. Consequently, in a given circuit, the arc voltage drop remains fairly constant, regardless of the current magnitude. Larger currents reduce the impedance, but the voltage drop remains the same. Arcs extinguish themselves at the ac zero crossing, requiring a voltage much greater than the arc voltage drop to reignite.

6.1.3   Voltage-Clamping Devices

Voltage-clamping components are not subject to the power-follow problems common in crowbar systems. Clamping devices include selenium cells, zener diodes, and varistors of various types. Zener diodes, using improved silicon rectifier technology, provide an effective voltage clamp for the protection of sensitive electronic circuitry from transient disturbances. On the other hand, power dissipation for zener units is usually somewhat limited, compared with other suppression methods.

Selenium cells and varistors, although entirely different in construction, act similarly on a circuit exposed to a transient overvoltage. Figure 6.4 illustrates the variable nonlinear impedance exhibited by a voltage-clamping device. The figure also shows how these components can reduce transient overvoltages in a given circuit. The voltage-divider network established by the source impedance (Z_s) and the clamping-device impedance (Z_c) attenuates voltage excursions at the load. It should be understood that the transient suppressor depends upon the source impedance to aid the clamping effect. A protection device cannot be fully effective in a circuit that exhibits a low source impedance because the voltage-divider ratio is reduced proportionately. It also must be recognized that voltage-clamp components divert surge currents; they do not absorb them. Care must be taken to ensure that the diversion path does not create new problems for the ac power system.

A typical voltage-vs.-current curve for a voltage-clamping device is shown in Figure 6.5. When the device is exposed to a high-voltage transient, the impedance of the component changes from a high standby value to a low conductive value, clamping the voltage at a specified level.

A selenium thyrector device, intended for low- to medium-power applications, is shown in Figure 6.6. Voltage-clamping components are available in a variety of configurations, from high-power chassis-mount units to surface-mount devices designed for use on printed wiring boards. Table 6.1 compares the attributes of a variety of transient-suppression technologies.

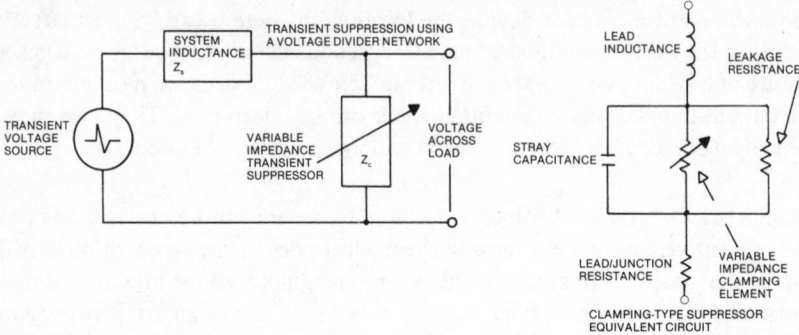

Figure 6.4   The mechanics of transient suppression using a voltage-clamping device.

Hybrid Suppression Circuits.   Any of the elements discussed so far in this chapter can be combined into a single hybrid unit to gain the best elements of each technology. A hybrid protector can be designed that will respond to several types of line disturbances. As shown in Figure 6.7, a gas tube commonly is combined with a silicon clamping device to provide two-stage suppression. When a transient hits, the gas tube will fire and crowbar the bulk of the energy. The clamping device will catch the leading edge of the transient that the gas tube may have missed. Proper matching of the gas tube and the clamping device is an important design parameter. If, for example, the tube is not presented with sufficient voltage to ensure that it fires, the silicon device will have to take the full surge energy, and may be destroyed in the process. The series impedance shown in the diagram can be a resistor or an inductor. It is sized to pass the full load current while still providing sufficient impedance to ensure that the tube will fire.

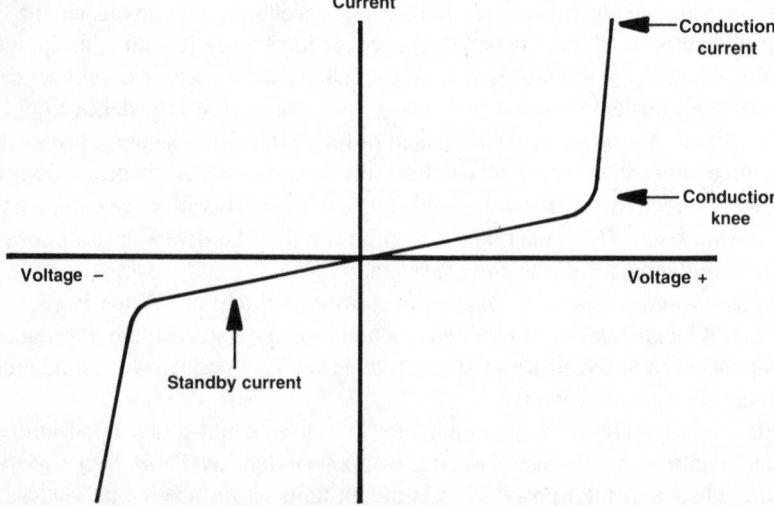

Figure 6.5   The voltage-vs.-current curve for a bipolar voltage-clamping device. The component is designed to be essentially invisible in the circuit until the applied positive or negative potential reaches or exceeds the *conduction knee* of the device.

Figure 6.6   Selenium thyrector voltage-clamping device.

A multistage suppressor is shown in Figure 6.8. The circuit is composed primarily of metal-oxide varistors (MOVs). V_{R1} and V_{R2} are the primary devices, and they absorb most of the overvoltage transients. Inductors L1 and L2 work with V_{R3} to exercise tight control over voltage excursions.

Table 6.1   Comparison of Various Transient Suppression Technologies

Voltage-current characteristics	Device type	Leakage	Follow current	Clamping voltage	Energy capability	Capacitance	Response time	Cost
V ↑ Clamping voltage — — — — Working voltage → I Transient current	Ideal device	Zero to low	No	Low	High	Low or high	Fast	Low
V ↑ — — — — Working voltage → I Transient current	Metal-oxide varistor	Low	No	Moderate to low	High	Moderate to high	Fast	Low
V ↑ Maximum current limit — — — — Working voltage → I Transient current	Zener diode	Low	No	Low	Low	Low	Fast	High
V ↑ Peak voltage (ignition) Working voltage — — — — → I Transient current	Gas discharge tube	Zero	Yes	High ignition voltage / Low clamp	High	Low	Slow	Low to high

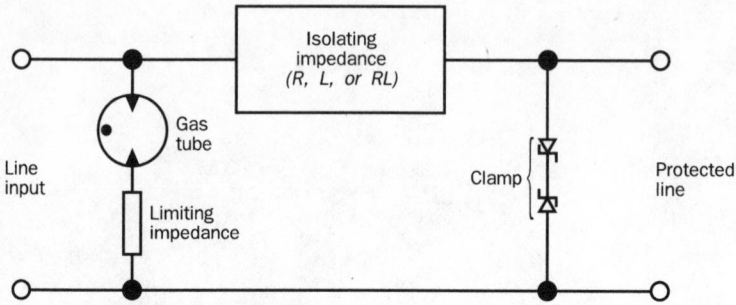

Figure 6.7   A two-stage hybrid protection circuit.

The response time of voltage-clamping devices is significantly affected by lead inductance. For this reason, lead lengths for all devices in a protection unit should be kept to a minimum.

6.1.4 Selecting Protection Components

Selecting a transient-suppression device for a given application is a complicated procedure that must take into account the following factors:

- The steady-state working voltage, including normal tolerances of the circuit.
- The transient energy to which the device is likely to be exposed.
- The voltage-clamping characteristics required in the application.
- Circuit-protection devices (such as fuses or circuit breakers) present in the system.
- The consequences of protection-device failure in a short-circuit mode.
- The sensitivity of load equipment to transient disturbances.

Most transient-suppression equipment manufacturers offer detailed application hand-books. Consult such reference data whenever planning to use a protection device. The

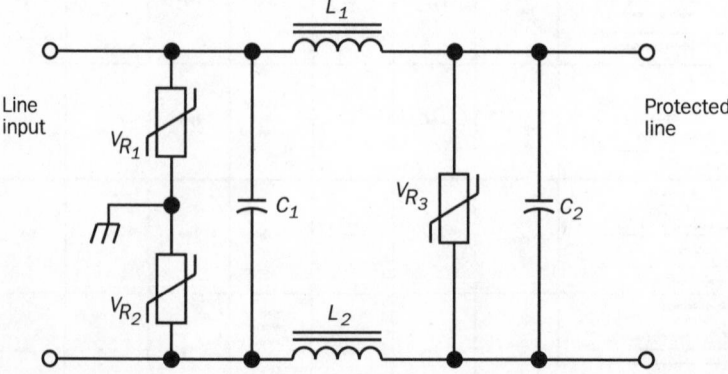

Figure 6.8   A multistage hybrid protection circuit utilizing MOVs, inductors, and capacitors.

specifications and ratings of suppression components are not necessarily interchange-able from one manufacturer to another. Carefully examine all variables in any planned addition of transient-suppression devices to a piece of equipment or ac power-distribution system. Make allowances for operation of the circuit under all reasonable conditions.

6.1.5 Performance Testing

The development of discrete protection devices such as those previously discussed has made it possible to design transient-suppression capabilities into virtually any piece of electronic equipment. Methods used in years past, including RC snubbers, common-variety isolation transformers, and RFI filter assemblies, did little to eliminate the threat posed by ac line transients. The following oscilloscope photos demonstrate graphically the improvement that the new generation of transient suppressors has made over previous methods.

Four devices were checked for performance using a transient waveform generator test set. The test instrument generates the waveform shown in Figure 6.9. The waveform is approximately 800 V P-P and 30 μs in duration. The waveform is synchronized to the ac line, allowing easy observation on an oscilloscope. The test pulse is designed to simulate the transient waveform typically generated by SCR controllers, contractor load switching, and large motor mode changes. The test setup is shown in Figure 6.10.

Figure 6.11 shows the 800 V P-P test waveform applied to a common isolation transformer. As the oscilloscope trace shows, the output transient is greater in magnitude than the input transient. This ringing commonly is found in general-purpose power transformers. New, high-performance isolation transformers, specifically designed to remove noise and transients from the ac line, perform much better than the unit used in this test. On the basis of this measurement, however, engineers should be cautioned against using a general-purpose transformer for isolation if transient suppression is not provided on the primary and secondary windings of the device.

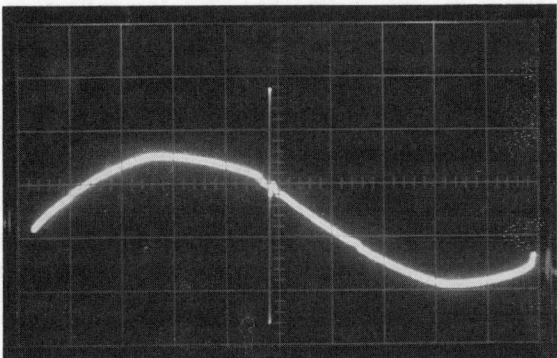

Figure 6.9   A transient waveform used to check the performance of ac power isolation and transient-suppression components. The spike is 800 V P-P and is positioned just past the 90° point on the ac wave. The oscilloscope photo scale = 135 V/div.

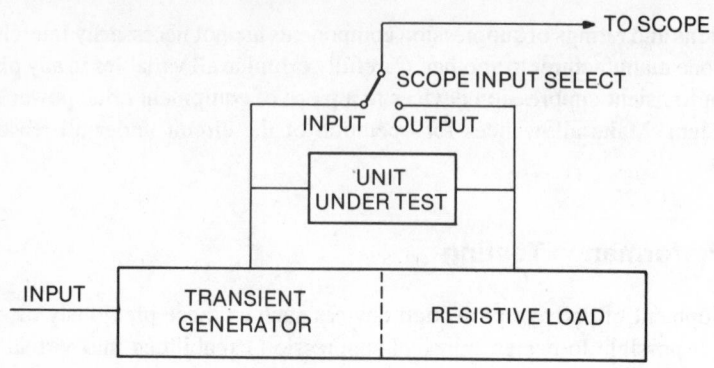

Figure 6.10   Test setup for measuring the performance of discrete transient-suppression devices.

Figure 6.12 shows the test waveform applied to a general-purpose RFI filter. As the oscilloscope photo shows, there is almost no attenuation of the pulse. The performance of most RFI filters varies considerably with changes in the frequency of the applied waveform.

Capacitors (in the range of 0.001 to 0.1 μF) are sometimes used for transient and noise suppression in electronic equipment, generally placed across the ac input terminals of the unit. Figure 6.13 shows the attenuation gained when a 0.25 μF capacitor is placed across the 800 V P-P spike. The capacitor reduces the transient by half, centered on the sine wave. This is contrasted with the MOV, which clips on maximum amplitude, positive and negative. An examination of the figure shows a brief period of oscillation following the transient. This oscillation is the result of the interaction of the capacitor and the inductance in the circuit. Placing a series resistor of approximately 100 Ω reduces the oscillation to a negligible amount, but it also significantly reduces the effectiveness of the capacitor in snubbing the transient, as shown in Figure 6.14.

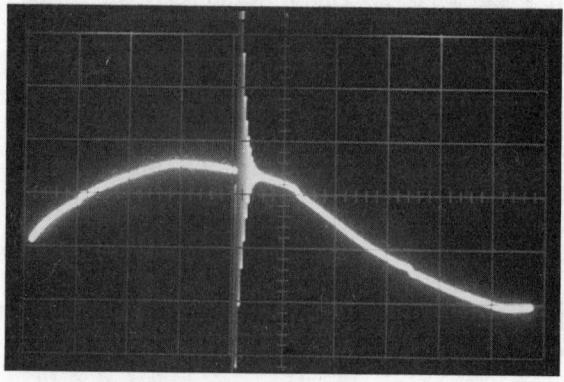

Figure 6.11   The effect of the test transient on a common isolation transformer. Note that the ringing caused by the spike exceeds the original disturbance.

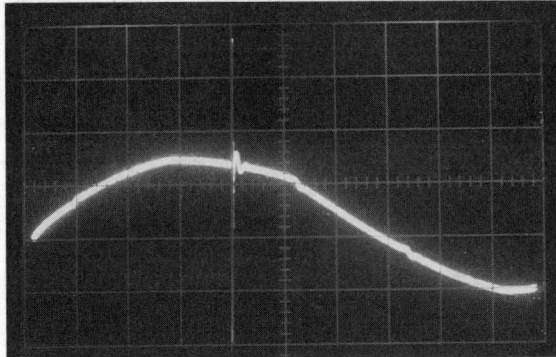

Figure 6.12   The performance of a common RFI filter when subjected to the 800 V P-P test waveform. Note there is little attenuation of the transient.

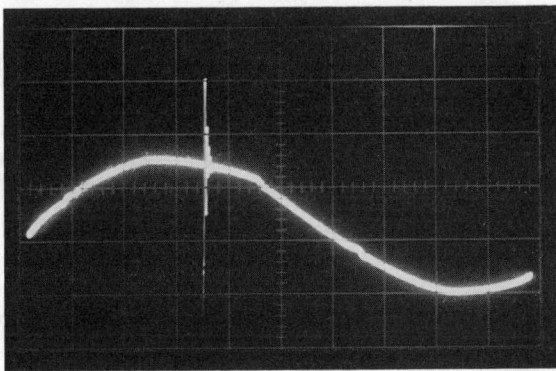

Figure 6.13   The transient-suppression performance of a 0.25 μF capacitor. Although the spike is reduced substantially, a small amount of oscillation can be seen in the display.

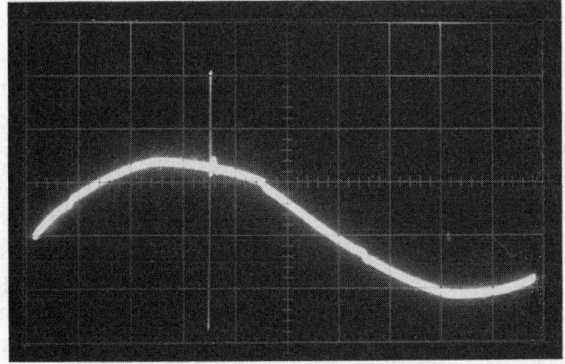

Figure 6.14   The effect of adding a 100 Ω series resistor to the 0.25 μF capacitor measured in Figure 6.13. Note that the resistor eliminates the oscillation, but it also reduces the transient-suppression effectiveness of the capacitor.

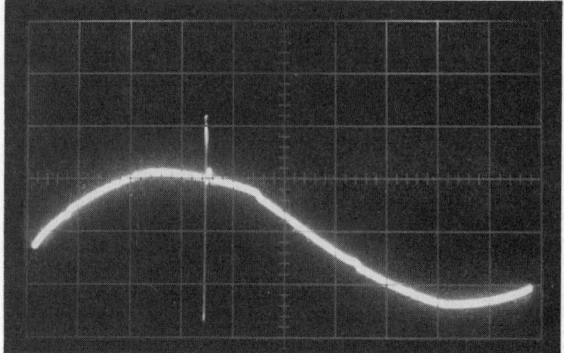

Figure 6.15   The transient-suppression performance of a MOV. Note the good clamping action and freedom from oscillation or ringing provided by the varistor.

The MOV is probably the most common discrete transient-suppression device. Figure 6.15 shows the performance of the varistor under the application of the 800 V P-P transient waveform. The device clips the transient at about 200 percent of the applied steady-state voltage. Although this level of protection may not be sufficient to prevent disallowed logic states in computer equipment or overloads in sensitive systems, it probably would prevent damage to the equipment on-line. This conclusion assumes that most power supplies in commercial/industrial equipment are designed for long-term operation at about 50 percent of their maximum capability. Although the MOV performs well by many standards, note the undershoot that occurs in the waveform. Varistors (and other similar devices) clip on maximum amplitude, without regard to the position of the disturbance on the waveform.

6.2   FACILITY PROTECTION

Most transient disturbances that a facility will experience enter the plant through the utility company ac power line. Effective transient suppression, therefore, begins with proper installation of the ac power-system wiring. If possible, arrange with the local utility to have a separate transformer feed the facility. The initial cost may be higher, but it will reduce the chances that transient disturbances from nearby operations will affect sensitive equipment. Do not allow the placement of noisy loads on the facility power line. Devices such as arc-welders, heavy electric motors, elevators, and other large loads can create an electrical environment that is prone to equipment malfunctions. It should be noted, however, that because transient disturbances, by definition, are high-frequency events, they will capacitively couple from the primary to the secondary of a typical utility company transformer. Simply installing a dedicated transformer that is not equipped with a Faraday shield (utility transformers generally do not have such shields) will not, in itself, protect equipment from damage. Installation of a dedicated utility transformer, however, will reduce the likelihood of problems and permit the establishment of a facility ground system independent of other users.

6.2.1 Facility Wiring

Insist that all ac wiring within the facility be performed by an experienced electrical contractor, and always fully within the local electrical code. Confirm that all wiring is sized properly for the load current. Table 6.2 lists the physical characteristics for various wire sizes. The current-carrying capability (ampacity) of single conductors in free air is listed in Table 6.3. The ampacity of conductors in a raceway or cable (three or fewer conductors) is listed in Table 6.4.

Synthetic insulation for wire and cable is classified into two broad categories: (1) *thermosetting*, and (2) *thermoplastic*. A wide variety of chemical mixtures can be found within each category. Most insulation is composed of compounds made from synthetic rubber polymers (thermosetting) and from synthetic materials (thermoplastics). Various materials are combined to provide specific physical and electrical

Table 6.2 Physical Characteristics of Standard Sizes of Copper Cable (at 25°C)

Wire Size (AWG/MCM)	Area (cmil)	Number of Conductors	Diameter Each Conductor (in)	DC Resistance Ω/1,000 ft Copper	Aluminum
12	6,530	1	0.0808	1.62	2.66
10	10,380	1	0.1019	1.018	1.67
8	16,510	1	0.1285	0.6404	1.05
6	26,240	7	0.0612	0.410	0.674
4	41,740	7	0.0772	0.259	0.424
3	52,620	7	0.0867	0.205	0.336
2	66,360	7	0.0974	0.162	0.266
1	83,690	19	0.0664	0.129	0.211
0	105,600	19	0.0745	0.102	0.168
00	133,100	19	0.0837	0.0811	0.133
000	167,800	19	0.0940	0.0642	0.105
0000	211,600	19	0.1055	0.0509	0.0836
250	250,000	37	0.0822	0.0431	0.0708
300	300,000	37	0.0900	0.0360	0.0590
350	350,000	37	0.0973	0.0308	0.0505
400	400,000	37	0.1040	0.0270	0.0442
500	500,000	37	0.1162	0.0216	0.0354
600	600,000	61	0.0992	0.0180	0.0295
700	700,000	61	0.1071	0.0154	0.0253
750	750,000	61	0.1109	0.0144	0.0236
800	800,000	61	0.1145	0.0135	0.0221
900	900,000	61	0.1215	0.0120	0.0197
1000	1,000,000	61	0.1280	0.0108	0.0177

Table 6.3   Permissible Ampacities of Single Conductors in Free Air

Wire Size	Copper Wire with		Aluminum Wire with	
	R, T, TW Insulation	RH, RHW, TH, THW Insulation	R, T, TW Insulation	RH, RHW, TH, THW Insulation
12	25	25	20	20
10	40	40	30	30
8	55	65	45	55
6	80	95	60	75
4	105	125	80	100
3	120	145	95	115
2	140	170	110	135
1	165	195	130	155
0	195	230	150	180
00	225	265	175	210
000	260	310	200	240
0000	300	360	230	280
250	340	405	265	315
300	375	445	290	350
350	420	505	330	395
400	455	545	355	425
500	515	620	405	485
600	575	690	455	545
700	630	755	500	595
750	655	785	515	620
800	680	815	535	645
900	730	870	580	700
1000	780	935	625	750

properties. Thermosetting compounds are characterized by their ability to be stretched, compressed, or deformed within reasonable limits under mechanical stress, and then to return to their original shape when the stress is removed. Thermoplastic insulation materials are best known for their electrical characteristics and relatively low cost. Thermoplastics permit insulation thickness to be reduced while maintaining good electrical properties.

Many different types of insulation are used for electric conductors. The operating conditions determine the type of insulation used. Insulation types are identified by abbreviations established in the National Electrical Code (NEC). The most popular types are:

- R: Rubber, rated for 140°F
- RH: Heat-resistant rubber, rated for 167°F

Table 6.4 Permissible Ampacities of Conductors in a Raceway or Cable (three or fewer conductors total)

Wire Size	Copper Wire with		Aluminum Wire with	
	R, T, TW Insulation	RH, RHW, TH, THW Insulation	R, T, TW Insulation	RH, RHW, TH, THW Insulation
12	20	20	15	15
10	30	30	25	25
8	40	45	30	40
6	55	65	40	50
4	70	85	55	65
3	80	100	65	75
2	95	115	75	90
1	110	130	85	100
0	125	150	100	120
00	145	175	115	135
000	165	200	130	155
0000	195	230	155	180
250	215	255	170	205
300	240	285	190	230
350	260	310	210	250
400	280	335	225	270
500	320	380	260	310
600	355	420	285	340
700	385	460	310	375
750	400	475	320	385
800	410	490	330	395
900	435	520	355	425
1000	455	545	375	445

- RHH: Heat-resistant rubber, rated for 194°F
- RHW: Moisture- and heat-resistant rubber, rated for 167°F
- T: Thermoplastic, rated for 140°F
- THW: Moisture- and heat-resistant thermoplastic, rated for 167°F
- THWN: Moisture- and heat-resistant thermoplastic with nylon, rated for 194°F

6.2.2 Utility Service Entrance

Figure 6.16 shows a typical service entrance, with the neutral line from the utility company tied to ground and to a ground rod at the meter panel. Where permitted by

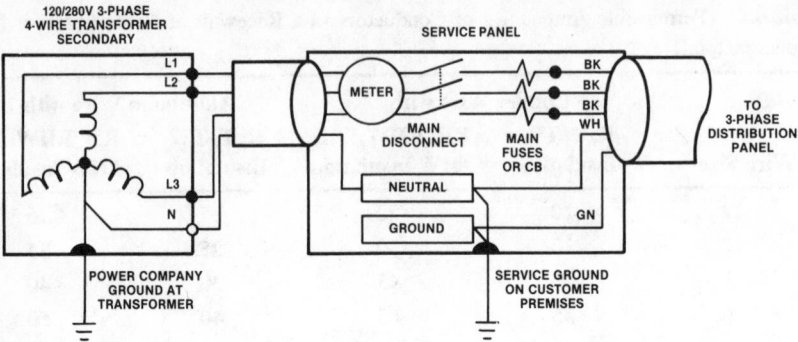

Figure 6.16   Connection arrangement for a three-phase utility company service panel.

the local code, this should be the only point at which neutral is tied to ground in the ac distribution system.

Figure 6.17 shows a three-phase power-distribution panel. Note that the neutral and ground connections are kept separate. Most ac distribution panels give the electrical contractor the ability to lift the neutral from ground by removing a short-circuiting screw in the breaker-panel chassis. Where permitted by local code, insulate the neutral lines from the cabinet. Bond the ground wires to the cabinet for safety. Always run a separate, insulated green wire for ground. Never rely on conduit or other mechanical structures to provide an ac system ground to electric panels or equipment.

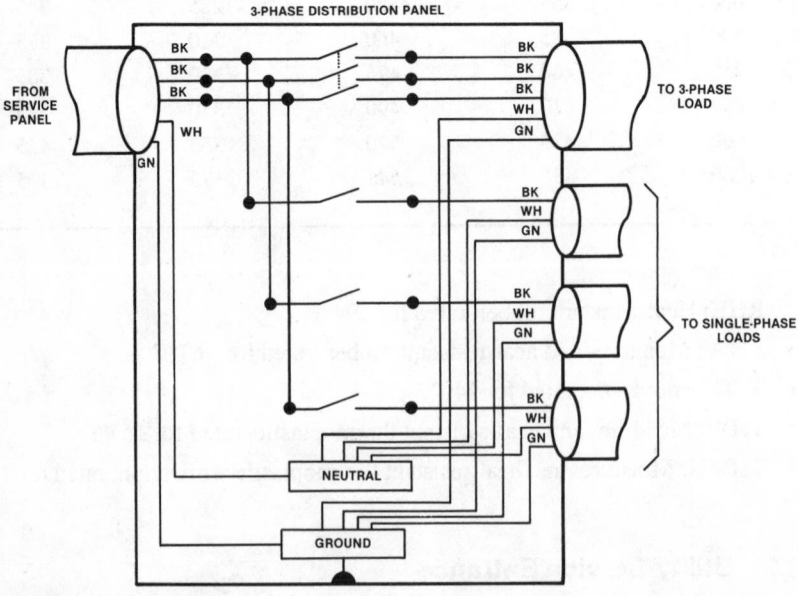

Figure 6.17   Connection arrangement of the neutral and green-wire ground system for a three-phase ac distribution panel.

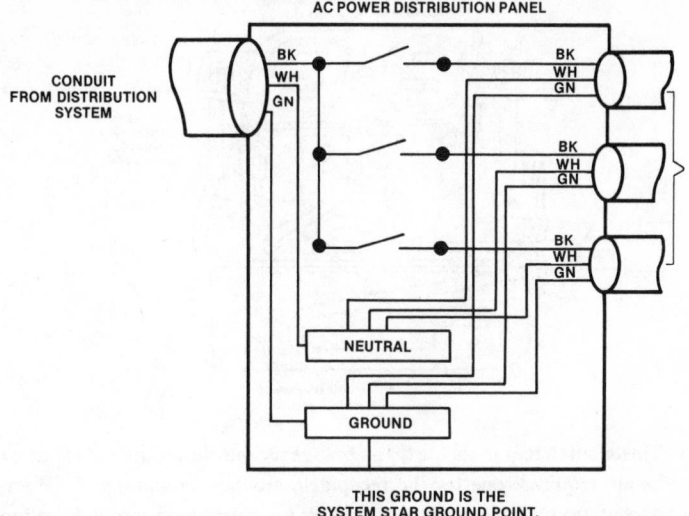

Figure 6.18 Connection arrangement of the neutral and green-wire ground system for a single-phase ac distribution panel.

A single-phase power-distribution panel is shown in Figure 6.18. Note that neutral is insulated from ground and that the insulated green ground wires are bonded to the panel chassis.

Conduit runs often are a source of noise. Corrosion of the steel-to-steel junctions can act as an RF detector. Conduit feeding sensitive equipment may contact other conduit runs powering noisy devices, such as elevators or air conditioners. Where possible (and permitted by the local code), eliminate this problem by using PVC pipe, Romex, or jacketed cable. If metal pipe must be used, send the noise to the power ground rods by isolating the green ground wire from the conduit with a ground-isolating (orange) receptacle. When using an orange receptacle, a second ground wire is required to bond the enclosure to the ground system, as shown in Figure 6.19. In a new installation, isolate the conduit from building metal structures or other conduit runs. Consult the local electrical code and an experienced electrical contractor before installing or modifying any ac power-system wiring. Make sure to also secure any necessary building permits for such work.

6.3 POWER-SYSTEM PROTECTION

Transient-protection methods for a commercial/industrial facility vary considerably, depending on the size and complexity of the plant, the sensitivity of equipment at the facility, and the extent of transient activity present on the primary power lines. Figure 6.20 shows one approach to transient suppression for a transmission facility. Lightning arresters are built into the 12 kV-to-208 V three-phase pole-mounted transformer. The service drop comes into the meter panel and is connected to a *primary lightning*

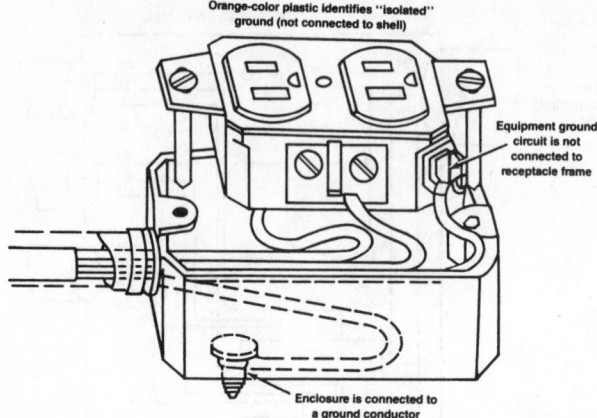

Figure 6.19   Installation requirements for an orange receptacle ac outlet. At least two ground conductor paths are required: one for the receptacle ground pin and one for the receptacle enclosure. (Adapted from: Federal Information Processing Standards Publication no. 94, *Guideline on Electrical Power for ADP Installations*, U.S. Department of Commerce, National Bureau of Standards, Washington, DC, 1983.)

arrester and a *primary varistor*. The circuit shown is duplicated three times for a three-wire wye (208 V phase-to-phase, 120 V phase-to-neutral) power system.

The primary arrester and varistor are placed at the service drop input point to protect the main circuit breaker and power-system wiring from high-voltage transients that are not clipped by the lightning arrester at the pole or by the varistors later in the circuit path. The primary varistor has a higher maximum clamp voltage than the varistors

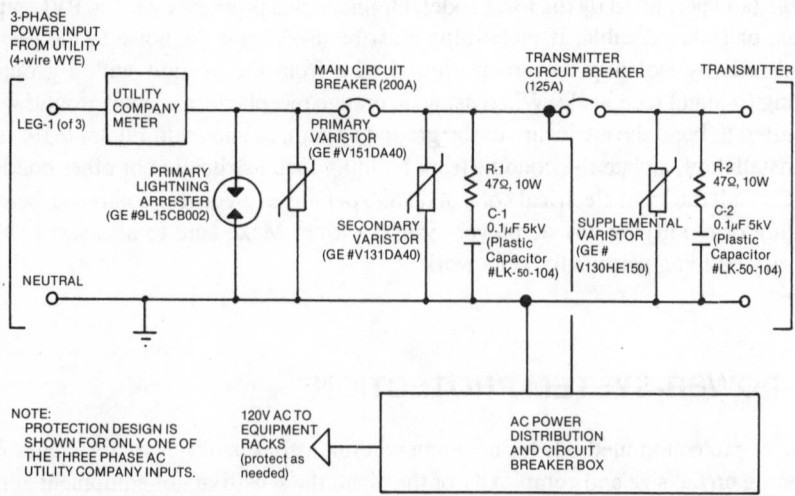

Figure 6.20   The application of transient-suppression components in a systemwide protection plan. Install such hardware with extreme care, and only after consultation with the local utility company and an electrical contractor.

located after the main breaker, causing the devices downstream to carry most of the clamp-mode current when a transient occurs (assuming low system inductance). If the main circuit breaker opens during a transient disturbance, the varistor at the service drop entrance will keep the voltage below a point that could damage the breaker or system wiring.

Placing overvoltage protection before the main service breaker may be considered only when the service drop transformer feeds a single load and when the transformer has transient protection of its own, including lightning arresters and primary-side fuses. Consult the local power company before placing any transient-suppression devices ahead of the main breaker.

Transient protection immediately after the main breaker consists of a *secondary varistor* and a capacitor between each leg and neutral. A 47 Ω 10 W series resistor protects the circuit if the capacitor fails. It also reduces the resonant effects of the capacitor and ac distribution-system inductance. The varistor clips overvoltages as previously described, and the resistor-capacitor network helps eliminate high-frequency transients on the line. The capacitor also places higher capacitive loading on the secondary of the utility company step-down transformer, reducing the effects of turn-on spikes caused by capacitive coupling between the primary and the secondary of the pole- or surface-mounted transformer.

As an extra measure of protection, a *supplemental varistor* and RC snubber are placed at the primary power input to the transmitter. Transient suppressors are placed as needed at the ac power-distribution panel and circuit-breaker box.

6.3.1 Staging

The transient-suppression system shown in Figure 6.20 uses a technique known as *staging* of protection components. An equivalent circuit of the basic system is shown in Figure 6.21. The staging approach takes advantage of the series resistance and impedance of the ac wiring system of a facility to aid in transient suppression.

When appreciable inductance or resistance exists in an ac distribution system, the protection components located at the utility company service drop entrance (the primary suppressors) will carry most of the suppressed-surge current in the event of

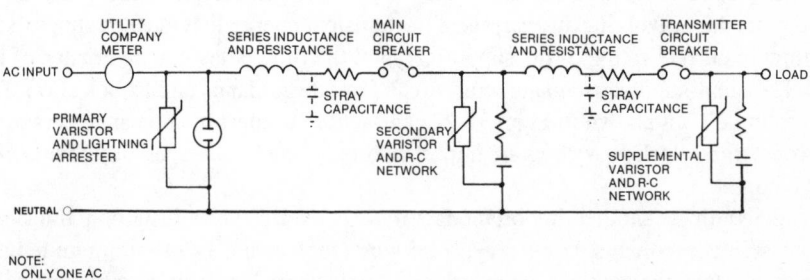

Figure 6.21   The use of ac system series inductance and resistance to aid transient suppressors in controlling line disturbances. This technique is known as *staging*.

a lightning strike or major transient disturbance. The varistors and RC networks downstream (the secondary and supplemental suppressors) are rated for clamp voltages lower than the primary protection devices. With the assistance of the ac circuit series resistance and impedance, the secondary and supplemental devices exercise tight control over voltage excursions.

The staged arrangement also protects the system from exposure caused by a transient-suppression device that becomes ineffective. The performance of an individual suppression component is more critical in a system that is protected at any one point than it is in a system protected at several points. The use of staged suppression also helps prevent transients generated by load equipment from being distributed to other sections of a facility, because suppressors can be located near offending machines.

Do not place transient suppressors of the same type in parallel to gain additional power-handling capability. Even suppressors that are identical in part number have specified tolerances; devices placed in parallel will not share the suppressed-surge current equally.

6.3.2 Design Cautions

Install transient suppressors at the utility service entrance with extreme care and only after consulting an experienced electrical contractor and the local utility company engineering department. Protection-device failure is rare, but it can occur, causing damage to the system unless the consequences of the failure are taken into account. Before installing a surge-limiting device, examine what would happen if the device failed in a short circuit (which is generally the case). Check for proper fusing on the protected lines, and locate transient-limiting devices in sealed enclosures to prevent damage to other equipment or injury to people if device failure occurs.

In the failure mode, current through the protection device is limited only by the applied voltage and source impedance. High currents can cause the internal elements of the component to melt and to eventually result in an open circuit. However, high currents often cause the component package to rupture, expelling package material in both solid and gaseous forms. A transient suppressor must be fused if the line on which it is operating has a circuit-breaker (or fuse) rating beyond the point that would provide protection against package rupture of the suppressor. Selecting the fuse is a complicated procedure involving an analysis of the transient energy that must be suppressed, the rupture current rating of the suppressor, and the time-delay characteristics of the fuse. Transient-suppressor manufacturers can provide guidance on fuse selection. The monitoring circuit shown in Figure 6.22 can be used to alert maintenance personnel to an open fuse. Such provisions are important for continued safe operation of sensitive load equipment.

Lead length is another important factor to consider when installing transient-suppression components. Use heavy, solid wire (such as no. 12 of minimum length) to connect protection devices to the ac lines. Avoid sharp bends. If possible, maintain a minimum bending radius of 8 in for interconnecting wires. Long leads act as inductors in the presence of high-frequency transients and as resistors when high-current surges are being clamped.

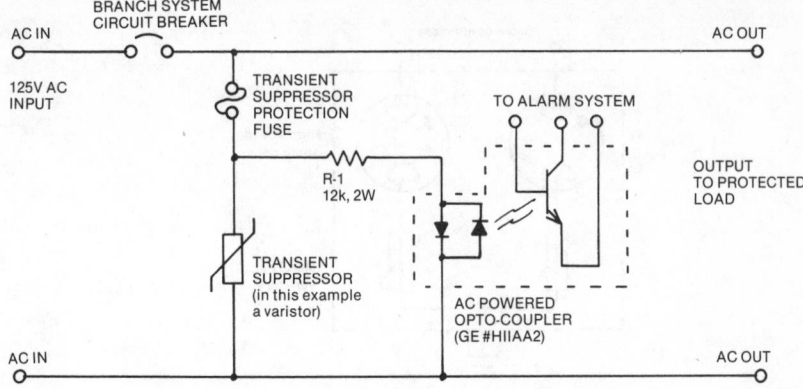

Figure 6.22 An open-fuse alarm circuit for a fused transient suppressor.

Give careful attention to proper heat sink design when installing transient-suppression devices. Some suppressors require an external heat sink to meet their published specifications. If an adequate heat sink is not provided, the result may be premature device failure.

Specifications. Transient-suppression components fail when subjected to events beyond their peak current/energy ratings. They also can fail when operated at steady-state voltages beyond their recommended values. Examine the manufacturer's product literature for each discrete protection device under consideration. Many companies have applications engineering departments that can assist in matching their product lines to specific requirements.

Consider using hybrid protection devices that provide increased product lifetime. For example, a varistor normally exhibits some leakage current. This leakage can lead to device heating and eventual failure. Hybrid devices are available that combine a varistor with a gas-filled spark-gap device to hold the leakage current to zero during standby operation, extending the expected product lifetime. During a transient, the spark-gap fires and the varistor clamps the pulse in the normal way. Such a device is shown in Figure 6.23. Hybrid devices of this type are immune to power-follow problems discussed in Section 6.1.2.

6.3.3 Single-Phasing

Any load using a three-phase ac power source is subject to the problem of *single-phasing*, the loss of one of the three legs from the primary power-distribution system. Single-phasing is usually a utility company problem, caused by a downed line or a blown pole-mounted fuse. The loss of one leg of a three-phase line results in a particularly dangerous situation for three-phase motors, which will overheat and sometimes fail. Figure 6.24 shows a simple protection scheme that has been used to protect industrial equipment from damage caused by single-phasing. At first glance, the system appears capable of easily handling the job, but operational problems can

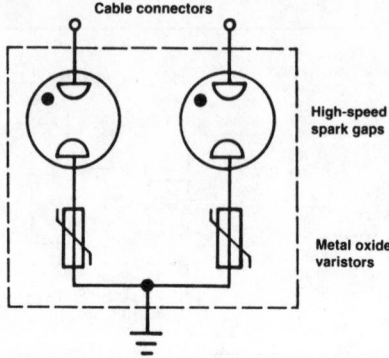

Figure 6.23   A hybrid voltage-protection device incorporating a gas-tube spark-gap and varistor in each suppression element. The design goal is to extend the life of the varistors.

result. The loss of one leg of a three-phase line rarely results in zero (or near-zero) voltages in the legs associated with the problem line. Instead, a combination of leakage currents caused by *regeneration* of the missing legs in inductive loads and the system load distribution usually results in voltages of some sort on the fault legs of the three-phase supply. It is possible, for example, to have phase-to-phase voltages of 220 V, 185 V, and 95 V on the legs of a three-phase, 208 V ac line experiencing a single-phasing problem. These voltages may change, depending upon what equipment is switched on at the facility.

Integrated circuit technology has provided a cost-effective solution to this common design problem. Phase-loss protection modules are available from several manufac-

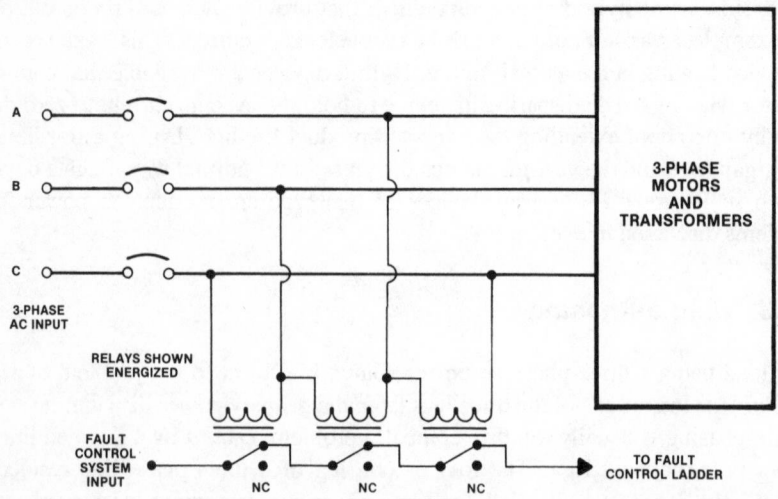

Figure 6.24   Phase-loss protection using relay logic.

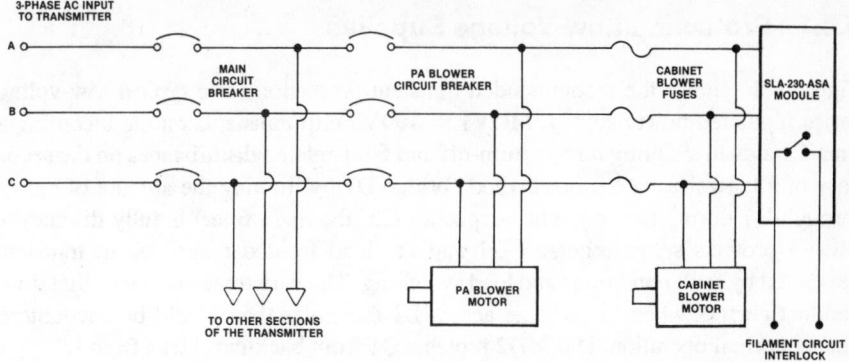

Figure 6.25   Phase-loss protection using a solid-state sensor. This circuit, used for a high-power transmitter, illustrates the importance of proper placement of the sensing device.

turers that provide a contact closure when voltages of proper magnitude and phase are present on the monitored line. The relay contacts can be wired into the logic control ladder of the protected load to prevent the application of primary ac power during a single-phasing condition. Figure 6.25 shows the recommended connection method for a high-power transmitter. Note that the input to the phase monitor module is taken from the final set of three-phase blower motor fuses. In this way, any failure inside the system that might result in a single-phasing condition is taken into account. The phase-loss protector shown in the figure includes a sensitivity adjustment for various nominal line voltages. The unit is small and relatively inexpensive.

Protection against single-phasing is particularly important when transient-suppression devices are placed at the utility company service entrance. The action of suppressors can cause one or more of the fuses at the service drop transformer to open, creating a fault condition. Positive protection against continued operation of the load under such circumstances is necessary to prevent equipment failure.

6.4 CIRCUIT-LEVEL APPLICATIONS

The transient-suppression provisions built into professional and industrial equipment have improved dramatically over the past few years. This trend is driven both by the increased availability of high-quality transient-suppression devices, and the increased need for clean power. Computer-based equipment is known for its intolerance of transient disturbances. As computers are integrated into an increasing number of facilities, clean power is a must.

The transient-suppression applications presented in this section are intended only as examples of ways that protection can be built into equipment to increase reliability. End-users should not attempt to modify existing equipment to provide increased transient-suppression capabilities. Such work is the domain of the equipment manufacturer. Transient suppression must be engineered into products during design and construction, not added on later in the field.

6.4.1   Protecting Low-Voltage Supplies

Figure 6.26 shows the recommended transient protection for a typical low-voltage series-regulated power supply. MOV1 to MOV3 clip transients on the incoming ac line. C1 aids in shunting turn-on/turn-off and fault-related disturbances on the secondary of T1. Resistor R1 protects diode bridge D1 by limiting the amount of current through D1 during turn-on, when capacitor C2 (the main filter) is fully discharged. MOV4 protects series regulator Q1 and the load from damage due to transients generated by fault conditions and load switching. The varistor is chosen so that it will conduct current when the voltage across L1 is greater than would be encountered during normal operation. Diode D2 protects Q1 from back-emf kicks from L1.

Three-terminal integrated circuit voltage regulator U1 is protected from excessive back-current because of a short circuit on its input side by diode D3. Capacitors C3 to C7 provide filtering and protect against RF pickup on the supply lines. Diode D4 protects U1 from back-emf kicks from an inductive load, and zener diode D5 protects the load from excessive voltage in case U1 fails (possibly impressing the full input voltage onto the load). D5 also protects U1 from overvoltages caused by transients generated through inductive load switching or fault conditions. D6 performs a similar function for the input side of U1.

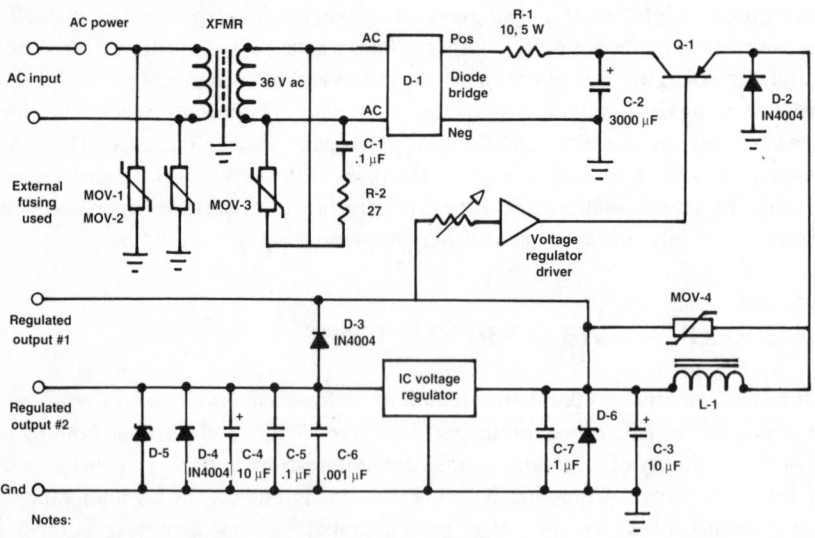

Notes:
1. MOV-1 and MOV-2 are GE MOV #V130LA20C.
2. MOV-3 is a GE MOV #V68ZA10.
3. Component values not shown are voltage-dependent.
   Types are selected according to system design.

Figure 6.26   Transient-overvoltage protection for a low-voltage power supply. A circuit such as this will survive well in the field despite frequent transient disturbances.

6.4.2 Protecting High-Voltage Supplies

A number of circuit configurations may be used for a high-voltage power supply. The most common incorporates a three-phase delta-to-wye transformer feeding a full-wave rectifier bridge, as shown in Figure 6.27. This arrangement provides high efficiency and low ripple content. With a well-balanced ac input line, the ripple component of the dc output is 4.2 percent, at a frequency 6 times the ac input frequency (360 Hz for a 60 Hz input). The dc output voltage is approximately 25 percent higher than the phase voltage, and each arm of the six-element rectifier must block only the phase voltage. The rms current through each rectifier element is 57 percent of the total average dc current of the load. The rectifier peak current is approximately equal to the value of the average dc output current. The typical power factor presented to the ac line is 95 percent.

Figure 6.28 illustrates the application of device staging to protect a high-voltage power supply from transient disturbances. As detailed previously for ac distribution systems, staging takes advantage of the series resistance and inductance of interconnecting wiring to assist in suppressing transient disturbances. The circuit includes two sets of varistors, primary and secondary units. The secondary set is rated for a lower clamp voltage and, together with the primary varistor groups, exercises tight control over disturbances entering the supply from the ac utility input.

Additional transient-suppression devices (CR1 to CR3) and three sets of RC snubbers (R1/C1 to R3/C3) clip transients generated by the power transformer during retarded-phase operation. On the transformer secondary, three groups of RC snubbers (R4/C4a-b to R6/C6a-b) provide additional protection to the load from turn-on/turn-off spikes and transient disturbances on the utility line.

Figure 6.29 shows the secondary side of a high-voltage power supply incorporating transient-overvoltage protection. RC networks are placed across the secondary windings of the high-voltage transformer, and a selenium thyrector (CR7) is placed across the choke. CR7 is essentially inactive until the voltage across the device exceeds a predetermined level. At the *trip point*, the device will break over into a conducting

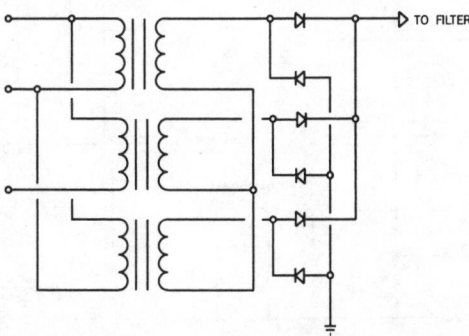

Figure 6.27 Three-phase full-wave bridge rectifier circuit.

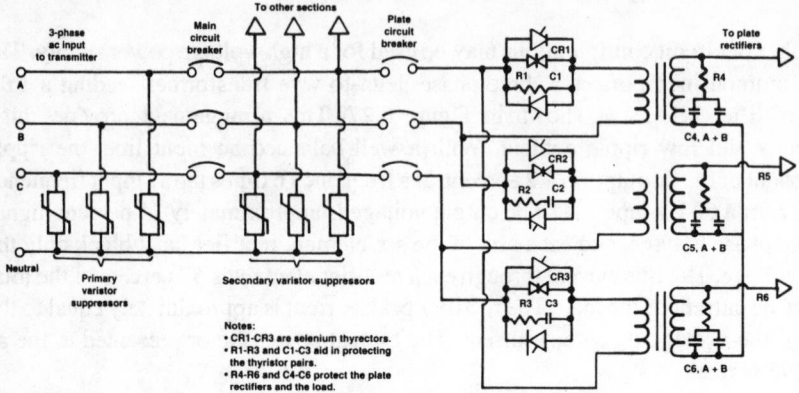

Figure 6.28 The use of ac system series inductance and resistance to aid transient suppressors in controlling line disturbances in a high-power transmitter.

state, shunting the transient overvoltage. CR7 is placed in parallel with L1 to prevent damage to other components in the system in the event of a sudden decrease in current drawn by the load. A sudden drop in load current will cause the stored energy of L1 to be discharged into the power supply and load circuits in the form of a high potential pulse. Such a transient can damage or destroy filter, feedthrough, or bypass capacitors. It also can damage wiring or cause arcing. CR7 prevents these problems by dissipating the stored energy in L1 as heat.

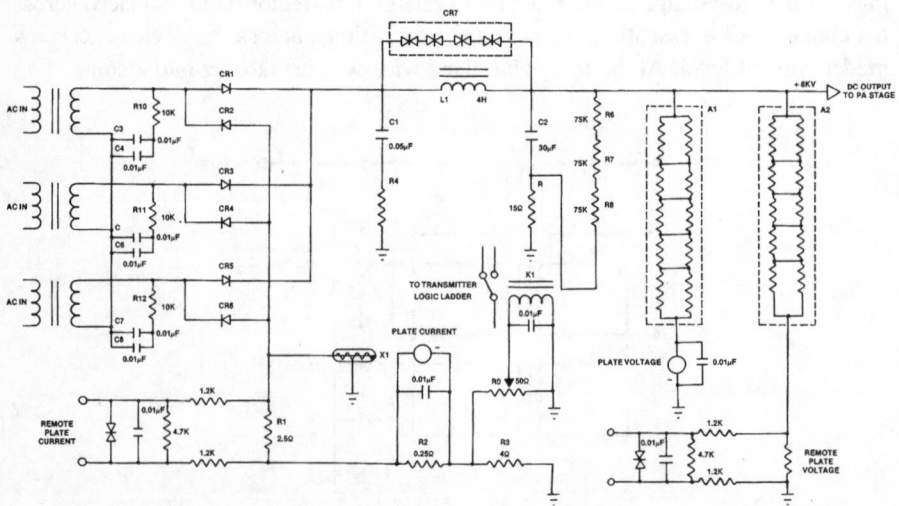

Figure 6.29 High-voltage power supply incorporating transient-suppression devices.

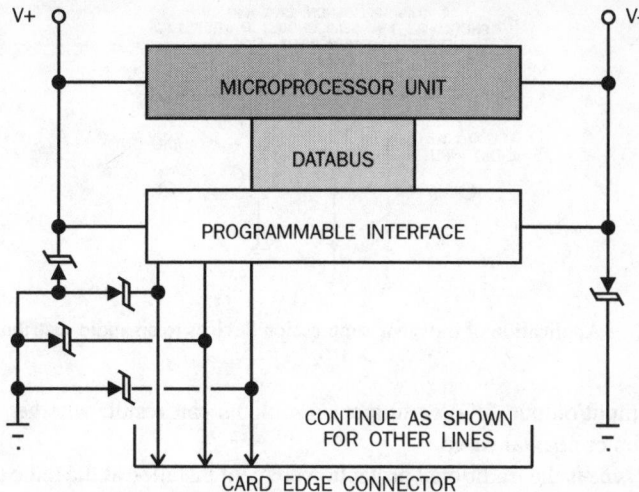

Figure 6.30   Application of transient-suppression devices to microcomputer circuits.

6.4.3 Protecting Logic Circuits

For maximum protection of microcomputer equipment, transient suppression must be designed into individual circuit boards. Figure 6.30 illustrates a typical application of on-the-board transient suppression. Multiple voltage-clamping devices are included in a single DIP package, making it possible to conveniently include protection on individual printed wiring boards. Figure 6.31 shows the application of transient suppressors in a voltage-follower circuit, common in many data acquisition systems. Note the use of suppression devices at the power-supply pins of the circuits shown in the figures.

6.4.4 Protecting Telco Lines

Voice and data communications networks are increasing in scope and complexity. Communication lines that travel from the telephone company (telco) central office to customer sites require transient protection. Furthermore, lines that travel from one building to another, or from one floor to another, also may pose a transient threat to

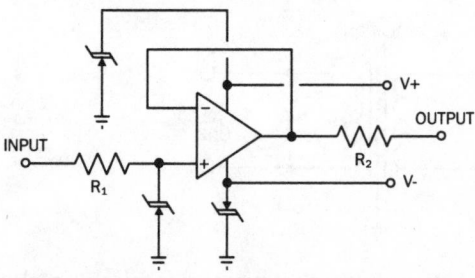

Figure 6.31   Application of transient-suppression devices to an analog voltage sampling circuit.

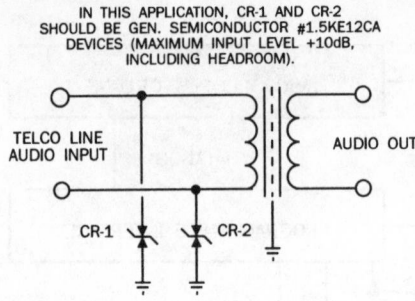

Figure 6.32   Application of transient-suppression devices to an audio distribution circuit.

computer input/output (I/O) equipment. Problems can result, whether the lines are twisted pairs or coaxial cable.

The *gas tube* is the traditional protection element installed at the telco central office (CO). The primary purpose of the gas tube (and its carbon predecessor) is to protect CO personnel from injury and CO equipment from damage in the event of a lightning flash to exposed lines or accidental contact with high-voltage utility company cables. Protection devices usually are included at the telco service entrance point on the customer's premises. Telco providers do their best to ensure that disturbances do not reach customers, but the final responsibility for transient protection lies with the equipment user.

Solid-state voltage-clamping devices generally are used to protect audio and data lines. The transient-clipping devices shown in Figure 6.32 are selected based on the typical audio voltage levels (including headroom) used on the loop. Figure 6.33 illustrates another protection arrangement, which prevents the introduction of noise into audio or data lines because of a common-mode imbalance that may result from transient suppressors being tied to ground. The use of a low-capacitance suppressor ensures minimum capacitive loading on the circuit.

For balanced telco lines, critical transient considerations include both the above-ground voltage of the two conductors (the common-mode voltage) and the voltage between the two conductors (the normal-mode voltage). When individual clamping devices are used on each conductor, as shown in Figure 6.32, one device will inevitably

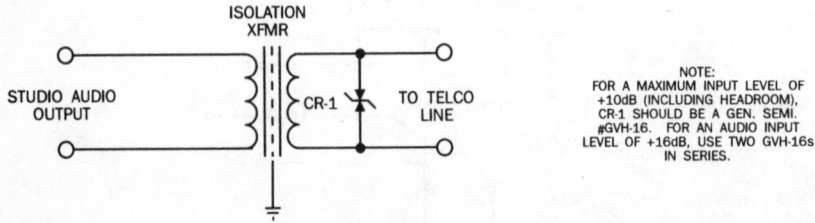

Figure 6.33   Application of a transient-suppression device to a telephone company audio or data line.

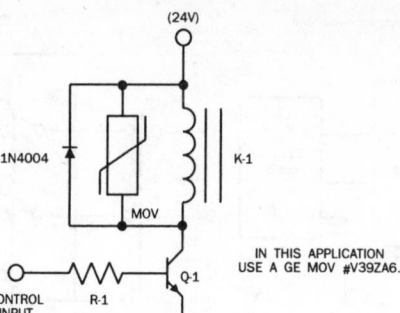

Figure 6.34   Transient suppression of a switched inductive load.

clamp before the other. This action can create a significant voltage differential that can damage sensitive equipment on the line. A common solution is the three-element gas tube. The device has a single gas chamber with two gaps, one for each side of the line. When one side reaches the ionization potential, both sides fire simultaneously to ground.

6.4.5   Inductive Load Switching

Any transistor that switches an inductive load must be provided with transient protection. Figure 6.34 shows the most common approach. Protection also is required for switches that control an appreciable amount of power, as illustrated in Figure 6.35. The use of a transient suppressor across switch or relay contacts will greatly extend the life of the switching elements.

6.4.6   Device Application Cautions

Building transient-suppression capability into a product is not as easy or straightforward as it might appear. Misapplication of a suppressor can reduce equipment reliability, not increase it. For example, Figure 6.36 (a) shows a transient-suppression arrangement that should be avoided. The relay contacts of K1 are protected by three transient suppressors. Although this is an acceptable application of the devices, the possibility of suppressor failure always must be considered. The usual failure mode is a short circuit. This being the case, a failure of any two of the three devices shown

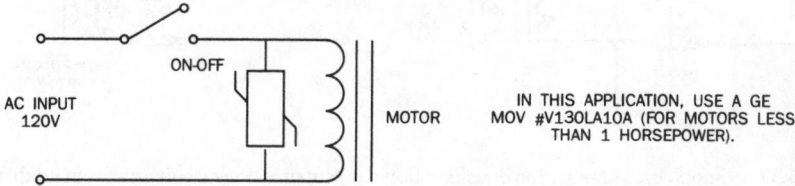

Figure 6.35   Transient suppression of switch arcing.

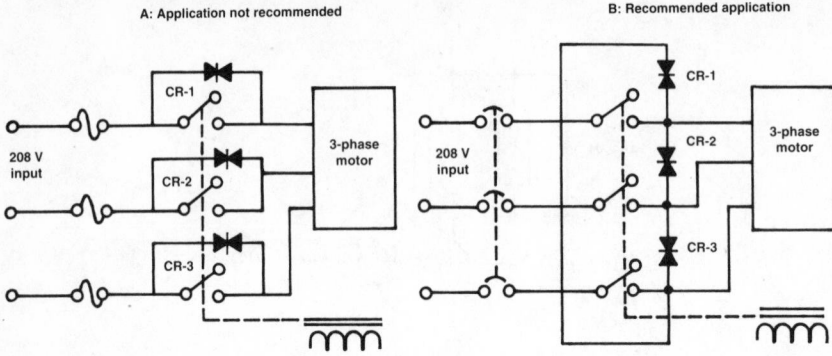

Figure 6.36 Suppressor application considerations: (a) suppression circuit arrangement that should be avoided; (b) alternate configuration that provides fail-safe operation. The application of any transient-suppression component must be considered carefully, keeping in mind the various modes of operation and the possibility of protection-device failure.

will cause a single-phasing condition, probably damaging the motor. A better arrangement is shown in Figure 6.36 (b). Device failure in this configuration will open the circuit breaker, shutting down the system but not destroying the motor.

Figure 6.37 (a) illustrates another example of inappropriate transient-suppressor application. As shown in the diagram, a protection device is placed across the *plate-on* button of a high-voltage power supply. Although this arrangement will extend the life of the switch contacts (especially if the button is switching 120 V ac or 208 V ac), failure of the device could prevent the high-voltage supply from being turned off in the event of an overload. Serious equipment damage could result. Figure 6.37 (b) shows the correct approach, in which the protection device is placed across the relay coil. Failure of the device in this configuration would shut down the system and prevent any further damage.

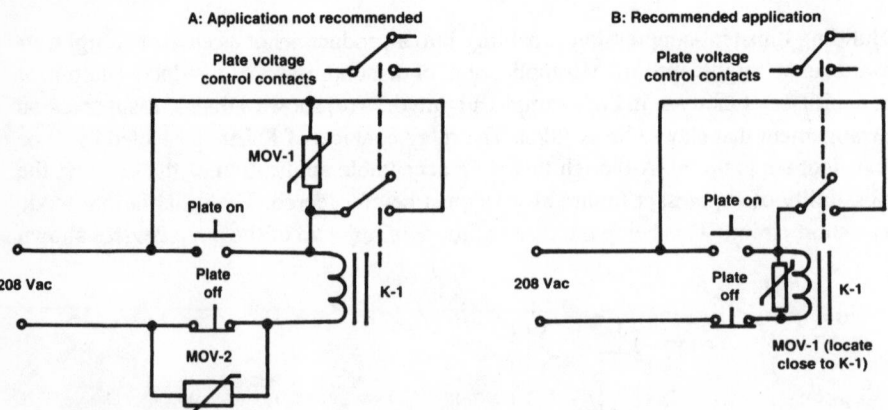

Figure 6.37 Suppressor application considerations: (a) suppression circuit arrangement that should be avoided; (b) alternate configuration that provides fail-safe operation.

6.5 BIBLIOGRAPHY

Belling, Michael: "Judging Surge Suppressor Performance," *Microservice Management* magazine, Intertec Publishing, Overland Park, KS, March 1990.

Block, Roger: "The Grounds for Lightning and EMP Protection," PolyPhaser Corporation, Gardnerville, NV, 1987.

"EMP Protection Device Applications Handbook," English Electric Valve publication.

Fardo, S., and D. Patrick: *Electrical Power Systems Technology*, Prentice-Hall, Englewood Cliffs, NJ, 1985.

Federal Information Processing Standards Publication no. 94, *Guideline on Electrical Power for ADP Installations*, U.S. Department of Commerce, National Bureau of Standards, Washington, DC, 1983.

Fink, D., and D. Christiansen: *Electronics Engineers' Handbook*, 3rd ed., McGraw-Hill, New York, 1989.

McPartland, Joseph, and Brian McPartland: *National Electrical Code Handbook*, 20th ed., McGraw-Hill, New York, 1990.

MOV Varistor Data and Applications Manual, General Electric Company, Auburn, NY, 1976.

Schneider, John: "Surge Protection and Grounding Methods for AM Broadcast Transmitter Sites," *Proceedings of the SBE National Convention and Broadcast Engineering Conference*, Society of Broadcast Engineers, Indianapolis, 1987.

Technical staff, "Update: Transient Suppressors," *Electronic Products*, June 1988.

Whitaker, Jerry: *Radio Frequency Transmission Systems: Design and Operation*, McGraw-Hill, New York, 1990.

———: *Maintaining Electronic Systems*, CRC Press, Boca Raton, FL, 1991.

7

FACILITY GROUNDING

Portions of this chapter were adapted from: Roger Block, "The Grounds for Lightning and EMP Protection," PolyPhaser Corporation, Gardnerville, NV, 1987.

7.1 INTRODUCTION

The attention given to the design and installation of a facility ground system is a key element in the day-to-day reliability of the plant. A well-designed and -installed ground network is invisible to the engineering staff. A marginal ground system, however, will cause problems on a regular basis. Grounding schemes can range from simple to complex, but any system serves three primary purposes:

- Provides for operator safety.
- Protects electronic equipment from damage caused by transient disturbances.
- Diverts stray radio frequency energy from sensitive audio, video, control, and computer equipment.

Most engineers view grounding mainly as a method to protect equipment from damage or malfunction. However, the most important element is operator safety. The 120 V or 208 V ac line current that powers most equipment can be dangerous — even deadly — if handled improperly. Grounding of equipment and structures provides protection against wiring errors or faults that could endanger human life.

Proper grounding is basic to protection against ac line disturbances. This applies whether the source of the disturbance is lightning, power-system switching activities, or faults in the distribution network. Proper grounding is also a key element in preventing radio frequency interference in transmission or computer equipment. A facility with a poor ground system may experience RFI problems on a regular basis. Implementing an effective ground network is not an easy task. It requires planning, quality components, and skilled installers. It is not inexpensive. However, proper grounding is an investment that will pay dividends for the life of the facility.

Any ground system consists of two key elements: (1) the earth-to-grounding electrode interface outside the facility, and (2) the ac power and signal-wiring systems inside the facility.

7.2 ESTABLISHING AN EARTH GROUND

The grounding electrode is the primary element of any ground system. The electrode can take many forms. In all cases, its purpose is to interface the electrode (a conductor) with the earth (a semiconductor). Grounding principles have been refined to a science. Still, however, many misconceptions exist regarding grounding. An understanding of proper grounding procedures begins with the basic earth-interface mechanism.

7.2.1 Grounding Interface

The grounding electrode (or ground rod) interacts with the earth to create a hemisphere-shaped volume, as illustrated in Figure 7.1. The size of this volume is related to the size of the grounding electrode. The length of the electrode has a much greater effect than the diameter. Studies have demonstrated that the earth-to-electrode resistance from a driven ground rod increases exponentially with the distance from that rod. At a given point, the change becomes insignificant. It has been found that for maximum effectiveness of the earth-to-electrode interface, each ground rod requires a hemisphere-shaped volume with a diameter that is approximately 2.2 times the rod length.

The constraints of economics and available real estate place practical limitations on the installation of a ground system. It is important, however, to keep the 2.2 rule in mind because it allows the facility design engineer to use the available resources to the best advantage. Figure 7.2 illustrates the effects of locating ground rods too close (less than 2.2 times the rod length). An overlap area is created that effectively wastes some of the earth-to-electrode capabilities of the two ground rods. Research has shown, for example, that two 10-ft ground rods driven only 1 ft apart provide about the same resistivity as a single 10-ft rod.

There are two schools of thought with regard to ground-rod length. The first approach states that extending ground-rod length beyond about 10 ft is of little value for most types of soil. The reasoning behind this conclusion is presented in Figure 7.3, where ground resistance is plotted as a function of ground-rod length. Beyond 10 ft in length, a point of diminishing returns is reached. The second school of thought concludes that optimum earth-to-electrode interface is achieved with long (40 ft or greater) rods, driven to penetrate the local water table. When planning this type of installation, consider the difficulty that may be encountered when attempting to drive long ground rods. The foregoing discussion assumes that the soil around the grounding

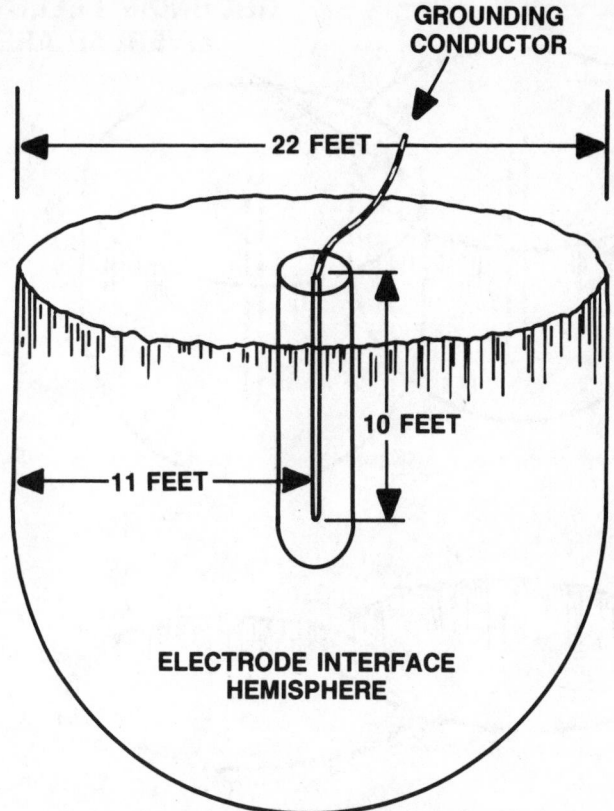

GROUNDING CONDUCTOR

22 FEET

10 FEET

11 FEET

ELECTRODE INTERFACE HEMISPHERE

Figure 7.1   The effective earth-interface hemisphere resulting from a single driven ground rod. The 90 percent effective area of the rod extends to a radius of approximately 1.1 times the length of the rod. (Adapted from: Roy Carpenter, "Improved Grounding Methods for Broadcasters," *Proceedings* of the SBE National Convention, Society of Broadcast Engineers, Indianapolis, 1987.)

electrode is reasonably uniform in composition. Depending upon the location, however, this may not be the case.

Horizontal grounding electrodes provide essentially the same resistivity as an equivalent-length vertical electrode. As Figure 7.4 demonstrates, the difference between a 10-ft vertical and a 10-ft horizontal ground rod is negligible (275 Ω vs. 250 Ω). This comparison includes the effects of the vertical connection element from the surface of the ground to the horizontal rod. Taken by itself, the horizontal ground rod provides an earth-interface resistivity of approximately 308 Ω when buried at a depth of 36 in.

Ground rods come in many sizes and lengths. The more popular sizes are 1/2, 5/8, 3/4, and 1 in. The 1/2-in size is available in steel with stainless-clad, galvanized, or copper-clad rods. All-stainless-steel rods also are available. Ground rods can be purchased in unthreaded or threaded (sectional) lengths. The sectional sizes are

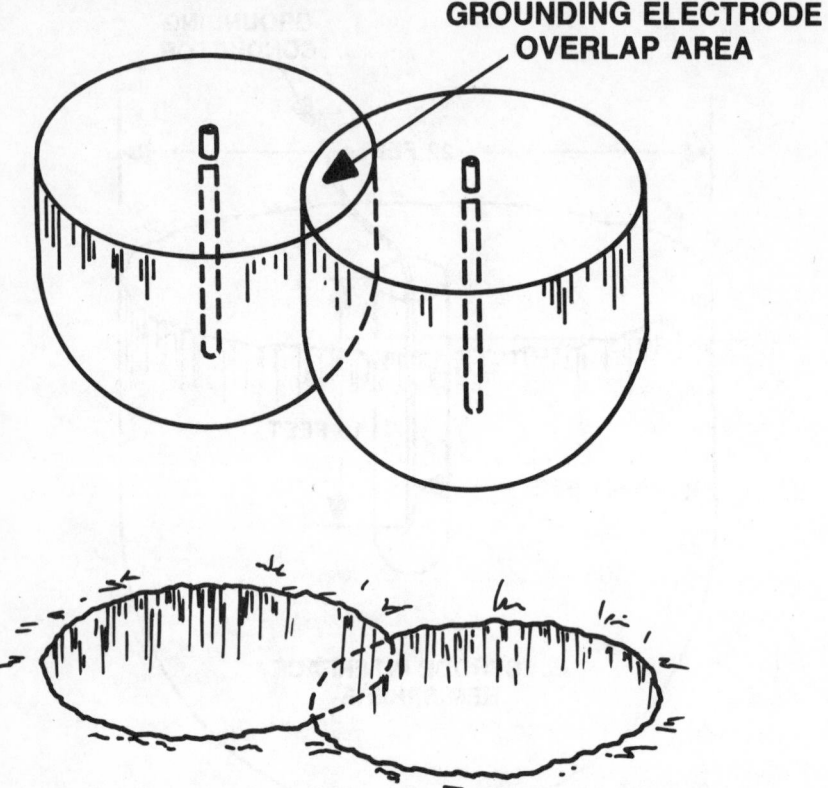

Figure 7.2   The effect of overlapping earth interface hemispheres by placing two ground rods at a spacing less than 2.2 times the length of either rod. The overlap area represents wasted earth-to-grounding electrode interface capability. (Adapted from: Roy Carpenter, "Improved Grounding Methods for Broadcasters," *Proceedings* of the SBE National Convention, Society of Broadcast Engineers, Indianapolis, 1987.)

typically 9/16-in or 1/2-in rolled threads. Couplers are made from the same materials as the rods. Couplers can be used to join 8- or 10-ft-length rods together. A 40-ft ground rod is driven one 10-ft section at a time.

The type and size of ground rod used is determined by how many sections are to be connected and how hard or rocky the soil is. Copper-clad 5/8-in x 10-ft rods are probably the most popular. Rod diameter has minimal effect on final ground imped-ance. Copper cladding is designed to prevent rust. The copper is not primarily to provide better conductivity. Although the copper certainly provides a better conductor interface to earth, the steel that it covers is also an excellent conductor when compared with ground conductivity. The thickness of the cladding is important only insofar as rust protection is concerned.

Soil Resistivity.   Wide variations in soil resistivity can be found within a given geographic area, as documented in Table 7.1. The wide range of values shown results from differences in moisture content, mineral content, and temperature.

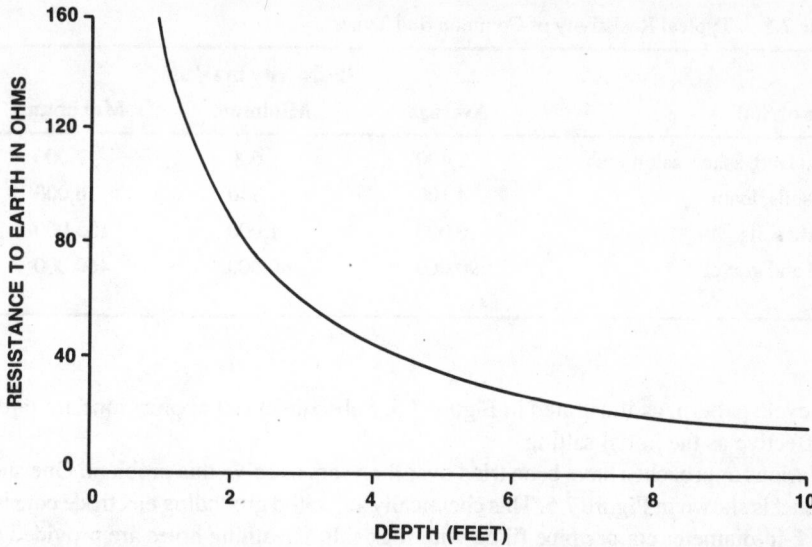

Figure 7.3   Charted grounding resistance as a function of ground-rod length. Ground-rod length in excess of 10 ft produces diminishing returns. The chart applies to a 1-in-diameter rod. (Adapted from: Roy Carpenter, "Improved Grounding Methods for Broadcasters," *Proceedings* of the SBE National Convention, Society of Broadcast Engineers, Indianapolis, 1987.)

7.2.2 Chemical Ground Rods

A chemically activated ground system is an alternative to the conventional ground rod. The idea behind the chemical ground rod is to increase the earth-to-electrode interface by conditioning the soil surrounding the rod. Experts have known for many years that the addition of ordinary table salt (NaCl) to soil will reduce the resistivity of the earth-to-ground electrode interface. With the proper soil moisture level (4 to 12 percent), *salting* can reduce soil resistivity from 10,000 Ω/m to less than 100 Ω/m. Salting the area surrounding a ground rod (or group of rods) follows a predictable

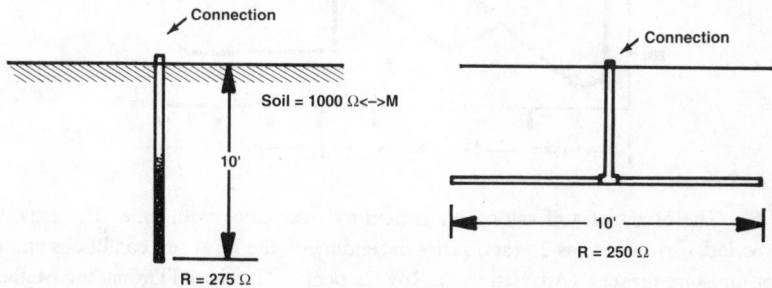

Figure 7.4   The effectiveness of vertical ground rods compared with horizontal ground rods. (Adapted from: Roy Carpenter, "Improved Grounding Methods for Broadcasters," *Proceedings* of the SBE National Convention, Society of Broadcast Engineers, Indianapolis, 1987.)

Table 7.1 Typical Resistivity of Common Soil Types

Type of Soil	Resistivity in Ω/cm		
	Average	Minimum	Maximum
Filled land, ashes, salt marsh	2,400	600	7,000
Top soils, loam	4,100	340	16,000
Hybrid soils	16,000	1,000	135,000
Sand and gravel	90,000	60,000	460,000

life-cycle pattern, as illustrated in Figure 7.5. Subsequent salt applications are rarely as effective as the initial salting.

Various approaches have been tried over the years to solve this problem. One such product is shown in Figure 7.6. This chemically activated grounding electrode consists of a 2-in-diameter copper pipe filled with rock salt. Breathing holes are provided on the top of the assembly, and seepage holes are located at the bottom. The theory of operation is simple. Moisture is absorbed from the air (when available) and is then absorbed by the salt. This creates a solution that seeps out of the base of the device and conditions the soil in the immediate vicinity of the rod.

Another approach is shown in Figure 7.7. This device incorporates a number of ports (holes) in the assembly. Moisture from the soil (and rain) is absorbed through the ports. The metallic salts subsequently absorb the moisture, forming a saturated

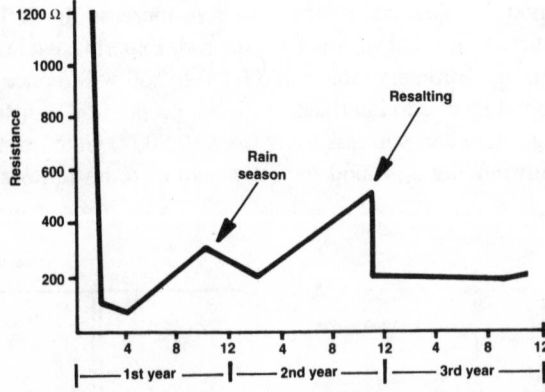

Figure 7.5 The effect of soil salting on ground-rod resistance with time. The expected resalting period, shown here as 2 years, varies depending on the local soil conditions and the amount of moisture present. (Adapted from: Roy Carpenter, "Improved Grounding Methods for Broadcasters," *Proceedings* of the SBE National Convention, Society of Broadcast Engineers, Indianapolis, 1987.)

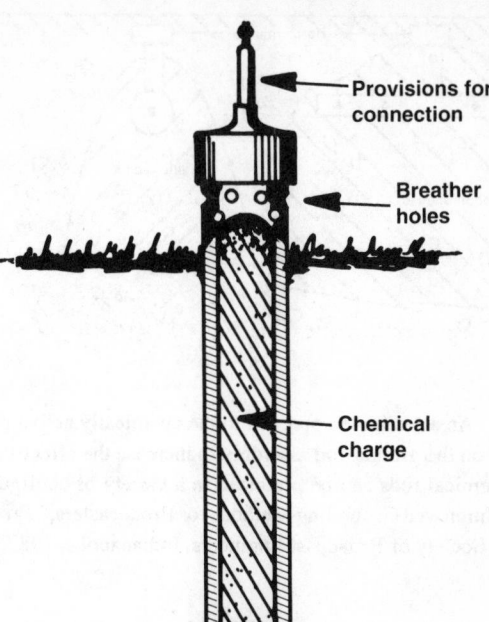

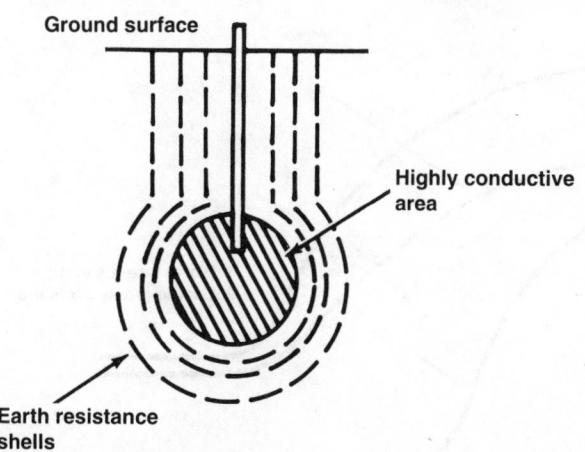

Figure 7.6 An air-breathing chemically activated ground rod: (a, *top*) breather holes at the top of the device permit moisture penetration into the chemical charge section of the rod; (b, *bottom*) a salt solution seeps out of the bottom of the unit to form a conductive shell. (Adapted from: Roy Carpenter, "Improved Grounding Methods for Broadcasters," *Proceedings* of the SBE National Convention, Society of Broadcast Engineers, Indianapolis, 1987.)

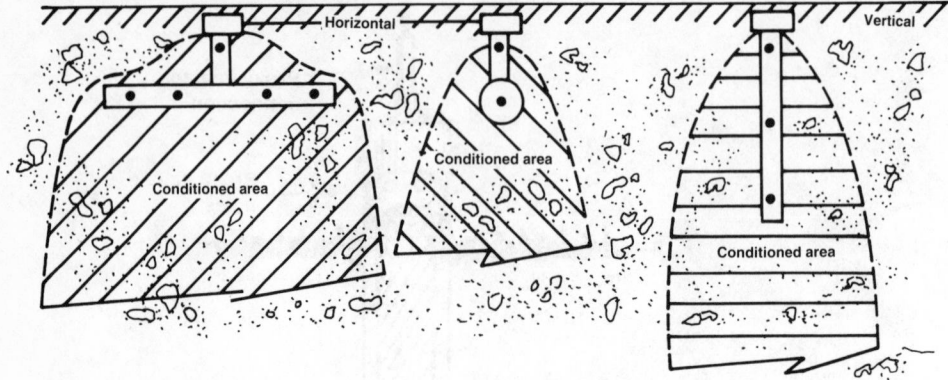

Figure 7.7   An alternative approach to the chemically activated ground rod. Multiple holes are provided on the ground-rod assembly to increase the effective earth-to-electrode interface. Note that chemical rods can be produced in a variety of configurations. (Adapted from: Roy Carpenter, "Improved Grounding Methods for Broadcasters," *Proceedings* of the SBE National Convention, Society of Broadcast Engineers, Indianapolis, 1987.)

solution that seeps out of the ports and into the earth-to-electrode hemisphere. Tests have shown that if the moisture content is within the required range, earth resistivity can be reduced by as much as 100:1. Figure 7.8 shows the measured performance of a typical chemical ground rod in three types of soil.

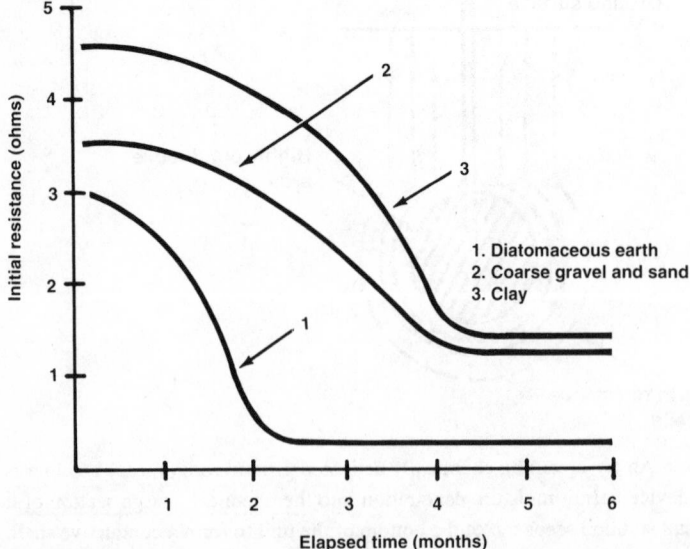

Figure 7.8   Measured performance of a chemical ground rod. (Adapted from: Roy Carpenter, "Improved Grounding Methods for Broadcasters," *Proceedings* of the SBE National Convention, Society of Broadcast Engineers, Indianapolis, 1987.)

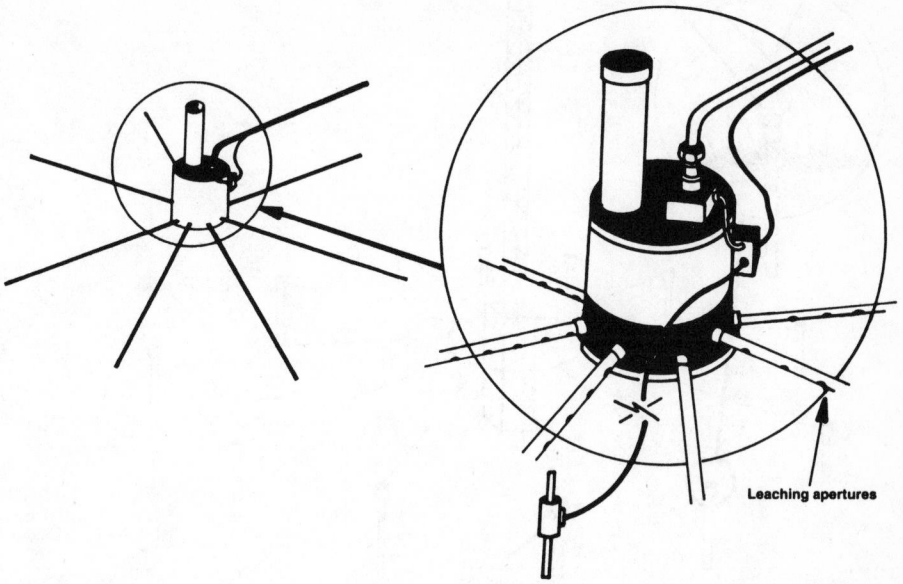

Leaching apertures

Figure 7.9 Hub and spoke counterpoise ground system. (Adapted from: Roy Carpenter, "Improved Grounding Methods for Broadcasters," *Proceedings* of the SBE National Convention, Society of Broadcast Engineers, Indianapolis, 1987.)

Implementations of chemical ground-rod systems vary depending on the application. Figure 7.9 illustrates a counterpoise ground consisting of multiple leaching apertures connected in a spoke fashion to a central hub. The system is serviceable in that additional salt compound can be added to the hub at required intervals to maintain the effectiveness of the ground. Figure 7.10 shows a counterpoise system made up of individual chemical ground rods interconnected with radial wires buried below the surface.

7.2.3 Ufer Ground System

Driving ground rods is not the only method of achieving a good earth-to-electrode interface. The concept of the *Ufer ground* has gained interest because of its simplicity and effectiveness. The Ufer approach (named for its developer), however, must be designed into a new structure. It cannot be added on later. The Ufer ground takes advantage of the natural chemical- and water-retention properties of concrete to provide an earth ground. Concrete retains moisture for 15 to 30 days after a rain. The material has a ready supply of ions to conduct current because of its moisture-retention properties, mineral content, and inherent pH. The large mass of any concrete foundation provides a good interface to ground.

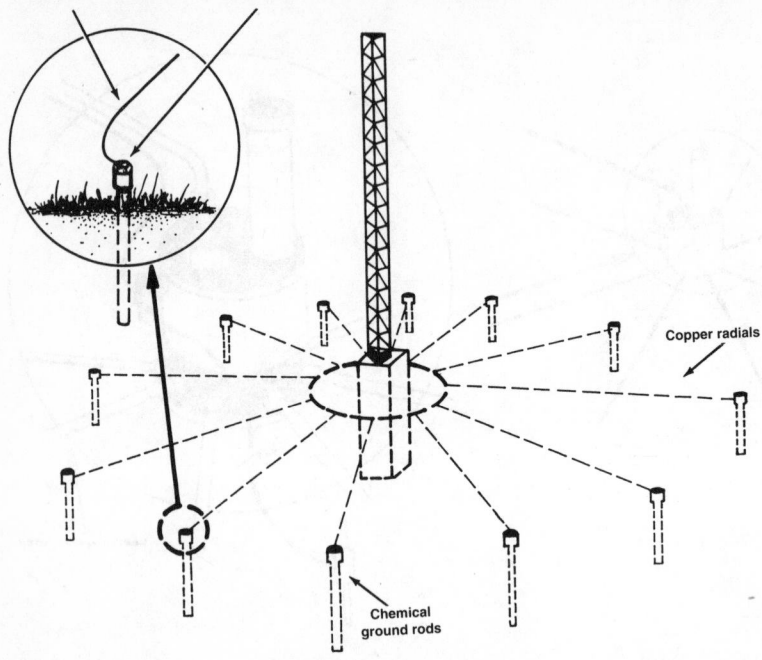

Figure 7.10 Tower grounding scheme using buried copper radials and chemical ground rods. (Adapted from: Roy Carpenter, "Improved Grounding Methods for Broadcasters," *Proceedings of the SBE National Convention*, Society of Broadcast Engineers, Indianapolis, 1987.)

A Ufer system, in its simplest form, is made by routing a solid-copper wire (no. 4 gauge or larger) within the foundation footing forms before concrete is poured. Figure 7.11 shows one such installation. The length of the conductor run within the concrete is important. Typically a 20-ft run (10 ft in each direction) provides a 5 Ω ground in 1000 Ω/m soil.

As an alternative, steel reinforcement bars (rebar) can be welded together to provide a rigid, conductive structure. A ground lug is provided to tie equipment to the ground system in the foundation. The rebar must be welded, not tied, together. If it is only tied, the resulting poor connections between rods can result in arcing during a current surge. This can lead to deterioration of the concrete in the affected areas.

The design of a Ufer ground is not to be taken lightly. Improper installation can result in a ground system that is subject to problems. The grounding electrodes must be kept a minimum of 3 in from the bottom and sides of the concrete to avoid the possibility of foundation damage during a large lightning surge. If an electrode is placed too near the edge of the concrete, a surge could turn the water inside the concrete to steam and break the foundation apart.

The Ufer approach also can be applied to guy-anchor points or the tower base, as illustrated in Figure 7.12. Welded rebar or ground rods sledged in place after the rebar cage is in position may be used. By protruding below the bottom concrete surface, the

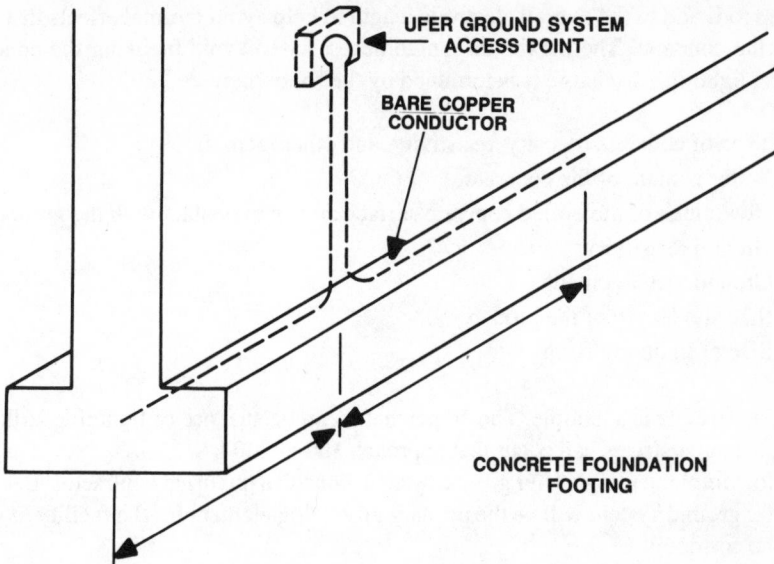

Figure 7.11   The basic concept of a Ufer ground system, which relies on the moisture-retentive properties of concrete to provide a large earth-to-electrode interface. Design of such a system is critical. Do not attempt to build a Ufer ground without the assistance of an experienced contractor. (Adapted from: Roger Block, "The Grounds for Lightning and EMP Protection," PolyPhaser Corporation, Gardnerville, NV, 1987.)

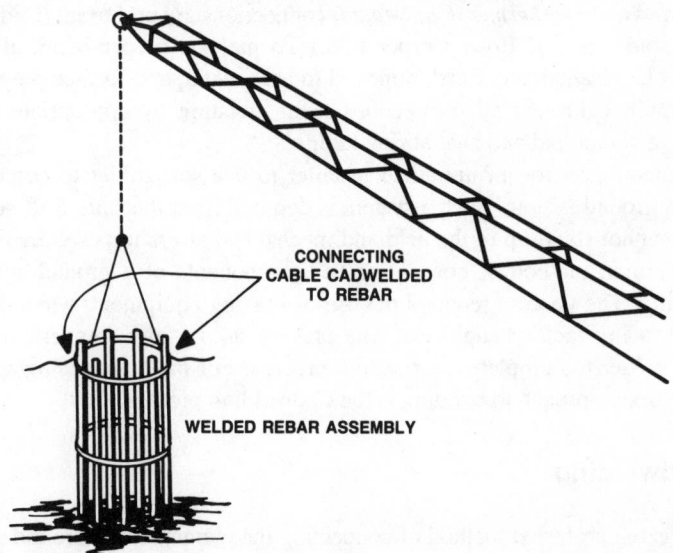

Figure 7.12   The Ufer ground system as applied to a transmission-tower base or guy-wire anchor point. When using this type of ground system, bond all rebar securely to prevent arcing in the presence of large surge currents. (Adapted from: Roger Block, "The Grounds for Lightning and EMP Protection," PolyPhaser Corporation, Gardnerville, NV, 1987.)

ground rods add to the overall electrode length to help avoid thermal effects that may crack the concrete. The maximum length necessary to avoid breaking the concrete under a lightning discharge is determined by the following:

- Type of concrete (density, resistivity, and other factors)
- Water content of the concrete
- How much of the buried concrete surface area is in contact with the ground
- Ground resistivity
- Ground water content
- Size and length of the ground rod
- Size of lightning flash

The last variable is a gamble. The 50 percent mean occurrence of lightning strikes is 18 A, but superstrikes can occur that approach 100 to 200 kA.

Before implementing a Ufer ground system, consult a qualified contractor. Because the Ufer ground system will be the primary grounding element for the facility, it must be done correctly.

7.3 BONDING GROUND-SYSTEM ELEMENTS

A ground system is only as good as the methods used to interconnect the component parts. Do not use soldered-only connections outside the equipment building. Crimped/brazed and *exothermic* (*Cadwelded*) connections are preferred. (Cadweld is a registered trademark of Erico Corporation.) To make a proper bond, all metal surfaces must be cleaned, any finish removed to bare metal, and surface preparation compound applied. Protect all connections from moisture by appropriate means, usually sealing compound and heat-shrink tubing.

It is not uncommon for an untrained installer to use soft solder to connect the elements of a ground system. Such a system is doomed from the start. Soft-soldered connections cannot stand up to the acid and mechanical stress imposed by the soil. The most common method of connecting the components of a ground system is silver-soldering. The process requires the use of brazing equipment, which may be unfamiliar to many facility engineers. The process uses a high-temperature/high-conductivity solder to complete the bonding process. For most grounding systems, however, the best approach to bonding is the Cadwelding process.

7.3.1 Cadwelding

Cadwelding is the preferred method of connecting the elements of a ground system. Molten copper is used to melt connections together, forming a permanent bond. This process is particularly useful in joining dissimilar metals. In fact, if copper and galvanized cable must be joined, Cadwelding is the only acceptable means. The completed connection will not loosen or corrode and will carry as much current as the

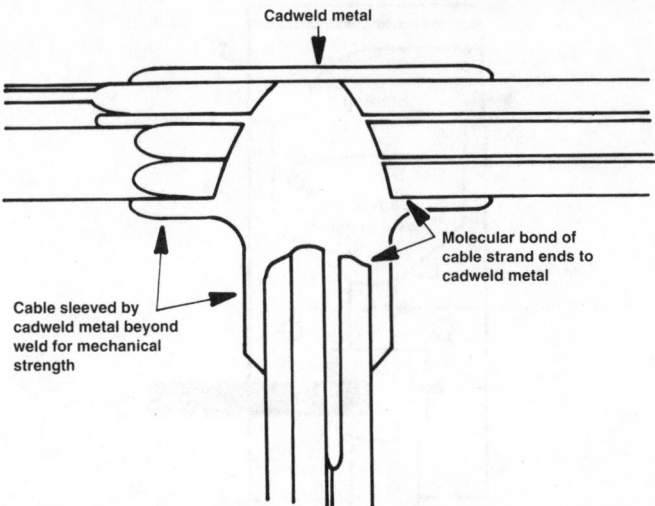

Cadweld metal

Molecular bond of
cable strand ends to
cadweld metal

Cable sleeved by
cadweld metal beyond
weld for mechanical
strength

Figure 7.13   The Cadweld bonding process. (Adapted from: Roger Block, "The Grounds for Lightning and EMP Protection," PolyPhaser Corporation, Gardnerville, NV, 1987.)

cable connected to it. Figure 7.13 illustrates the bonding that results from the Cadweld process.

Cadwelding is accomplished by dumping powdered metals (copper oxide and aluminum) from a container into a graphite crucible and igniting the material by means of a flint lighter. Reduction of the copper oxide by the aluminum produces molten copper and aluminum oxide slag. The molten copper flows over the conductors, bonding them together. The process is illustrated in Figure 7.14. Figure 7.15 shows a typical Cadweld mold. A variety of special-purpose molds are available to join different-size cables and copper strap. Figure 7.16 shows the bonding process for a copper-strap-to-ground-rod interface.

[a]   [b]   [c]

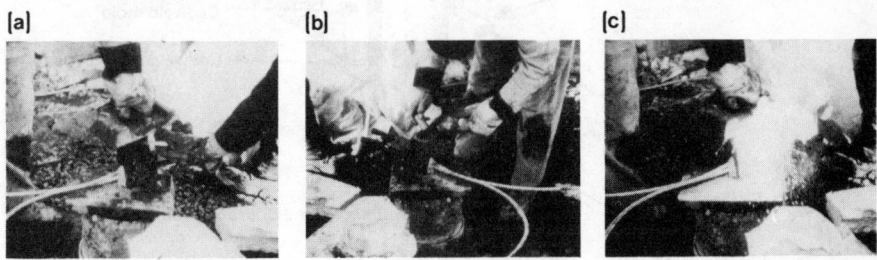

Figure 7.14   Cadwelding is the preferred method of joining the elements of a ground system. This photo sequence illustrates the procedure: (a) the powdered copper oxide and aluminum compound are added to the Cadweld mold after the conductors have been mechanically joined; (b) final preparation of the bond before igniting; (c) the chemical reaction that bonds the materials together.

224   AC POWER SYSTEMS

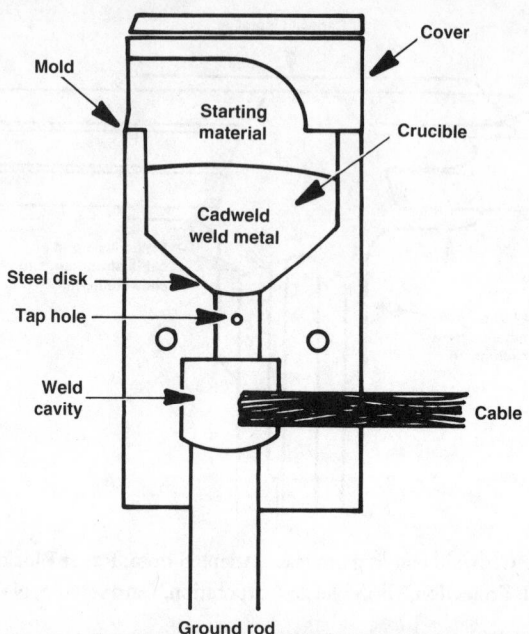

Figure 7.15   Typical Cadweld mold for connecting a cable to a ground rod. (Adapted from: Roger Block, "The Grounds for Lightning and EMP Protection," PolyPhaser Corporation, Gardnerville, NV, 1987.)

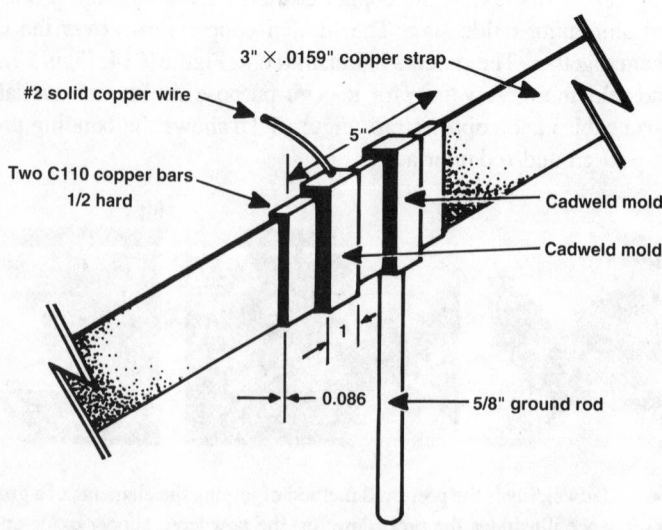

Figure 7.16   Cadweld mold for connecting a copper strap to a ground rod. (Adapted from: Roger Block, "The Grounds for Lightning and EMP Protection," PolyPhaser Corporation, Gardnerville, NV, 1987.)

7.3.2 Ground-System Inductance

Conductors interconnecting sections or components of an earth ground system must be kept as short as possible to be effective. The inductance of a conductor is a major factor in its characteristic impedance to surge energy. For example, consider a no. 6 AWG copper wire 10 m in length. The wire has a dc resistance of 0.013 Ω and an inductance of 10 μH. For a 1000 A lightning surge with a 1 μs rise time, the resistive voltage drop will be 13 V, but the reactive voltage drop will be 10 kV. Furthermore, any bends in the conductor will increase its inductance and further decrease the effectiveness of the wire. Bends in ground conductors should be gradual. A 90° bend is electrically equivalent to a 1/4-turn coil. The sharper the bend, the greater the inductance.

Because of the fast rise time of most lightning discharges and power-line transients, the *skin effect* plays an important role in ground-conductor selection. When planning a facility ground system, view the project from an RF standpoint.

7.3.3 Grounding Tower Elements

Guyed towers are better than self-supporting towers at dissipating lightning surge currents. This is true, however, only if the guy anchors are grounded properly. Use of the Ufer technique is one way of effectively grounding the anchors. For anchors not provided with a Ufer ground during construction, other, more conventional, techniques may be used.

Never rely on the turnbuckles as a path for lightning energy. The current resulting from a large flash probably will weld the turnbuckles in position. If the turnbuckles are provided with a safety loop of guy cable (as they should be), the loop may be damaged where it contacts the guys and turnbuckle. Figure 7.17 shows the preferred method of grounding guy wires: Tie them together above the loop and turnbuckles. Do not make these connections with copper wire, even if they are Cadwelded. During periods of precipitation, water shed from the top copper wire will carry ions that may react with the lower galvanized (zinc) guy wires. This reaction washes off the zinc coating, allowing rust to develop.

The best way to make the connection is with all-galvanized materials. This includes the grounding wire, clamps, and ground rods. It may not be possible to use all galvanized materials because, at some point, a connection to copper conductors will be required. *Battery action* caused by the dissimilar metal junction may allow the zinc to act as a sacrificed anode. The zinc eventually will disappear into the soil, leaving a bare steel conductor that can fall victim to rust.

Ruling out an all-galvanized system, the next best scheme uses galvanized wire (guy-wire material) to tie the guy wires together. Just above the soil, Cadweld the galvanized wire to a copper conductor that penetrates below grade to the perimeter ground system. The height above grade for the connection is determined by the local snowfall or flood level. The electric conductivity of snow, although low, can cause battery action from the copper through the snow to the zinc. The Cadwelded joint should be positioned above the usual snow or flood level.

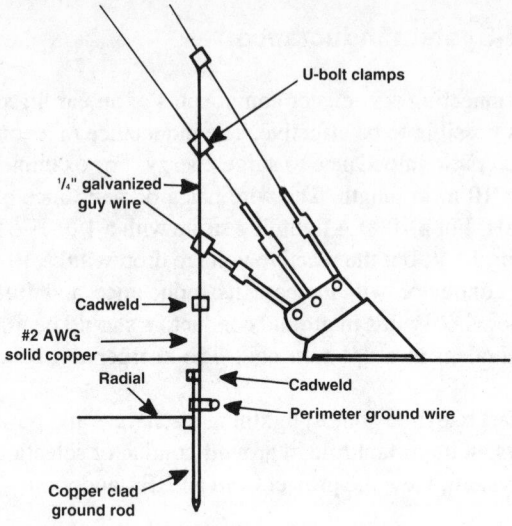

U-bolt clamps

¼" galvanized
guy wire

Cadweld

#2 AWG
solid copper

Radial

Cadweld

Perimeter ground wire

Copper clad
ground rod

Figure 7.17   Recommended guy-anchor grounding procedure. (Adapted from: Roger Block, "The Grounds for Lightning and EMP Protection," PolyPhaser Corporation, Gardnerville, NV, 1987.)

7.3.4   Ground-Wire Dressing

Figure 7.18 illustrates the proper way to bond the tower base ground leads to the buried ground system. Dress the leads close to the tower from the lowest practical structural element at the base. Keep the conductors as straight and short as possible. Avoid any sharp bends. Attach the ground wires to the tower only at one or more existing bolts (or holes). Do not drill any holes into the tower. Do not loosen any bolts to make the ground-wire attachment. Use at least two 3- to 4-in copper straps between the base of the tower and the buried ground system. Position the straps next to the concrete pier of the tower base. For towers more than 200 ft in height, use four copper straps, one on each side of the pier.

Figure 7.19 illustrates the proper way to bond guy wires to the buried ground system. The lead is dressed straight down from the topmost to the lowest guy. It should conform as close to vertical as possible, and be dressed downward from the lower side of each guy wire after connecting to each wire. To ensure that no arcing will occur through the turnbuckle, a connection from the anchor plate to the perimeter ground circle is recommended. No. 2 gauge copper wire is recommended. This helps minimize the unavoidable inductance created by the conductor being in the air. Interconnect leads that are suspended in air must be dressed so that no bending radius is less than 8 in.

A *perimeter* ground — a circle of wire connected at several points to ground rods driven into the earth — should be installed around each guy-anchor point. The perimeter system provides a good ground for the anchor, and when tied together with the tower base radials, acts to rapidly dissipate lightning energy in the event of a flash. Tower base radials are buried wires interconnected with the tower base ground that extend away from the center point of the structure.

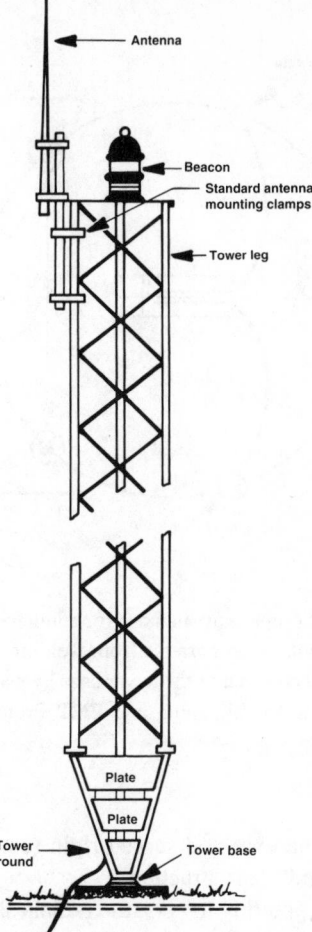

Figure 7.18   Ground-conductor dressing for the base of a guyed tower. (Adapted from: Roger Block, "The Grounds for Lightning and EMP Protection," PolyPhaser Corporation, Gardner-ville, NV, 1987.)

The required depth of the perimeter ground and the radials depends upon soil conductivity. Generally speaking, however, about 8 in below grade is sufficient. In soil with good conductivity, the perimeter wire may be as small as no. 10 gauge. Because no. 2 gauge is required for the segment of conductor suspended in air, it may be easier to use no. 2 throughout. An added advantage is that the same size Cadweld molds may be used for all bonds.

7.3.5   Facility Ground Interconnection

Any radial that comes within 2 ft of a conductive structure must be tied into the ground system. Bury the interconnecting wire, if possible, and approach the radial at a 45°

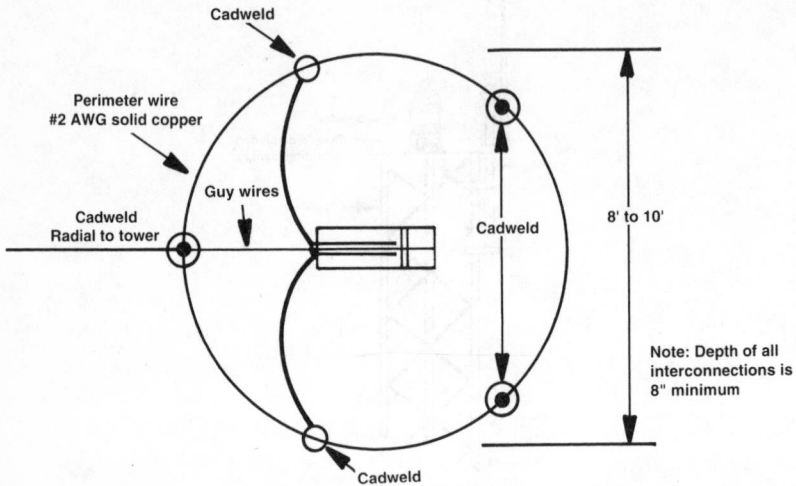

Figure 7.19 Top view of proper guy-anchor grounding techniques. A properly dressed and installed ground wire prevents surge currents from welding turnbuckles and damaging safety loops. The perimeter ground connects to the tower base by way of a radial wire. (Adapted from: Roger Block, "The Grounds for Lightning and EMP Protection," PolyPhaser Corporation, Gardnerville, NV, 1987.)

angle, pointing toward the expected surge origin (usually the tower). Cadweld the conductor to the radial and to the structure.

For large-base, self-supporting towers, the radials should be split among each leg pad, as shown in Figure 7.20. The radials may be brought up out of the soil (in air) and each attached at spaced locations around the foot pad. Some radials may have to be tied together first and then joined to the foot pad. Remember, if space between the radial lines can be maintained, less mutual inductance (coupling) will exist, and the system surge impedance will be lower.

It is desirable to have a continuous one-piece ring with rods around the equipment building. Connect this ring to one, and only one, of the tower radials, thus forming no radial loops. See Figure 7.21. Bury the ring to the same depth as the radials to which it interconnects. Connect power-line neutral to the ring. *Warning*: Substantial current may flow when the power-line neutral is connected to the ring. Follow safety procedures when making this connection.

Install a ground rod (if the utility company has not installed one) immediately outside the generator or utility company vault, and connect this rod to the equipment building perimeter ground ring. Route a no. 1/0 insulated copper cable from the main power panel inside the generator vault to the ground rod outside the vault. Cut the cable to length, strip both ends, and tie one end to the power-line neutral at the main power panel in the generator vault. Connect the other end to the ground rod. *Warning*:

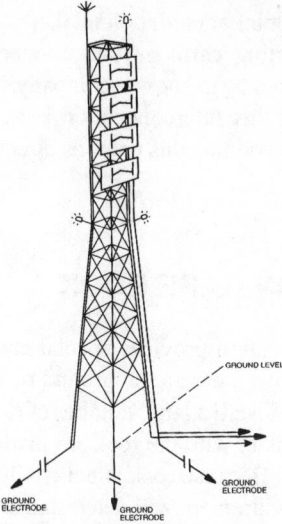

Figure 7.20   Interconnecting a self-supporting tower to the buried ground system.

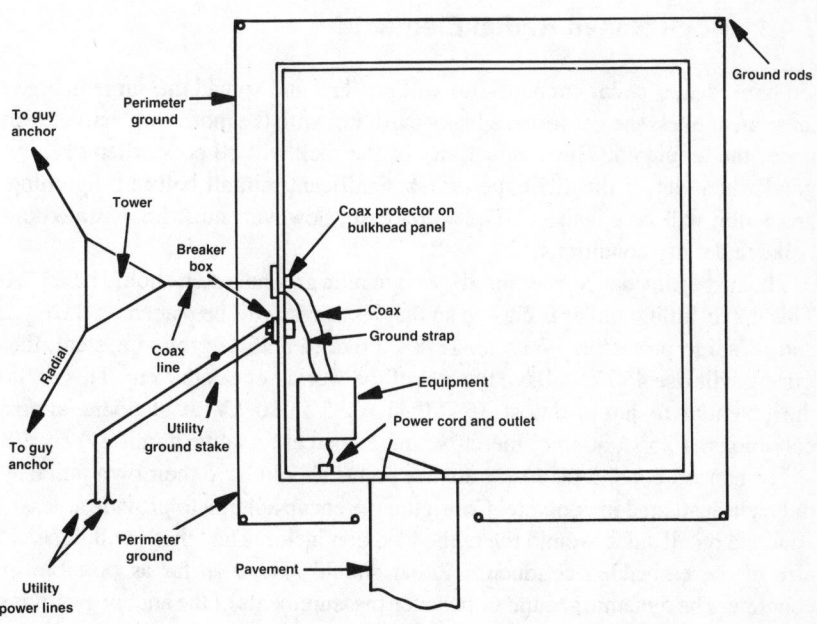

Figure 7.21   Interconnecting the metal structures of a facility to the ground system. (Adapted from: Roger Block, "The Grounds for Lightning and EMP Protection," PolyPhaser Corporation, Gardnerville, NV, 1987.)

Use care when making this connection. Hazardous voltage may exist between the power-line neutral and any point at earth potential.

Do not remove any existing earth ground connections to power-line neutral, particularly if they are installed by the power company. To do so may violate the local electrical code. The goal of this interconnection is to minimize noise that may be present on the neutral, and to conduct this noise as directly as possible outside to earth ground.

7.4   GROUNDING ON BARE ROCK

A bare rock mountaintop location provides special challenges to the facility design engineer. There is no soil, thus there are no ground rods. Radials are the only means to develop a ground system. Install a large number of radials, laid straight, but not too taut. The portions not in contact with the rock are in air and form an inductance that will choke the surge current. Because rock is not conductive when it is dry, keep the radials short. Only a test measurement will determine how short the radials should be. A conventional earth-resistance tester will tell only half the story (besides, ground rods cannot be placed in rock for such a measurement). A dynamic ground tester offers the only way to obtain the true surge impedance of the system.

7.4.1   Rock-Based Radial Elements

On bare rock, a radial counterpoise will conduct and spread the surge charge over a large area. In essence, it forms a leaky capacitor with the more conductive earth on or under the mountain. The conductivity of the rock will be poor when dry, but quite good when wet. If the site experiences significant rainfall before a lightning flash, protection will be enhanced. The worst case, however, must be assumed: an early strike under dry conditions.

The surge impedance, measured by a dynamic ground tester, should be 25 Ω or less. This upper-limit number is chosen so that less stress will be placed on the equipment and its surge protectors. With an 18 kA strike to a 25 Ω ground system, the entire system will rise 450 kV above the rest of the world at peak current. This voltage has the potential to jump almost 15.750 in (0.35 in/10 kV at standard atmospheric conditions of 25°C, 30 in of mercury, and 50 percent relative humidity).

For nonsoil conditions, tower anchor points should have their own radial systems or be encapsulated in concrete. Configure the encapsulation to provide at least 3 in of concrete on all sides around the embedded conductor. The length will depend on the size of the embedded conductor. Rebar should extend as far as possible into the concrete. The dynamic ground impedance measurements of the anchor grounds should each be less than 25 Ω.

The size of the bare conductor for each tower radial (or for an interconnecting wire) will vary, depending on soil conditions. On rock, a bare no. 1/0 or larger wire is recommended. Flat, solid-copper strap would be better, but may be blown or ripped if not covered with soil. If some amount of soil is available, no. 6 cable should be sufficient. Make the interconnecting radial wires continuous, and bury them as deep as possible; however, the first 6 to 10 in will have the most benefit. Going below 18 in will not be cost-effective, unless in a dry, sandy soil where the water table can be reached and ground-rod penetration is shallow. If only a small amount of soil exists, use it to cover the radials to the extent possible. It is more important to cover radials in the area near the tower than at greater distances. If, however, soil exists only at the outer distances and cannot be transported to the inner locations, use the soil to cover the outer portions of the radials.

If soil is present, install ground rods along the radial lengths. Spacing between ground rods is affected by the depth that each rod is driven; the shallower the rod, the closer the allowed spacing. Because the ultimate depth a rod may be driven cannot always be predicted by the first rod driven, use a maximum spacing of 15 ft when selecting a location for each additional rod. Drive rods at building corners first (within 24 in but not closer than 6 in to a concrete footer unless that footer has an encapsulated Ufer ground), then fill in the space between the corners with additional rods.

Drive the ground rods in place. Do not auger; set in place, then back fill. The soil compactness is never as great on augured-hole rods when compared with driven rods. The only exception is when a hole is augured or blasted for a ground rod or rebar and then back-filled in concrete. Because concrete contains lime (alkali base) and is porous, it absorbs moisture readily, giving it up slowly. Electron carriers are almost always present, making the substance a good conductor.

If a Ufer ground is not being implemented, the radials may be Cadwelded to a subterranean ring, with the ring interconnected to the tower foot pad via a minimum of three no. 1/0 wires spaced at 120° angles and Cadwelded to the radial ring.

7.5 TRANSMISSION-SYSTEM GROUNDING

Nature can devastate a communications site. Lightning can seriously damage an unprotected facility with little or no warning, leaving an expensive and time-consuming repair job. The first line of defense is proper grounding of the communications system.

7.5.1 Transmission Line

All coax and waveguide lines must include grounding kits for bonding the transmission line at the antenna. On conductive structures, this may be accomplished by bonding the tail of the grounding kit to the structure itself. Remove all nonconductive paint and corrosion before attachment. Do not drill holes, and do not loosen any existing tower member bolts. Antenna clamping hardware may be used, or an all-stainless-steel hose clamp of the appropriate size may be substituted. The location of the tower-top ground is not as critical as the bottom grounding kit.

On nonconductive structures, a no. 1/0 or larger wire must be run down the tower. Bond the transmission-line grounding kit to this down-run. Keep the wire as far away from all other conductive runs (aviation lights, coax, and waveguide) as possible. Separation of 2 ft is preferred; 18 in is the minimum. If any other ground lines, conduit, or grounded metallic-structure members must be traversed that are closer than 18 in, they too must be grounded to the down-lead ground line to prevent flashover.

At the point where the coax or waveguide separates from the conductive structure (metal tower), a coax or waveguide grounding kit must be installed. Secure the connection to a large vertical structure member with a small number of joints. This requirement is not critical if the cables emanate from a center tower face. Attach to a structural member as low as possible on the tower.

Dress the grounding kit tails in a nearly straight 45° downward angle to the tower member. On nonconductive structures, a metal busbar must be used. Ground the bar to one or more of the vertical downconductors, and as close to ground level as possible.

Coaxial cables, lighting conduit, and other lines on the tower must be secured properly to the structure. Figure 7.22 illustrates several common attachment methods.

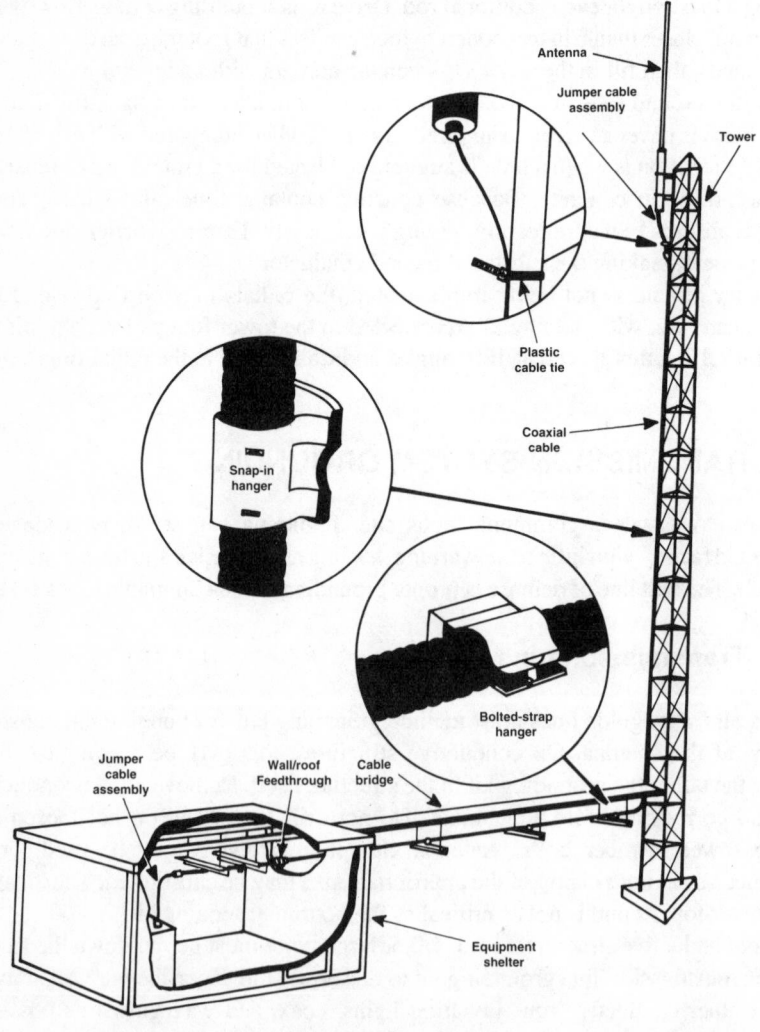

Figure 7.22 Transmission-line mounting and grounding procedures for a communications site.

7.5.2 Cable Considerations

Ground-strap connections must withstand weathering and maintain low electrical resistance between the grounded component and earth. Corrosion impairs ground-strap performance. Braided-wire ground straps should not be used in outside installations. Through capillary action resembling that of a wick, the braid conveys water, which accelerates corrosion. Eventually, advancing corrosion erodes the ground-strap cable. Braid also may act as a duct that concentrates water at the bond point. This water speeds corrosion, which increases the electrical resistance of the bond. A jacketed, seven-strand copper wire strap (no. 6 to no. 2) is recommended for transmission-line grounding at the tower.

7.5.3 Satellite Antenna Grounding

Most satellite receiving/transmitting antenna piers are encapsulated in concrete. Consideration should be given, therefore, to implementing a Ufer ground for the satellite dish. A 4-in-diameter pipe, submerged 4 to 5 ft in an 18-in-diameter (augered)

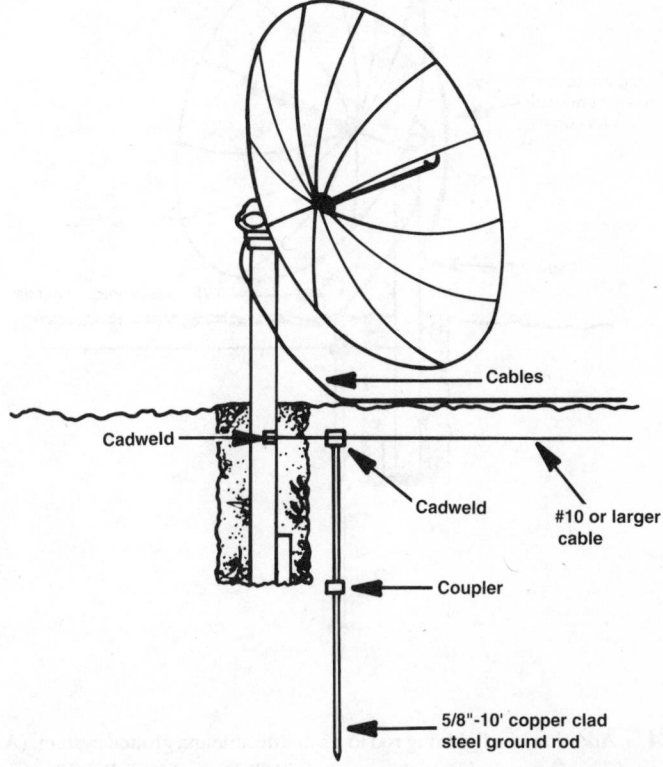

Figure 7.23 Grounding a satellite receiving antenna. (Adapted from: Roger Block, "The Grounds for Lightning and EMP Protection," PolyPhaser Corporation, Gardnerville, NV, 1987.)

hole will provide a good start for a Ufer-based ground system. It should be noted that an augered hole is preferred because digging and repacking the soil around the pier will create higher ground resistance. In areas of good soil conductivity (100 Ω/m or less) this basic Ufer system may be adequate for the antenna ground. Figure 7.23 shows the preferred method: a hybrid Ufer/ground-rod and radial system. A cable connects the mounting pipe (Ufer ground) to a separate driven ground rod. The cable then is connected to the facility ground system. In areas of poor soil conductivity, additional ground rods are driven at increments (2.2 times the rod length) between the satellite dish and the facility ground system. Run all cables underground for best performance. Make the interconnecting copper wire no. 10 size or larger; bury the wire at least 8 in below finished grade. Figure 7.24 shows the addition of a lightning rod to the satellite dish.

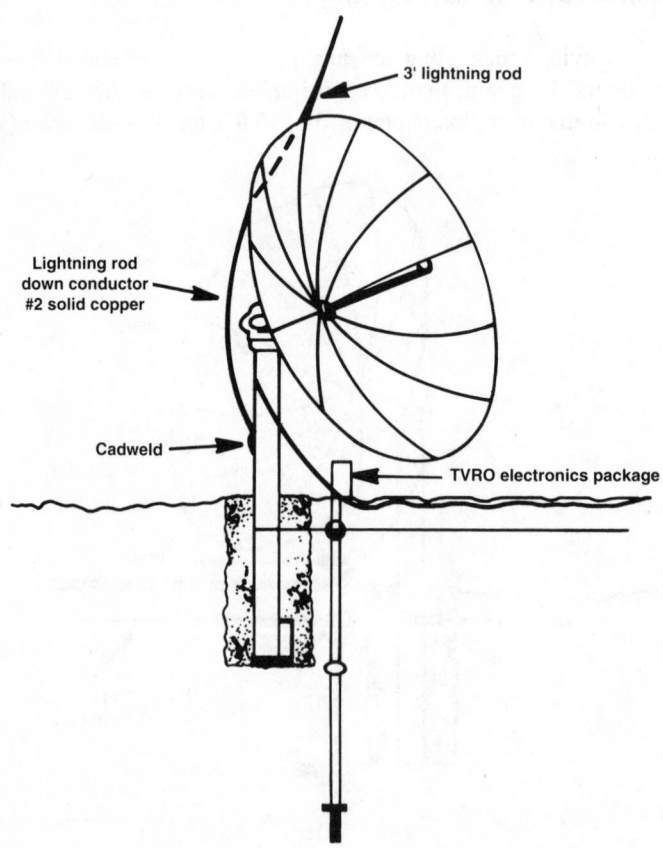

Figure 7.24   Addition of a lightning rod to a satellite antenna ground system. (Adapted from: Roger Block, "The Grounds for Lightning and EMP Protection," PolyPhaser Corporation, Gardnerville, NV, 1987.)

7.6   DESIGNING A BUILDING GROUND SYSTEM

After the required grounding elements have been determined, they must be connected together into a unified system. Many different approaches may be taken, but the goal is the same: Establish a low-resistance, low-inductance path to surge energy. Figure 7.25 shows a building ground system using a combination of ground rods and buried bare-copper radial wires. This design is appropriate when the building is large or located in an urban area. This approach also may be used when the facility is located in a highrise building that requires a separate ground system. Most newer office buildings have ground systems designed into them. If a comprehensive building ground system is provided, use it. For older structures (constructed of wood or brick), a separate ground system will be required.

Figure 7.26 shows another approach in which a perimeter ground strap is buried around the building and ground rods are driven into the earth at regular intervals (2.2 times the rod length). The ground ring consists of a one-piece copper conductor that is bonded to each ground rod.

If a transmission or microwave tower is located at the site, connect the tower ground system to the main ground point via a copper strap. The width of the strap must be at least 1 percent of the length and, in any event, not less than 3 in wide. The building ground system is not a substitute for a tower ground system, no matter what the size

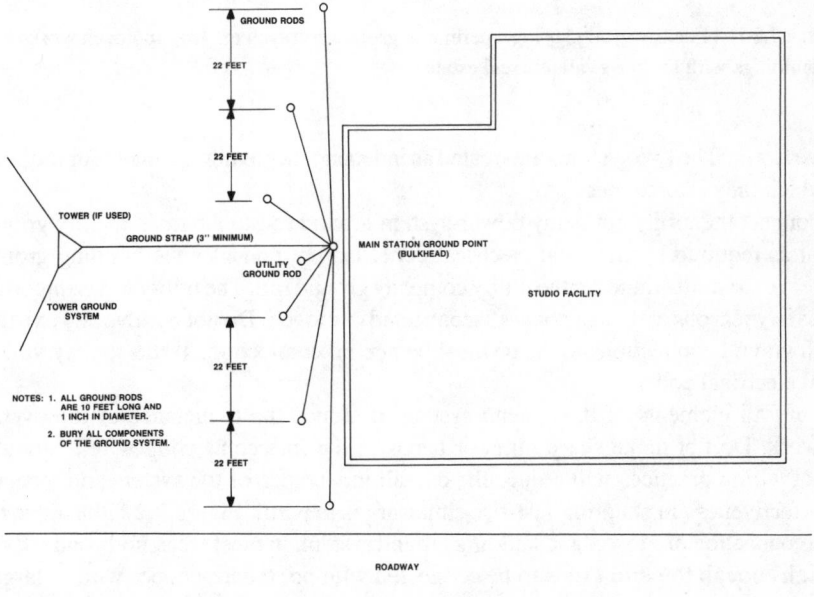

Figure 7.25   A facility ground system using the hub-and-spoke approach. The available real estate at the site will dictate the exact configuration of the ground system. If a tower is located at the site, the tower ground system is connected to the building ground as shown.

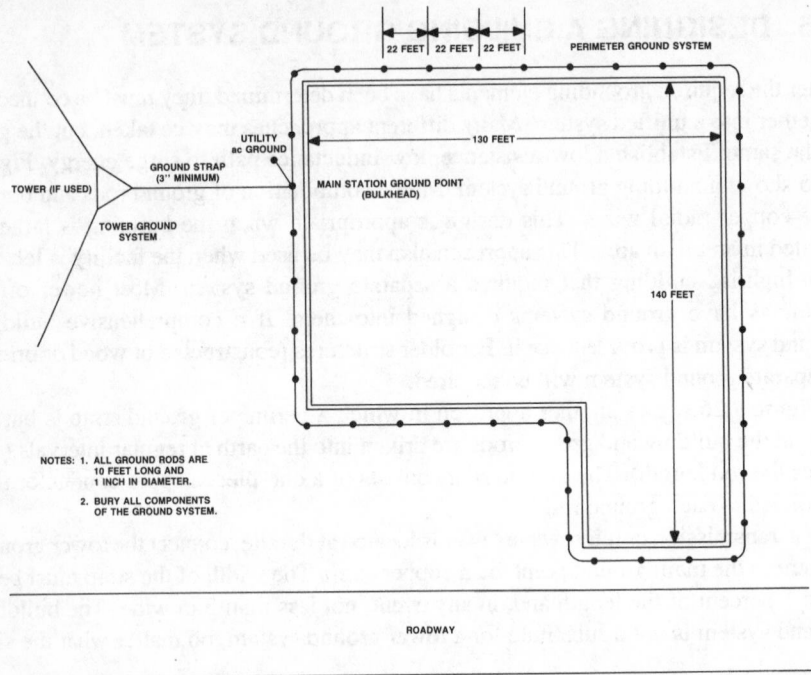

Figure 7.26   Facility ground using a perimeter ground-rod system. This approach works well for buildings with limited available real estate.

of the tower. The two systems are treated as independent elements, except for the point at which they interconnect.

Connect the utility company power-system ground rod to the main facility ground point as required by the local electrical code. Do not consider the building ground system to be a substitute for the utility company ground rod. The utility rod is important for safety reasons and must not be disconnected or moved. Do not remove any existing earth ground connections to the power-line neutral connection. To do so may violate local electrical code.

Bury all elements of the ground system to reduce the inductance of the overall network. Do not make sharp turns or bends in the interconnecting wires. Straight, direct wiring practices will reduce the overall inductance of the system and increase its effectiveness in shunting fast-rise-time surges to earth. Figure 7.27 illustrates the interconnection of a tower and building ground system. In most areas, soil conductivity is high enough to permit rods to be connected with no. 6 bare-copper wire or larger. In areas of sandy soil, use copper strap. A wire buried in low-conductivity, sandy soil tends to be inductive and less effective in dealing with fast-rise-time current surges. As stated previously, make the width of the ground strap at least 1 percent of its overall length. Connect buried elements of the system as shown in Figure 7.28.

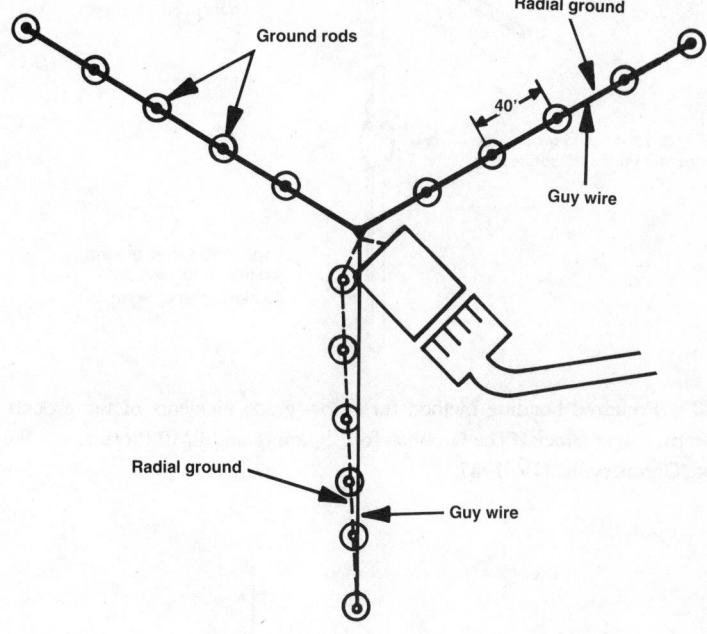

Figure 7.27   A typical guy-anchor and tower-radial grounding scheme. The radial ground is no. 6 copper wire. The ground rods are 5/8 in • 10 ft. (Adapted from: Roger Block, "The Grounds for Lightning and EMP Protection," PolyPhaser Corporation, Gardnerville, NV, 1987.)

For small installations with a low physical profile, a simplified grounding system can be implemented, as shown in Figure 7.29. A grounding plate is buried below grade level, and a ground wire ties the plate to the microwave tower mounted on the building.

7.6.1 Bulkhead Panel

The bulkhead panel is the cornerstone of an effective facility grounding system. The concept of the bulkhead is simple: Establish one reference point to which all cables entering and leaving the equipment building are grounded and to which all transient-suppression devices are mounted. Figure 7.30 shows a typical bulkhead installation

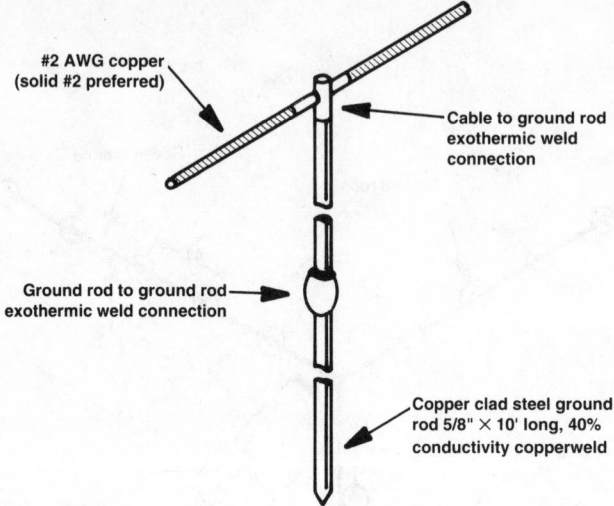

#2 AWG copper
(solid #2 preferred)

Cable to ground rod
exothermic weld
connection

Ground rod to ground rod
exothermic weld connection

Copper clad steel ground
rod 5/8" × 10' long, 40%
conductivity copperweld

Figure 7.28 Preferred bonding method for below-grade elements of the ground system. (Adapted from: Roger Block, "The Grounds for Lightning and EMP Protection," PolyPhaser Corporation, Gardnerville, NV, 1987.)

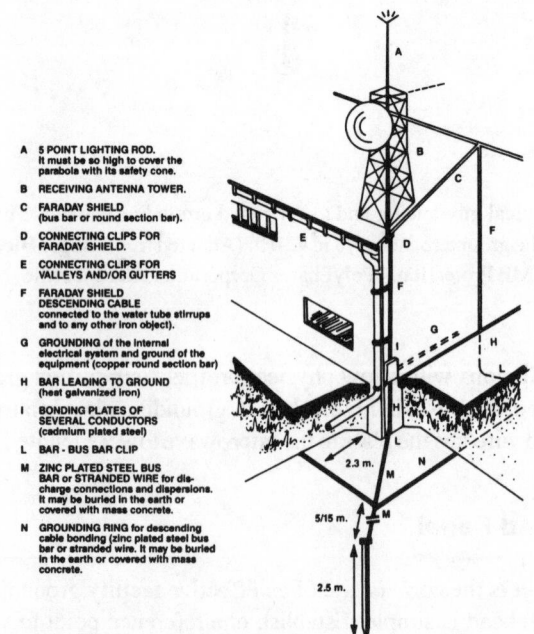

A   5 POINT LIGHTING ROD.
It must be so high to cover the
parabola with its safety cone.

B   RECEIVING ANTENNA TOWER.

C   FARADAY SHIELD
(bus bar or round section bar).

D   CONNECTING CLIPS FOR
FARADAY SHIELD.

E   CONNECTING CLIPS FOR
VALLEYS AND/OR GUTTERS

F   FARADAY SHIELD
DESCENDING CABLE
connected to the water tube stirrups
and to any other iron object).

G   GROUNDING of the internal
electrical system and ground of the
equipment (copper round section bar)

H   BAR LEADING TO GROUND
(heat galvanized iron)

I   BONDING PLATES OF
SEVERAL CONDUCTORS
(cadmium plated steel)

L   BAR - BUS BAR CLIP

M   ZINC PLATED STEEL BUS
BAR or STRANDED WIRE for discharge connections and dispersions.
It may be buried in the earth or
covered with mass concrete.

N   GROUNDING RING for descending
cable bonding (zinc plated steel bus
bar or stranded wire. It may be buried
in the earth or covered with mass
concrete.

2.3 m.

5/15 m.

2.5 m.

Figure 7.29 Grounding a small microwave transmission tower.

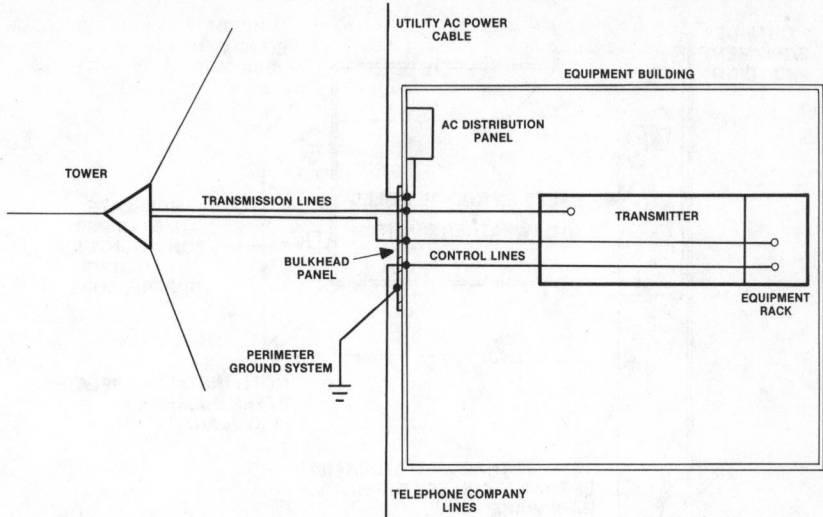

Figure 7.30   The basic design of a bulkhead panel for a facility. The bulkhead establishes the grounding reference point for the plant.

for a broadcast or communications facility. The panel size depends on the spacing, number, and dimensions of the coaxial lines, power cables, and other conduit entering or leaving the building.

To provide a weatherproof point for mounting transient-suppression devices, the bulkhead can be modified to accept a subpanel, as shown in Figure 7.31. The subpanel is attached so that it protrudes through an opening in the wall and creates a secondary plate on which transient suppressors are mounted and grounded. A typical cable/suppressor-mounting arrangement for a communications site is shown in Figure 7.32. To handle the currents that may be experienced during a lightning strike or large transient on the utility company ac line, the bottommost subpanel flange (which joins the subpanel to the main bulkhead) must have a total surface-contact area of at least 0.75 in^2 per transient suppressor.

Because the bulkhead panel will carry significant current during a lightning strike or ac line disturbance, it must be constructed of heavy material. The recommended material is 1/8-in C110 (solid copper) 1/2 hard. This type of copper stock weighs nearly 5 1/2 lb/ft^2 and sells for about $2.25/lb (U.S.). Installing a bulkhead is an expensive job, but one that will pay dividends for the life of the facility. Use 18-8 stainless-steel mounting hardware to secure the subpanel to the bulkhead.

Because the bulkhead panel establishes the central grounding point for all equipment within the building, it must be tied to a low-resistance (and low-inductance) perimeter ground system. The bulkhead establishes the *main facility ground point*, from which all grounds inside the building are referenced. A typical bulkhead installation for a small communications site is shown in Figure 7.33.

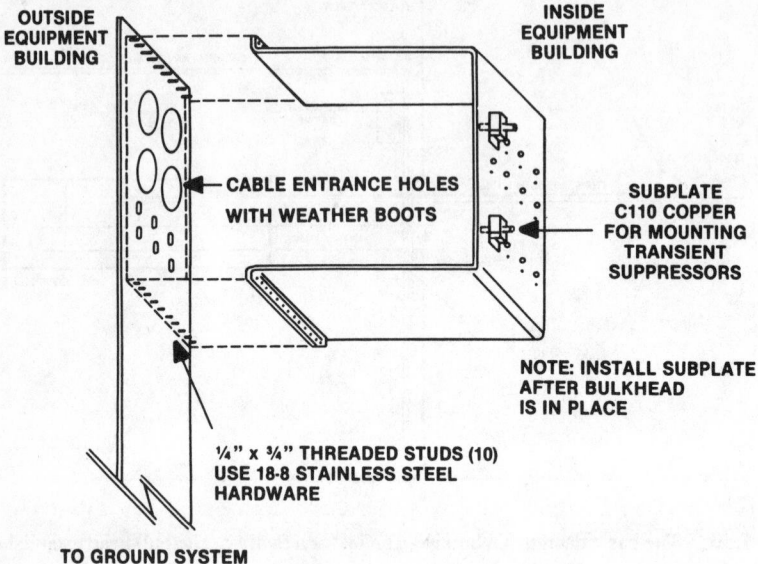

Figure 7.31 The addition of a subpanel to a bulkhead as a means of providing a mounting surface for transient-suppression components. To ensure that the bulkhead is capable of handling high surge currents, use the hardware shown. (Adapted from: Roger Block, "The Grounds for Lightning and EMP Protection," PolyPhaser Corporation, Gardnerville, NV, 1987.)

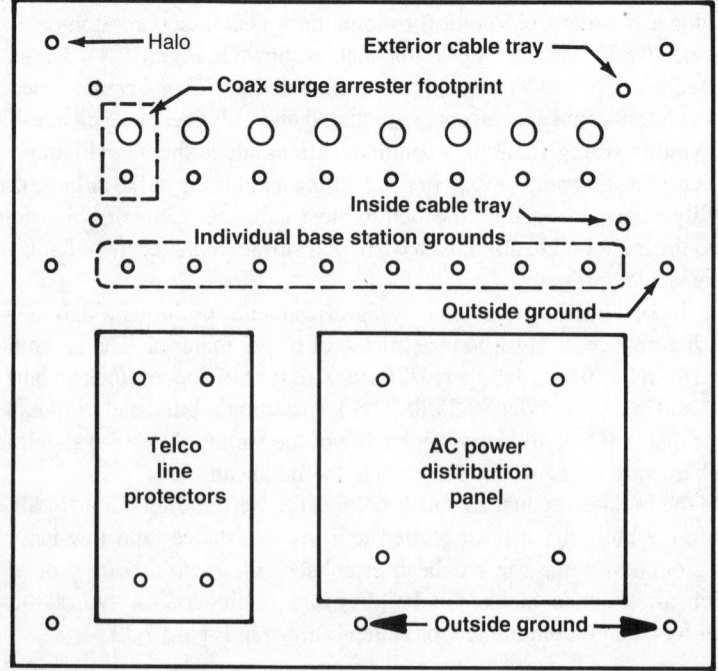

Figure 7.32 Mounting-hole layout for a communications site bulkhead subpanel.

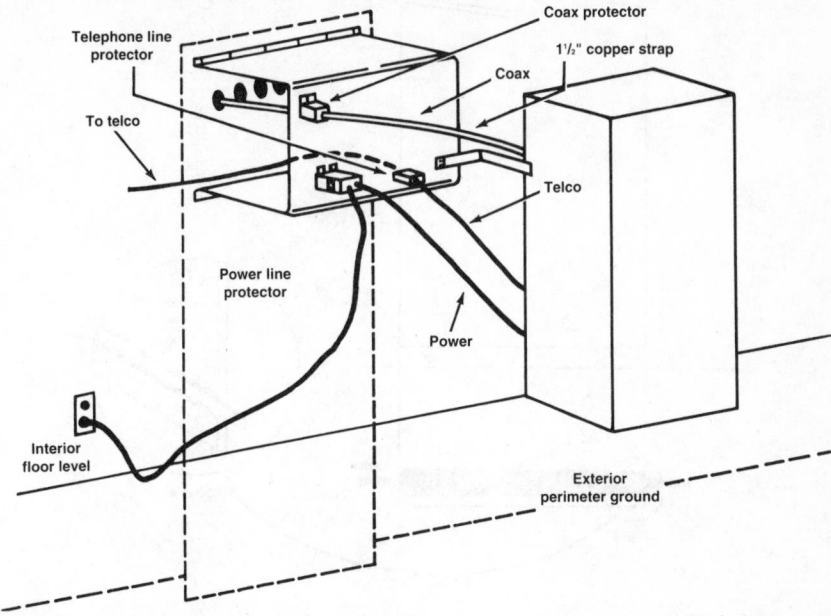

Figure 7.33   Bulkhead installation at a small communications site. (Adapted from: Roger Block, "The Grounds for Lightning and EMP Protection," PolyPhaser Corporation, Gardnerville, NV, 1987.)

7.6.2 Bulkhead Grounding

A properly installed bulkhead panel will exhibit lower impedance and resistance to ground than any other equipment or cable grounding point at the facility. Waveguide and coax line grounding kits should be installed at the bulkhead panel as well as at the tower. Dress the kit tails downward at a straight 45° angle using 3/8-in stainless-steel hardware to the panel. Position the stainless-steel lug at the tail end flat against a cleaned spot on the panel. Joint compound will be needed for aluminum and is recommended for copper panels.

Because the bulkhead panel will be used as the central grounding point for all the equipment inside the building, the lower the inductance to the perimeter ground system, the better. The best arrangement is to simply extend the bulkhead panel down the outside of the building, below grade, to the perimeter ground system. This will give the lowest resistance and the smallest inductive voltage drop. This approach is illustrated in Figure 7.34.

If cables are used to ground the bulkhead panel, secure the interconnection to the outside ground system along the bottom section of the panel. Use multiple no. 1/0 or larger copper wire or several solid-copper straps. If using strap, attach with stainless-steel hardware, and apply joint compound for aluminum bulkhead panels. Clamp and Cadweld, or silver-solder for copper/brass panels. If no. 1/0 or larger wire is used,

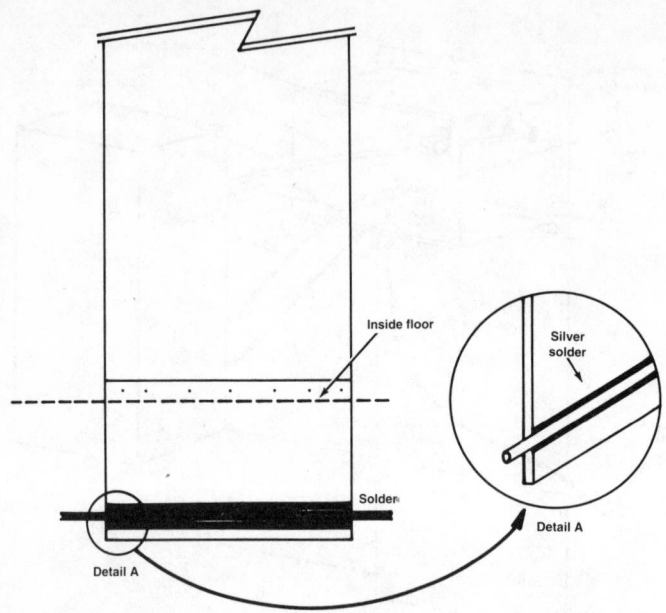

Figure 7.34   The proper way to ground a bulkhead panel and provide a low-inductance path for surge currents stripped from cables entering and leaving the facility. The panel extends along the building exterior to below grade. It is silver-soldered to a no. 2/0 copper wire that interconnects with the outside ground system. (Adapted from: Roger Block, "The Grounds for Lightning and EMP Protection," PolyPhaser Corporation, Gardnerville, NV, 1987.)

employ crimp lug and stainless-steel hardware. Measure the dc resistance. It should be no greater than 0.01 Ω between the ground system and the panel. Repeat this measurement on an annual basis.

If the antenna feed lines do not enter the equipment building via a bulkhead panel, treat them in the following manner:

1. Mount a feed-line ground bar on the wall of the building approximately 4 in below the feed-line entry point.
2. Connect the outer conductor of each feed line to the feed-line ground bar using an appropriate grounding kit.
3. Connect a no. 1/0 cable or 3- to 6-in-wide copper strap between the feed-line ground bar and the external ground system. Make the joint a Cadweld or silver-solder connection.
4. Mount coaxial arresters on the edge of the bar.
5. Weatherproof all connections.

7.6.3 Lightning Protectors

A variety of lightning arresters are available for use on coaxial transmission lines, utility ac feeds, and telephone cables. The protector chosen must be carefully matched to the requirements of the application. Do not use air gap protectors because these types are susceptible to air pollution, corrosion, temperature, humidity, and manufacturing tolerances. The turn-on speed of an air gap device is a function of all of the foregoing elements. A simple gas-tube-type arrestor is an improvement, but neither of these devices will operate reliably to protect shunt-fed cavities, isolators, or receivers that include static drain inductors to ground (which most have). Such voltage-sensitive crowbar devices are short-circuited out by the dc path to ground found in these circuits. The inductive change in current per unit time (Ldi/dt) voltage drop is usually not enough to fire the protector, but it may be sufficient to destroy the inductor and then the receiver front end. Instead, select a protector that does not have dc continuity on the coaxial center pin from input to output. Such units have a series capacitor between the gas tube and the equipment center pin that will allow the voltage to build up so that the arrester may fire properly.

7.6.4 Typical Installation

Figure 7.35 illustrates a common grounding arrangement for a remotely located grounded-tower (FM, TV, or 2-way radio) transmitter plant. The tower and guy wires are grounded using 10-ft-long copper-clad ground rods. The antenna is bonded to the tower, and the transmission line is bonded to the tower at the point where it leaves the structure and begins the horizontal run into the transmitter building. Before entering the structure, the line is bonded to a ground rod through a connecting cable. The transmitter itself is grounded to the transmission line and to the ac power-distribution system ground. This, in turn, is bonded to a ground rod where the utility feed enters the building. The goal of this arrangement is to strip all incoming lines of damaging overvoltages before they enter the facility. One or more lightning rods are mounted at the top of the tower structure. The rods extend at least 10 ft above the highest part of the antenna assembly.

Such a grounding configuration, however, has built-in problems that can make it impossible to provide adequate transient protection to equipment at the site. Look again at the example. To equipment inside the transmitter building, two grounds actually exist: the utility company ground and the antenna ground. One ground will have a lower resistance to earth, and one will have a lower inductance in the connecting cable or copper strap from the equipment to the ground system.

Using the Figure 7.35 example, assume that a transient overvoltage enters the utility company meter panel from the ac service line. The overvoltage is clamped by a protection device at the meter panel, and the current surge is directed to ground. But *which ground*, the utility ground or the antenna ground?

The utility ground surely will have a lower inductance to the current surge than the antenna ground, but the antenna probably will exhibit a lower resistance to ground than the utility side of the circuit. Therefore, the surge current will be divided between the two grounds, placing the transmission equipment in series with the surge suppressor and the antenna ground system. A transient of sufficient potential will damage the transmission equipment.

Transients generated on the antenna side because of a lightning discharge are no less troublesome. The tower is a conductor, and any conductor is also an inductor. A

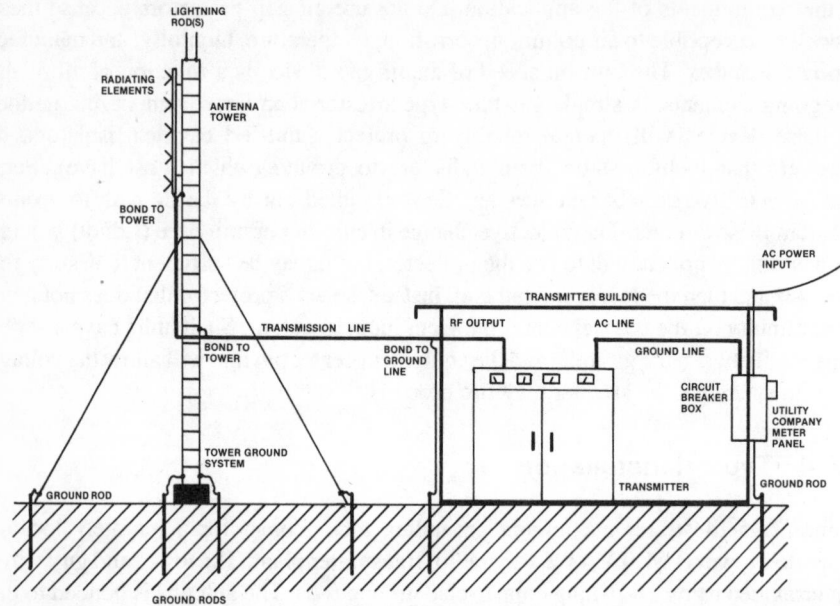

Figure 7.35 A common, but not ideal, grounding arrangement for a transmission facility using a grounded tower. A better configuration involves the use of a bulkhead panel through which all cables pass into and out of the equipment building.

typical 150-ft self-supporting tower may exhibit as much as 40 μH inductance. During a fast-rise-time lightning strike, an instantaneous voltage drop of 360 kV between the top of the tower and the base is not unlikely. If the coax shield is bonded to the tower 15 ft above the earth (as shown in the previous figure), 10 percent of the tower voltage drop (36 kV) will exist at that point during a flash. Figure 7.36 illustrates the mechanisms involved.

The only way to ensure that damaging voltages are stripped off all incoming cables (coax, ac power, and telephone lines) is to install a bulkhead entrance panel and tie all transient-suppression hardware to it. Configuring the system as shown in Figure 7.37 strips away all transient voltages through the use of a single-point ground. The bulkhead panel is the ground reference for the facility. With such a design, secondary surge current paths do not exist, as illustrated in Figure 7.38.

7.6.5 Checklist for Proper Grounding

A methodical approach is necessary in the design of a facility ground system. Consider the following points:

1. Install a bulkhead panel to provide mechanical support, electric grounding, and lightning protection for coaxial cables, power feeds, and telephone lines entering the equipment building.

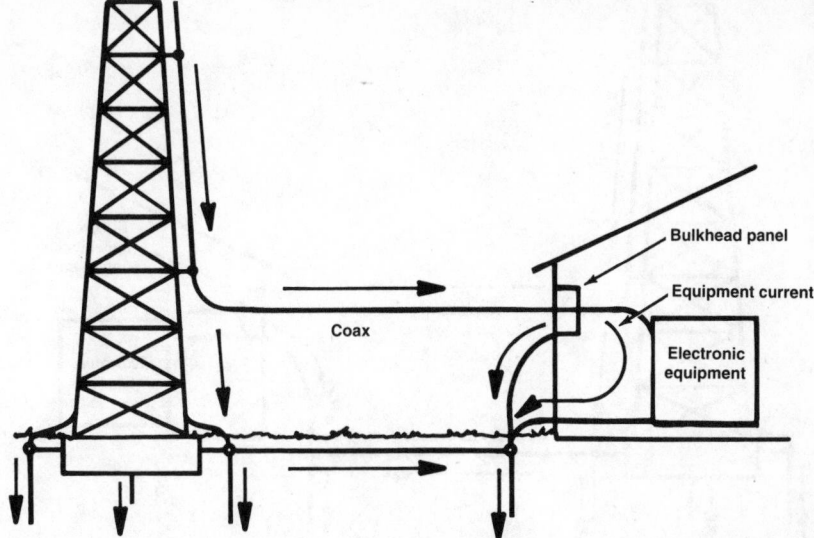

Figure 7.36   The equivalent circuit of the facility shown in Figure 7.35. Note the discharge current path through the electronic equipment.

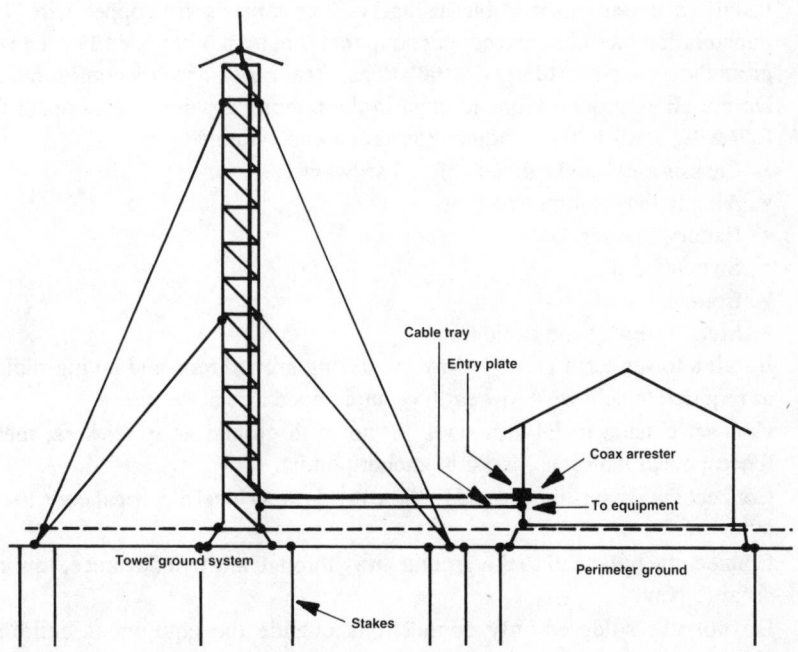

Figure 7.37   The preferred grounding arrangement for a transmission facility using a bulkhead panel. With this configuration, all damaging transient overvoltages are stripped off the coax, power, and telephone lines before they can enter the equipment building.

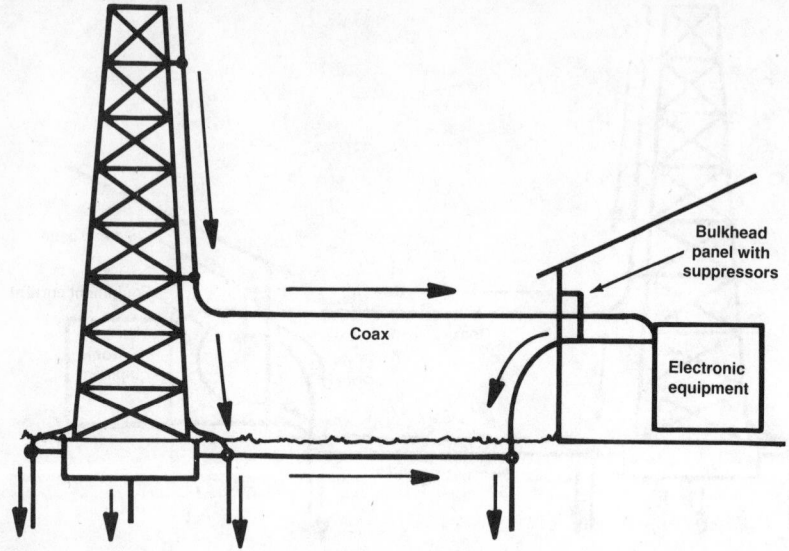

Coax

Bulkhead
panel with
suppressors

Electronic
equipment

Figure 7.38   The equivalent circuit of the facility shown in Figure 7.37. Discharge currents are prevented from entering the equipment building.

2.   Install an internal ground bus using no. 2 or larger solid-copper wire. (At transmission facilities, use copper strap that is at least 3 in wide.) Form a *star* grounding system. At larger installations, form a *star-of-stars* configuration. Do not allow ground loops to exist in the internal ground bus. Connect the following items to the building internal ground system:
     • Chassis racks and cabinets of all hardware
     • All auxiliary equipment
     • Battery charger
     • Switchboard
     • Conduit
     • Metal raceway and cable tray
3.   Install a tower earth ground array by driving ground rods and laying radials as required to achieve a low earth ground impedance at the site.
4.   Connect outside metal structures to the earth ground array (towers, metal fences, metal buildings, and guy-anchor points).
5.   Connect the power-line ground to the array. Follow local electrical code to the letter.
6.   Connect the bulkhead to the ground array through a low-inductance, low-resistance bond.
7.   Do not use soldered-only connections outside the equipment building. Crimped, brazed, and exothermic (Cadwelded) connections are preferable. For a proper bond, all metal surfaces must be cleaned, any finish removed to bare metal, and surface preparation compound applied (where necessary). Protect all connections from moisture by appropriate means (sealing compound and heat sink tubing).

7.7 BIBLIOGRAPHY

Block, Roger: "How to Ground Guy Anchors and Install Bulkhead Panels," *Mobile Radio Technology* magazine, Intertec Publishing, Overland Park, KS, February 1986.

———: "The Grounds for Lightning and EMP Protection," PolyPhaser Corporation, Gardnerville, NV, 1987.

Carpenter, R. B.: "Improved Grounding Methods for Broadcasters," *Proceedings* of the SBE National Convention, Society of Broadcast Engineers, Indianapolis, 1987.

Defense Civil Preparedness Agency, "EMP Protection for AM Radio Stations," Washington, DC, TR-61-C, May 1972.

Hill, Mark: "Computer Power Protection," *Broadcast Engineering* magazine, Intertec Publishing, Overland Park, KS, April 1987.

Little, Richard: "Surge Tolerance: How Does Your Site Rate?," *Mobile Radio Technology* magazine, Intertec Publishing, Overland Park, KS, June 1988.

Midkiff, John: "Choosing the Right Coaxial Cable Hanger," *Mobile Radio Technology* magazine, Intertec Publishing, Overland Park, KS, April 1988.

Mullinack, Howard G.: "Grounding for Safety and Performance," *Broadcast Engineering* magazine, Intertec Publishing, Overland Park, KS, October 1986.

Schneider, John: "Surge Protection and Grounding Methods for AM Broadcast Transmitter Sites," *Proceedings* of the SBE National Convention, Society of Broadcast Engineers, Indianapolis, 1987.

Sullivan, Thomas: "How to Ground Coaxial Cable Feedlines," *Mobile Radio Technology* magazine, Intertec Publishing, Overland Park, KS, April 1988.

Technical Reports LEA-9-1, LEA-0-10, and LEA-1-8, Lightning Elimination Associates, Santa Fe Springs, CA.

8

AC SYSTEM
GROUNDING PRACTICES

8.1 INTRODUCTION

Installing an effective ground system to achieve a good earth-to-grounding-electrode interface is only half the battle for a facility designer. The second, and equally important, element of any ground system is the configuration of grounding conductors inside the building. Many different methods can be used to implement a ground system, but some conventions always should be followed to ensure a low-resistance (and low-inductance) layout that will perform as required. Proper grounding is important whether or not the facility is located in a high-RF field.

8.1.1 Building Codes

The primary purpose of grounding electronic hardware is to prevent electric shock hazard. The National Electrical Code (NEC) and local building codes are designed to provide for the safety of the workplace. Local codes always should be followed. Occasionally, code sections are open to some interpretation. When in doubt, consult a field inspector. Codes constantly are being changed or expanded because new situations arise that were not anticipated when the codes were written. Sometimes, an interpretation will depend upon whether the governing safety standard applies to building wiring or to a factory-assembled product to be installed in a building. Underwriters Laboratories (UL) and other qualified testing organizations examine products at the request and expense of manufacturers or purchasers, and list products if the examination reveals that the device or system presents no significant safety hazard when installed and used properly.

Municipal and county safety inspectors generally accept UL and other qualified testing laboratory certification listings as evidence that a product is safe to install. Without a listing, the end-user may not be able to obtain the necessary wiring permits and inspection sign-off. On-site wiring must conform with local wiring codes. Most codes are based on the NEC. Electrical codes specify wiring materials, wiring devices, circuit protection, and wiring methods.

8.2 SINGLE-POINT GROUND

Single-point grounding is the basis of any properly designed facility ground network. Fault currents and noise should have only one path to the facility ground. Single-point grounds can be described as *star* systems in which radial elements circle out from a central hub. A star system is illustrated in Figure 8.1. Note that all equipment grounds are connected to a *main ground point*, which is then tied to the facility ground system. Multiple ground systems of this type can be cascaded as needed to form a *star-of-stars* facility ground system. The key element in a single-point ground is that each piece of equipment has one ground reference. Fault energy and noise then are efficiently drained to the outside earth ground system. The single-point ground is basically an extension of the bulkhead panel discussed in Section 7.6.1.

8.2.1 Facility Ground System

Figure 8.2 illustrates a star grounding system as applied to an ac power-distribution transformer and circuit-breaker panel. Note that a central ground point is established for each section of the system: one in the transformer vault and one in the circuit-breaker box. The breaker ground ties to the transformer vault ground, which is connected to the building ground system. Figure 8.3 shows single-point grounding applied to a data processing center. Note how individual equipment groups are formed

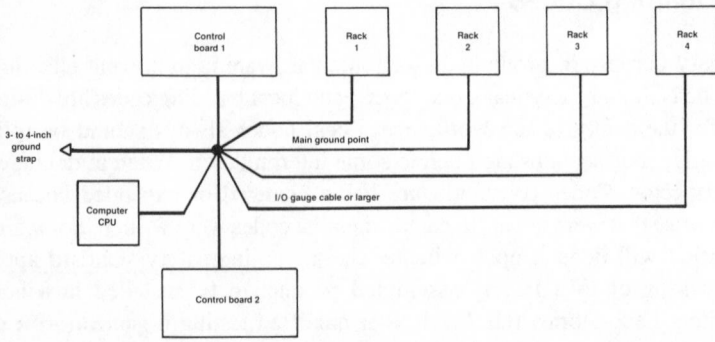

Figure 8.1   Typical facility grounding system. The *main facility ground point* is the reference from which all grounding is done at the plant. If a bulkhead entrance panel is used, it will function as the main ground point.

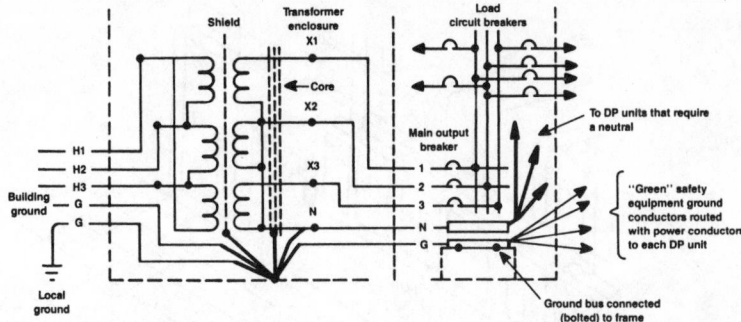

Figure 8.2 Single-point grounding applied to a power-distribution system. (Adapted from: Federal Information Processing Standards Publication no. 94, *Guideline on Electrical Power for ADP Installations*, U.S. Department of Commerce, National Bureau of Standards, Washington, DC, 1983.)

into a star grounding system, and how different groups are formed into a star-of-stars configuration. A similar approach can be taken for a data processing center using multiple modular power center (MPC) units, as shown in Figure 8.4. The terminal mounting wall is the reference ground point for the facility.

Figure 8.5 shows the recommended grounding arrangement for a typical broadcast or audio/video production facility. The building ground system is constructed using heavy-gauge copper wire (no. 4 or larger) if the studio is not located in an RF field, or a wide copper strap (3-in minimum) if the facility is located near an RF energy source. A common method of determining the required size of the ground strap (in inches) inside the building is to specify the minimum width of the strap as 1.5 percent of its length. For example, if the total grounding run from the perimeter ground system to the farthest piece of equipment is 350 ft, use a 5.25-in ground strap. For short runs in an RF field, do not use a ground strap that is less than 3 in wide.

Run the strap or cable from the perimeter ground to the main facility ground point. Branch out from the main ground point to each major piece of equipment and to the various equipment rooms. Establish a *local ground point* in each room or group of racks. Use a separate ground cable for each piece of equipment (no. 12-gauge or larger). Figure 8.6 shows the grounding plan for a communications facility. Equipment grounding is handled by separate conductors tied to the bulkhead panel (entry plate). A "halo" ground is constructed around the perimeter of the room. Cable trays are tied into the halo. All electronic equipment is grounded to the bulkhead to prevent ground-loop paths.

The ac line ground connection for individual pieces of equipment often presents a built-in problem for the system designer. If the equipment is grounded through the chassis to the equipment-room ground point, a ground loop may be created through the green-wire ground connection when the equipment is plugged in. The solution to this problem involves careful design and installation of the ac power-distribution system to minimize ground-loop currents, while at the same time providing the

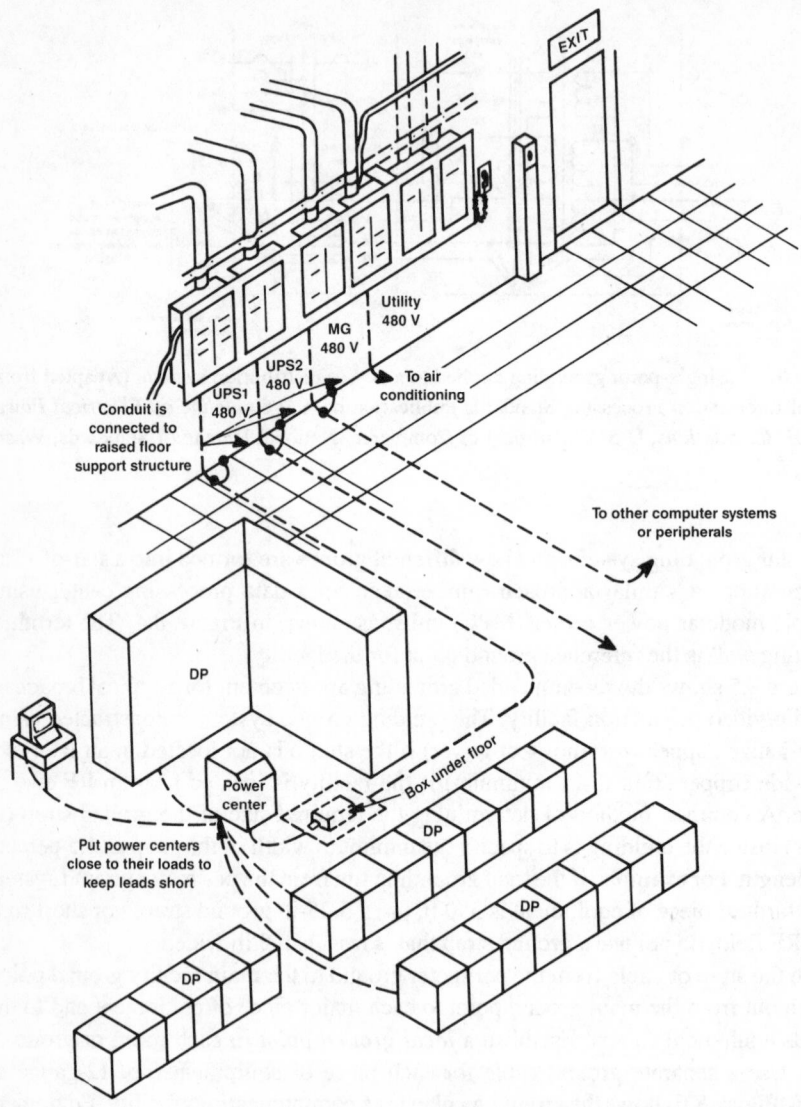

Figure 8.3   Configuration of a star-of-stars grounding system at a data processing facility. (Adapted from: Federal Information Processing Standards Publication no. 94, *Guideline on Electrical Power for ADP Installations*, U.S. Department of Commerce, National Bureau of Standards, Washington, DC, 1983.)

required protection against ground faults. Some equipment manufacturers provide a convenient solution to the ground-loop problem by isolating the signal ground from the ac and chassis ground. This feature offers the user the best of both worlds: the ability to create a signal ground system and ac ground system free of interaction and ground loops.

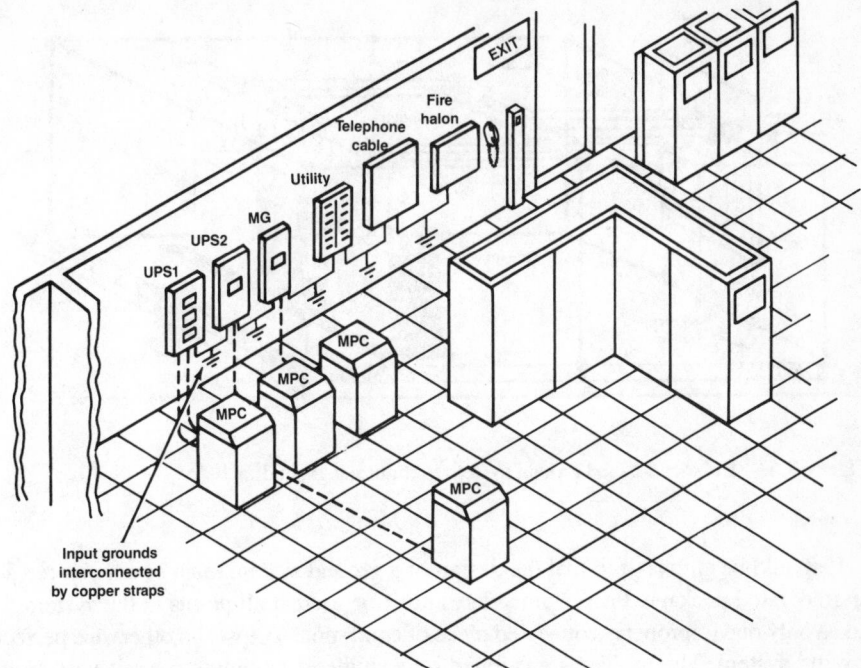

Figure 8.4 Establishing a star-based single-point ground system using multiple modular power centers. (Adapted from: Federal Information Processing Standards Publication no. 94, *Guideline on Electrical Power for ADP Installations*, U.S. Department of Commerce, National Bureau of Standards, Washington, DC, 1983.)

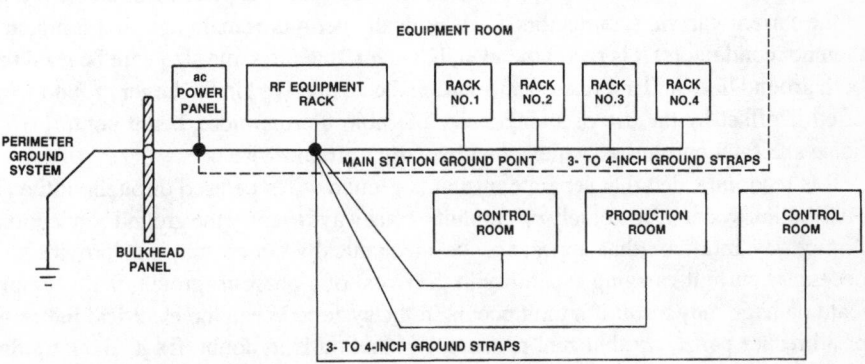

Figure 8.5 Typical grounding arrangement for individual equipment rooms at a communications facility. The ground strap from the main ground point establishes a *local ground point* in each room, to which all electronic equipment is bonded.

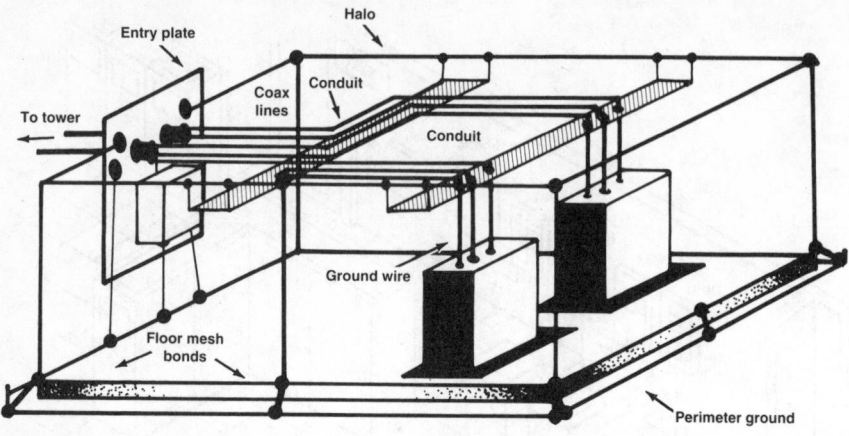

Figure 8.6   Bulkhead-based ground system including a grounding halo.

It should be emphasized that the design of a ground system must be considered as an integrated package. Proper procedures must be used at all points in the system. It takes only one improperly connected piece of equipment to upset an otherwise perfect ground system. The problems generated by a single grounding error can vary from trivial to significant, depending upon where in the system the error exists. This consideration leads, naturally, to the concept of ground-system maintenance for a facility. Check the ground network from time to time to ensure that no faults or errors have occurred. Any time new equipment is installed, or old equipment is removed from service, careful attention must be given to the possible effects that such work will have on the ground system.

Grounding Conductor Size.   The NEC and local electrical codes specify the minimum wire size for grounding conductors. The size varies, depending upon the rating of the current-carrying conductors. Code typically permits a smaller ground conductor than hot conductors. It is recommended, however, that the same size wire be used for both ground lines and hot lines. The additional cost involved in the larger ground wire often is offset by the use of a single size of cable. Furthermore, better control over noise and fault currents is achieved with a larger ground wire.

It is recommended that separate insulated ground wires be used throughout the ac distribution system. Do not rely on conduit or raceways to carry the ground connection. A raceway interface that appears to be mechanically sound may not provide the necessary current-carrying capability in the event of a phase-to-ground fault. Significant damage may result if a fault occurs in the system. When the electrical integrity of a breaker panel, conduit run, or raceway junction is in doubt, fix it. Back up the mechanical connection with a separate ground conductor of the same size as the current-carrying conductors. Loose joints have been known to shower sprays of sparks during phase-to-ground faults, creating a fire hazard. Secure the ground cable using appropriate hardware. Clean attachment points of any paint or dirt accumulation. Properly label all cables.

Structural steel, compared with copper, is a poor conductor at any frequency. At dc, steel has a resistivity 10 times that of copper. As frequency rises, the skin effect is more pronounced because of the magnetic effects involved. A no. 6 copper wire can have less RF impedance than a 12-in steel "I" beam. Furthermore, because of their bolted, piecemeal construction, steel racks and building members should not be depended upon alone for circuit returns.

8.2.2 Power-Center Grounding

A modular power center, commonly found in computer-room installations, provides a comprehensive solution to ac power-distribution and ground-noise considerations. Such equipment is available from several manufacturers, with various options and features. A computer power-distribution center generally includes an isolation transformer designed for noise suppression, distribution circuit breakers, power-supply cables, and a status-monitoring unit. The system concept is shown in Figure 8.7. Input

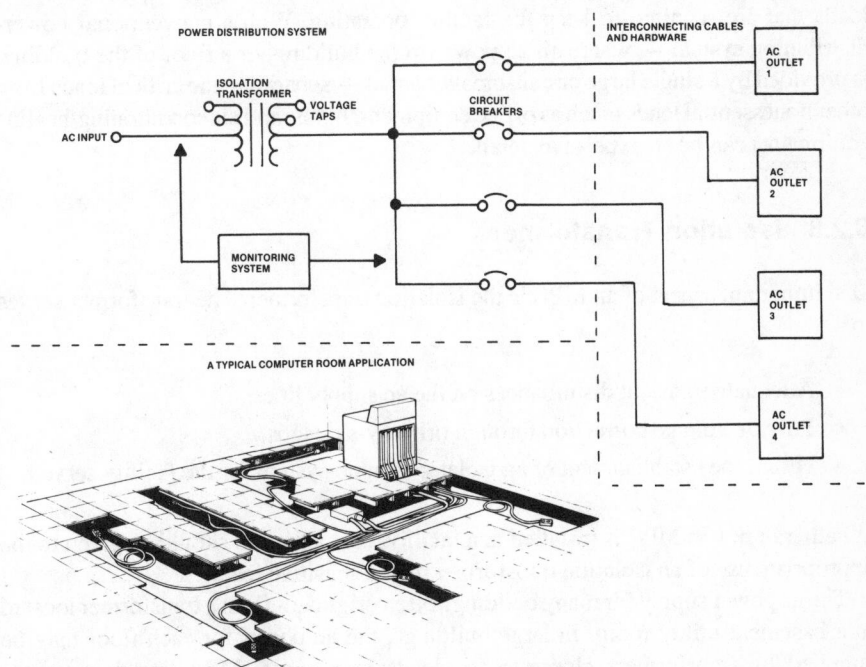

Figure 8.7   The basic concept of a computer-room modular power center. Both single- and multiphase configurations are available. When ordering an MPC, the customer can specify cable lengths and terminations, making installation quick and easy.

power is fed to an isolation transformer with primary taps to match the ac voltage required at the facility. A bank of circuit breakers is included in the chassis, and individual preassembled and terminated cables supply ac power to the various loads. A status-monitoring circuit signals the operator if any condition is detected outside normal parameters.

The ground system is an important component of the MPC. A unified approach, designed to prevent noise or circulating currents, is taken to grounding for the entire facility. This results in a clean ground connection for all equipment on-line.

The use of a modular power center can eliminate the inconvenience associated with rigid conduit installations. Distribution systems also are expandable to meet future facility growth. If the plant ever is relocated, the power center can move with it. MPC units usually are expensive. However, considering the costs of installing circuit-breaker boxes, conduit, outlets, and other hardware on-site by a licensed electrician, the power-center approach may be economically viable. The use of a power center also will make it easier to design a standby power system for the facility. Many computer-based operations do not have a standby generator on site. Depending on the location of the facility, it may be difficult or even impossible to install a generator to provide standby power in the event of a utility company outage. However, by using the power-center approach to ac distribution for computer and other critical load equipment, an uninterruptible power system may be installed easily to power only the loads that are required to keep the facility operating. With a conventional power-distribution system — where all ac power to the building, or a floor of the building, is provided by a single large circuit-breaker panel — separating the critical loads from other nonessential loads (such as office equipment, lights, and air conditioning/heating equipment) can be an expensive detail.

8.2.3 Isolation Transformers

One important aspect of an MPC is the isolation transformer. The transformer serves to:

- Attenuate transient disturbances on the ac supply lines.
- Provide voltage correction through primary-side taps.
- Permit the establishment of an isolated ground system for the facility served.

Whether or not an MPC is installed at a facility, consideration should be given to the appropriate use of an isolation transformer near a sensitive load.

The ac power supply for many buildings often originates from a transformer located in a basement utility room. In large buildings, the ac power for each floor may be supplied by transformers closer to the loads they serve. Most transformers are 208 Y/120 V three-phase. Many fluorescent lighting circuits operate at 277 V, supplied by a 408 Y/277 V transformer. Long feeder lines to DP systems and other sensitive loads raise the possibility of voltage fluctuations based on load demand and ground-loop-induced noise.

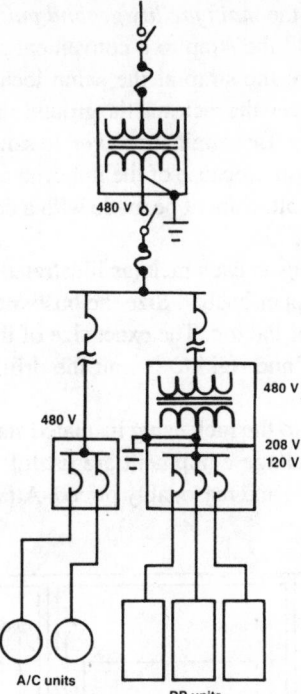

480 V

480 V

480 V

208 V
120 V

A/C units

DP units

Figure 8.8   Preferred power-distribution configuration for a data processing site. (Adapted from: Federal Information Processing Standards Publication no. 94, *Guideline on Electrical Power for ADP Installations*, U.S. Department of Commerce, National Bureau of Standards, Washington, DC, 1983.)

Figure 8.8 illustrates the preferred method of power distribution in a building. A separate dedicated isolation transformer is located near the DP equipment, providing good voltage regulation and permitting the establishment of an effective single-point star ground in the data processing center. Note that the power-distribution system voltage shown in the figure (480 V) is maintained at 480 V until it reaches the DP step-down isolation transformer. Use of this higher voltage provides more efficient transfer of electricity throughout the plant. At 480 V, the line current is about 43 percent of the current in a 208 V system for the same conducted power.

8.2.4   Grounding Equipment Racks

The installation and wiring of equipment racks must be planned carefully to avoid problems during day-to-day operations. Figure 8.9 shows the recommended approach. Bond adjacent racks together with 3/8- to 1/2-in-diameter bolts. Clean the contacting surfaces by sanding down to bare metal. Use lockwashers on both ends of the bolts. Bond racks together using at least six bolts per side (three bolts for each vertical rail).

Run a ground strap from the *main facility ground point*, and bond the strap to the base of each rack. Spot-weld the strap to a convenient spot on one side of the rear portion of each rack. Secure the strap at the same location for each rack used. A mechanical connection between the rack and the ground strap may be made using bolts and lockwashers, if necessary. Be certain, however, to sand down to bare metal before making the ground connection. Because of the importance of the ground connection, it is recommended that each attachment be made with a combination of crimping and silver-solder.

Install a vertical ground bus in each rack (as illustrated in Figure 8.9). Use about 1 1/2-in-wide, 1/4-in-thick copper busbar. Size the busbar to reach from the bottom of the rack to about 1 ft short of the top. The exact size of the busbar is not critical, but it must be sufficiently wide and rigid to permit the drilling of 1/8-in holes without deforming.

Mount the ground busbar to the rack using insulated standoffs. Porcelain standoffs commonly found in high-voltage equipment are useful for this purpose. Porcelain standoffs are readily available and reasonably priced. Attach the ground busbar to the

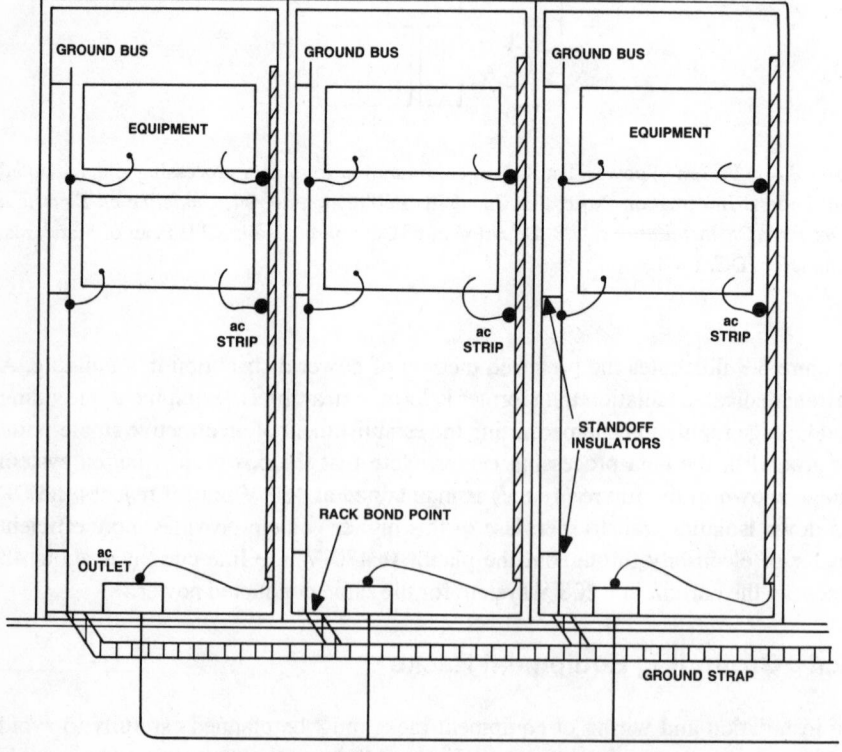

Figure 8.9   Recommended grounding method for equipment racks. To make assembly of multiple racks easier, position the ground connections and ac receptacles at the same location in all racks.

rack at the point that the facility ground strap attaches to the rack. Silver-solder the busbar to the rack and strap at the same location in each rack used.

Install an ac receptacle box at the bottom of each rack. Isolate the conduit from the rack. The easiest approach involves using an insulated bushing between the conduit and the receptacle box. With this arrangement, the ac outlet box can be mounted directly to the bottom of the rack near the point that the ground strap and ground busbar are bonded to the rack. An alternative approach involves the use of an orange-type receptacle. The orange-type outlet isolates the green-wire power ground from the receptacle box. Use insulated standoffs to mount the ac outlet box to the rack. Bring out the green-wire ground, and bond it to the rack near the point that the ground strap and ground busbar are silver-soldered to the rack. The goal of this configuration is to keep the green-wire ac and facility system grounds separate from the ac distribution conduit and metal portions of the building structure. Carefully check the local electrical code to ensure that such configurations are legal.

Although the foregoing procedure is optimum from a signal-grounding standpoint, it should be pointed out that under a ground fault condition, performance of the system may be unpredictable if high currents are being drawn in the current-carrying conductors supplying the load. Vibration of ac circuit elements resulting from the magnetic field effects of high-current-carrying conductors is insignificant as long as all conductors are within the confines of a given raceway or conduit. A ground fault will place return current outside of the normal path. If sufficiently high currents are being conducted, the consequences can be devastating. Sneak currents from ground faults have been known to destroy wiring systems that were installed exactly to code.

The fail-safe wiring method for equipment-rack ac power involves use of orange-type outlets, with the receptacle green-wire ground routed back to the breaker-panel star ground system. Insulate the receptacle box from the rack to prevent conduit-based noise currents from contaminating the rack ground system. Try to route the power conduit and facility ground cable or strap via the same path if such a compromise configuration is deemed necessary. As stated previously, keep metallic conduit and building structures insulated from the facility ground line, except at the bulkhead panel (main grounding point).

Mount a vertical ac strip inside each rack to power the equipment. Insulate the power strip from the rack using porcelain standoffs. Power equipment from the strip using standard three-prong grounding ac plugs. Do not defeat the safety ground connection. Equipment manufacturers use this ground to drain transient energy.

Mount equipment in the rack using normal metal mounting screws. If the location is in a high-RF field, clean the rack rails and equipment-panel connection points to ensure a good electrical bond. This is important because in a high-RF field, detection of RF energy can occur at the junctions between equipment chassis and the rack.

Connect a separate ground wire from each piece of equipment in the rack to the vertical ground busbar. Use no. 12 stranded copper wire (insulated) or larger. Connect the ground wire to the busbar by drilling a hole in the busbar at a convenient elevation near the equipment. Fit one end of the ground wire with an enclosed-hole solderless terminal connector (no. 10-sized hole or larger). Attach the ground wire to the busbar using no. 10 (or larger) hardware. Use an internal-tooth lockwasher between the busbar

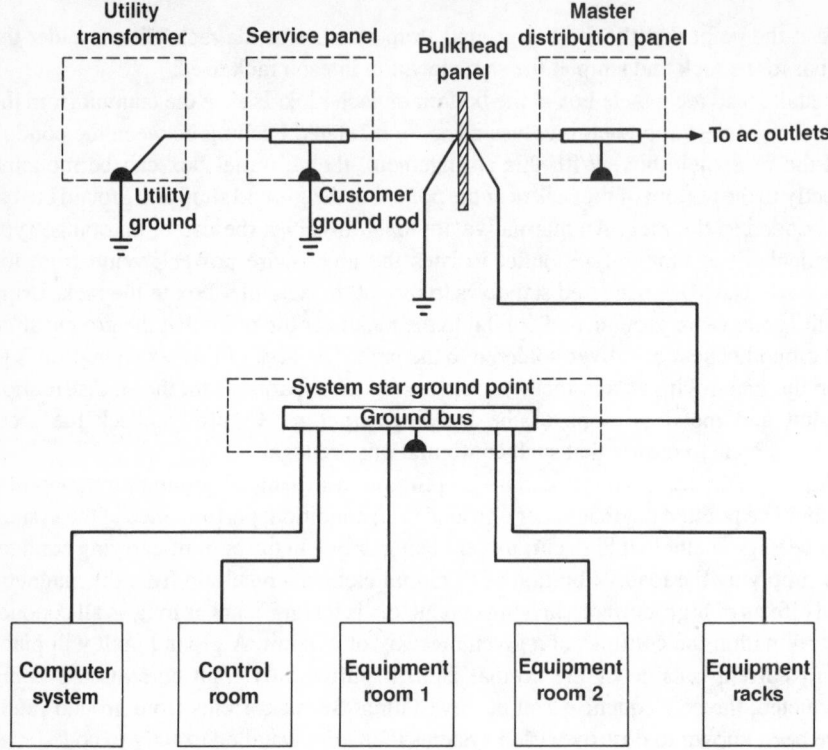

Figure 8.10   Equivalent ground circuit diagram for a medium-sized commercial/industrial facility.

and the nut. Fit the other end of the ground wire with a terminal that will be accepted by the equipment. If the equipment has an isolated signal ground terminal, tie it to the ground busbar.

Figure 8.10 shows each of the grounding elements discussed in this section integrated into one diagram. This approach fulfills the requirements of personnel safety and equipment performance.

Follow similar grounding rules for simple one-rack equipment installations. Figure 8.11 illustrates the grounding method for a single open-frame equipment rack. The vertical ground bus is supported by insulators, and individual jumpers are connected from the ground rail to each chassis.

8.3   GROUNDING SIGNAL-CARRYING CABLES

Proper ground-system installation is the key to minimizing noise currents on signal-carrying cables. Audio, video, and data lines are often subject to ac power noise currents and RFI. The longer the cable run, the more susceptible it is to disturbances. Unless care is taken in the layout and installation of such cables, unacceptable performance of the overall system may result.

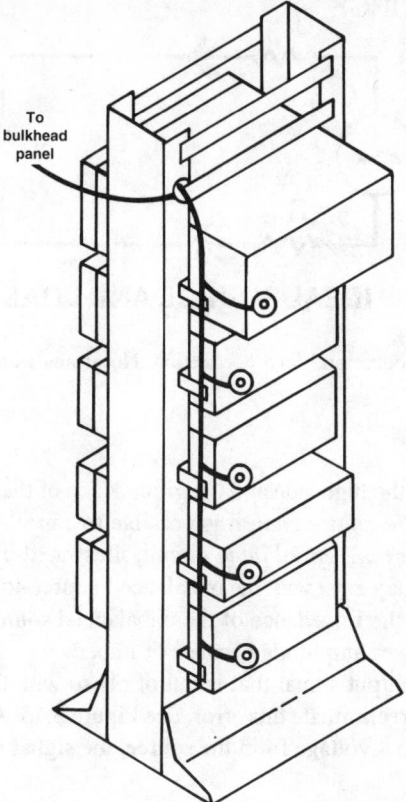

To
bulkhead
panel

Figure 8.11   Ground bus for an open-frame equipment rack.

8.3.1 Analyzing Noise Currents

Figure 8.12 shows a basic source and load connection. No grounds are present, and both the source and the load float. This is the optimum condition for equipment interconnection. Either the source or the load may be tied to ground with no problems, provided only one ground connection exists. *Unbalanced* systems are created when each piece of equipment has one of its connections tied to ground, as shown in Figure 8.13. This condition occurs if the source and load equipment have unbalanced (single-ended) inputs and outputs. This type of equipment utilizes chassis ground (or common) for one of the conductors. Problems are compounded when the equipment is separated by a significant distance.

As shown in Figure 8.13, a difference in ground potential causes current flow in the ground wire. This current develops a voltage across the wire resistance. The ground-noise voltage adds directly to the signal itself. Because the ground current is usually the result of leakage in power transformers and line filters, the 60 Hz signal gives rise to hum of one form or another. Reducing the wire resistance through a heavier ground conductor helps the situation, but cannot eliminate the problem.

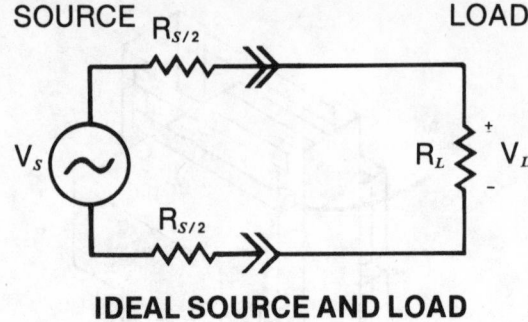

IDEAL SOURCE AND LOAD

Figure 8.12   A basic source and load connection. No grounds are indicated, and both the source and the load float.

By amplifying both the high side and the ground side of the source and subtracting the two to obtain a *difference signal*, it is possible to cancel the ground-loop noise. This is the basis of the *differential input* circuit, illustrated in Figure 8.14. Unfortunately, problems still may exist with the unbalanced-source-to-balanced-load system. The reason centers on the impedance of the unbalanced source. One side of the line will have a slightly lower amplitude because of impedance differences in the output lines. By creating an output signal that is out of phase with the original, a balanced source can be created to eliminate this error. See Figure 8.15. As an added benefit, for a given maximum output voltage from the source, the signal voltage is doubled over the unbalanced case.

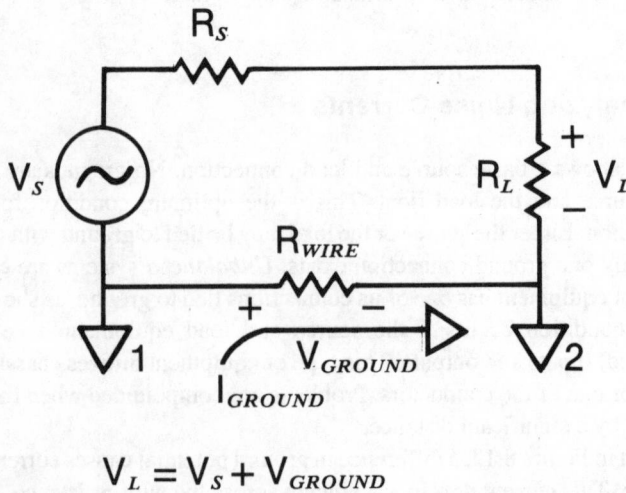

Figure 8.13   An unbalanced system in which each piece of equipment has one of its connections tied to ground.

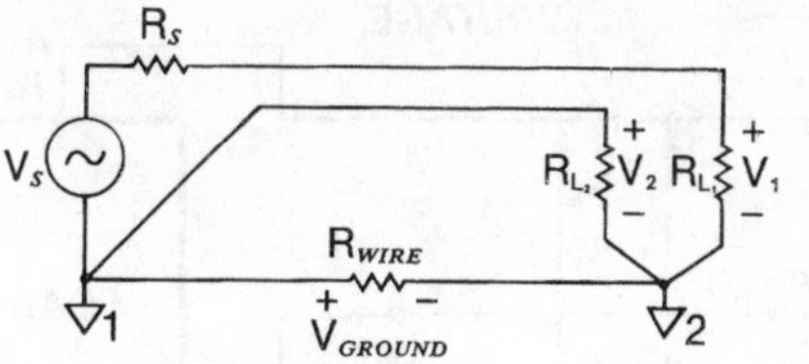

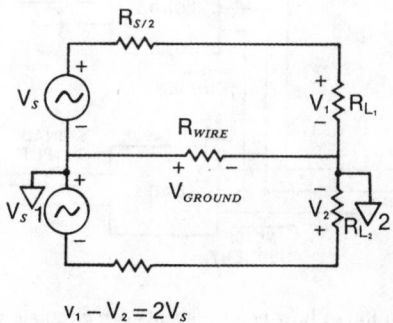

$$V_1 = V_S + V_{GROUND}$$
$$V_2 = V_{GROUND}$$
$$V_1 - V_2 = V_S$$

Figure 8.14   Ground-loop noise can be canceled by amplifying both the high side and the ground side of the source and subtracting the two signals.

8.3.2 Types of Noise

Two basic types of noise can appear on ac power, audio, video, and computer data lines within a facility: normal mode and common mode. Each type has a particular effect on sensitive load equipment. The normal-mode voltage is the potential difference that exists between pairs of power (or signal) conductors. This voltage also is referred to as the *transverse-mode* voltage. The common-mode voltage is a potential difference (usually noise) that appears between the power or signal conductors and the local ground reference. The differences between normal-mode and common-mode noise are illustrated in Figure 8.16.

The common-mode noise voltage will change depending upon what is used as the ground reference point. It is often possible to select a ground reference that has a minimum common-mode voltage with respect to the circuit of interest, particularly if the reference point and the load equipment are connected by a short conductor. Common-mode noise can be caused by electrostatic or electromagnetic induction.

$$V_1 - V_2 = 2V_S$$

Figure 8.15   A balanced source configuration where the inherent amplitude error of the system shown in Figure 8.14 is eliminated.

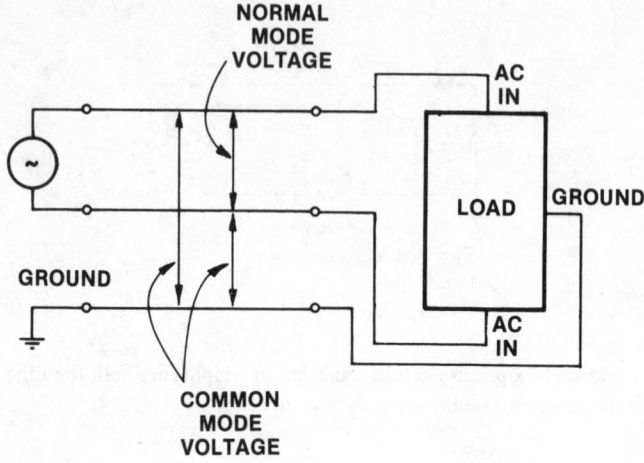

Figure 8.16 The principles of normal-mode and common-mode noise voltages as they apply to ac power circuits.

In practice, a single common-mode or normal-mode noise voltage is rarely found. More often than not, load equipment will see both common-mode and normal-mode noise signals. In fact, unless the facility wiring system is unusually well-balanced, the noise signal of one mode will convert some of its energy to the other mode.

Common-mode and normal-mode noise disturbances typically are caused by momentary impulse voltage differences among parts of a distribution system that have differing ground potential references. If the sections of a system are interconnected by a signal path in which one or more of the conductors are grounded at each end, the ground offset voltage can create a current in the grounded signal conductor. If noise voltages of sufficient potential occur on signal-carrying lines, normal equipment operation can be disrupted. See Figure 8.17.

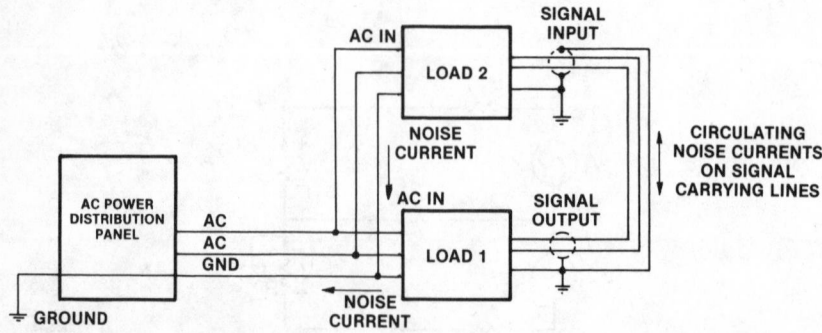

Figure 8.17 An illustration of how noise currents can circulate within a system because of the interconnection of various pieces of hardware.

8.3.3   Skin Effect

Low-level signal cables are particularly susceptible to high-frequency noise energy because of the *skin effect* of current-carrying conductors. When a conductor carries an alternating current, a magnetic field is produced that surrounds the wire. This field continually is expanding and contracting as the ac current wave increases from zero to its maximum positive value and back to zero, then through its negative half-cycle. The changing magnetic lines of force cutting the conductor induce a voltage in the conductor in a direction that tends to retard the normal flow of current in the wire. This effect is more pronounced at the center of the conductor. Thus, current within the conductor tends to flow more easily toward the surface of the wire. The higher the frequency, the greater the tendency for current to flow at the surface. The depth of current flow is a function of frequency, and is determined from the following equation:

$$d = \frac{2.6}{\sqrt{uf}}$$

Where:

d   = depth of current in mils
u   = permeability (copper = 1, steel = 300)
f   = frequency of signal in megahertz

It can be calculated that at a frequency of 100 kHz, current flow penetrates a conductor by 8 mils. At 1 MHz, the skin effect causes current to travel in only the top 2.6 mils in copper, and even less in almost all other conductors. Therefore, the series impedance of conductors at high frequencies is significantly higher than at ac power line frequencies. This makes low-level signal-carrying cables particularly susceptible to disturbances resulting from RFI.

8.3.4   Patch-Bay Grounding

Patch panels for audio, video, and data circuits require careful attention to planning to avoid built-in grounding problems. Because patch panels are designed to tie together separate pieces of equipment, often from remote areas of a facility, the opportunity exists for ground loops. The first rule of patch-bay design is to never use a patch bay to switch low-level (microphone) signals. If mic sources must be patched from one location to another, install a bank of mic-to-line amplifiers to raise the signal levels to 0 dBm before connection to the patch bay. Most video output levels are 1 V P-P, giving them a measure of noise immunity. Data levels are typically 5 V. Although these line-level signals are significantly above the noise floor, capacitive loading and series resistance in long cables can reduce voltage levels to a point that noise becomes a problem.

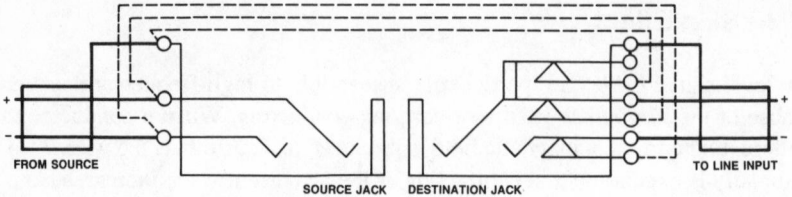

Figure 8.18   Patch-panel wiring for seven-terminal normalling jack fields. Use patch cords that connect ground (sleeve) at both ends.

Newer-design patch panels permit switching of ground connections along with signal lines. Figure 8.18 illustrates the preferred method of connecting an audio patch panel into a system. Note that the source and destination jacks are *normalled* to establish ground signal continuity. When another signal is plugged into the destination jack, the ground from the new source is carried to the line input of the destination jack. With such an approach, jack cords that provide continuity between sleeve (ground) points are required.

If only older-style, conventional jacks are available, use the approach shown in Figure 8.19. This configuration will prevent ground loops, but because destination shields are not carried back to the source when normalling, noise will be higher. Bus all destination jack sleeves together, and connect to the local (rack) ground. The wiring methods shown in Figures 8.18 and 8.19 assume balanced input and output lines with all shields terminated at the load (input) end of the equipment.

8.3.5   Input/Output Circuits

Common-Mode Rejection Ratio (CMRR) is the measure of how well an input circuit rejects ground noise. The concept is illustrated in Figure 8.20. The input signal to a differential amplifier is applied between the plus and minus amplifier inputs. The stage will have a certain gain for this signal condition, called the *differential gain*. Because the ground-noise voltage appears on both the plus and minus inputs simultaneously, it is common to both inputs.

The amplifier subtracts the two inputs, giving only the difference between the voltages at the input terminals at the output of the stage. The gain under this condition

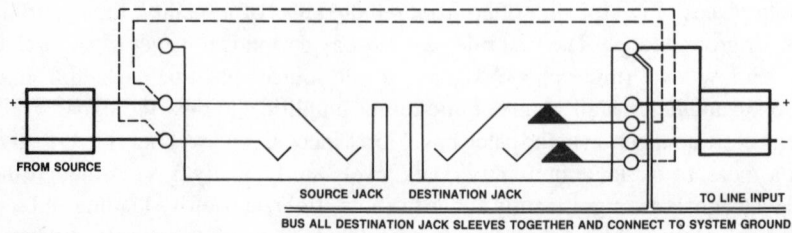

Figure 8.19   Patch-panel wiring for conventional normalling jack fields. Use patch cords that connect ground (sleeve) at both ends.

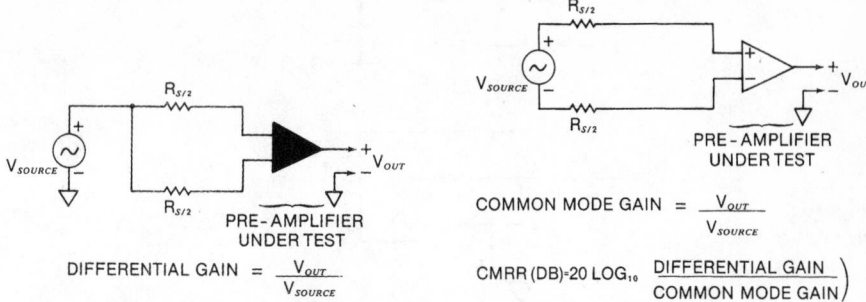

Figure 8.20   The concept of common-mode rejection ratio (CMRR) for an active-balanced input circuit: (a, *left*) differential gain measurement; (b, *right*) calculating CMRR.

should be zero, but in practice it is not. CMRR is the ratio of these two gains (the differential gain and the common-mode gain) in decibels. The larger the number, the better. For example, a 60 dB CMRR means that a ground signal common to the two inputs will have 60dB less gain than the desired differential signal. If the ground noise is already 40 dB below the desired signal level, the output noise level will be 100 dB below the desired signal level. If, however, the noise is already part of the differential signal, the CMRR will do nothing to improve it.

Active-balanced I/O circuits are the basis of nearly all professional audio interconnections (except for speaker connections), many video signals, and increasing numbers of data lines. A wide variety of circuit designs have been devised for active-balanced inputs. All have the common goal of providing high CMRR and adequate gain for subsequent stages. All also are built around a few basic principles.

Figure 8.21 shows the simplest and least expensive approach, using a single operational amplifier (op-amp). For a unity gain stage, all the resistors are the same value. This circuit presents an input impedance to the line that is different for the two sides. The positive input impedance will be twice that of the negative input. The CMRR is dependent on the matching of the four resistors and the balance of the source impedance. The noise performance of this circuit, which usually is limited by the resistors, is a tradeoff between low loading of the line and low noise.

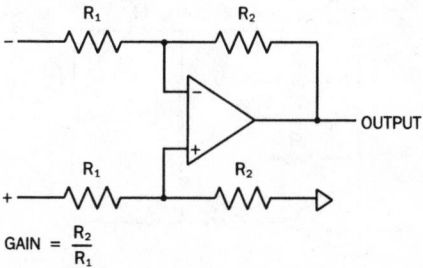

Figure 8.21   The simplest and least expensive active-balanced input op-amp circuit. Performance depends on resistor-matching and the balance of the source impedance.

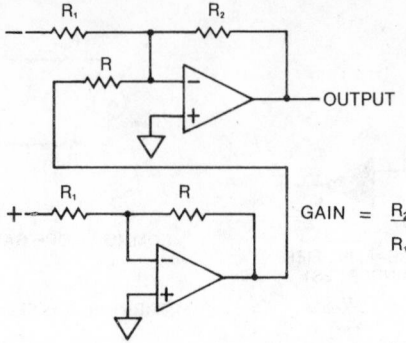

Figure 8.22   An active-balanced input circuit using two op-amps, one to invert the positive input terminal and the other to buffer the difference signal. Without adjustments, this circuit will provide about 50 dB CMRR.

Another approach, shown in Figure 8.22, uses a buffering op-amp stage for the positive input. The positive signal is inverted by the op-amp, then added to the negative input of the second inverting amplifier stage. Any common-mode signal on the positive input (which has been inverted) will cancel when it is added to the negative input signal. Both inputs have the same impedance. Practical resistor-matching limits the CMRR to about 50 dB. With the addition of an adjustment pot, it is possible to achieve 80 dB CMRR, but component aging will degrade this over time.

Adding a pair of buffer amplifiers before the summing stage results in an instrumentation-grade circuit, as shown in Figure 8.23. The input impedance is increased substantially, and any source impedance effects are eliminated. More noise is introduced by the added op-amp, but the resistor noise usually can be decreased by reducing impedances, causing a net improvement in system noise.

Early active-balanced output circuits used the approach shown in Figure 8.24. The signal is buffered to provide one phase of the balanced output. This signal then is inverted with another op-amp to provide the other phase of the output signal. The outputs are taken through two resistors, each of which is half of the desired source impedance. Because the load is driven from the two outputs, the maximum output voltage is double that of an unbalanced stage.

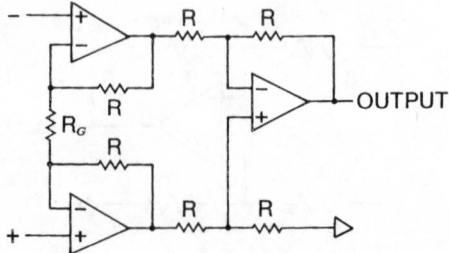

Figure 8.23   An active-balanced input circuit using three op-amps to form an instrumentation-grade circuit. The input signals are buffered, then applied to a differential amplifier.

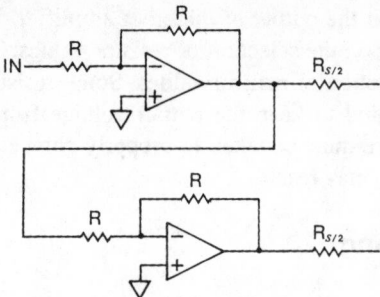

Figure 8.24 A basic active-balanced output circuit. This configuration works well when driving a single balanced load.

The Figure 8.24 circuit works reasonably well if the load is always balanced, but it suffers from two problems when the load is not balanced. If the negative output is shorted to ground by an unbalanced load connection, the first op-amp is likely to distort. This produces a distorted signal at the input to the other op-amp. Even if the circuit is arranged so that the second op-amp is grounded by an unbalanced load, the distorted output current probably will show up in the output from coupling through grounds or circuit-board traces. Equipment that uses this type of balanced stage often provides a second set of output jacks that are wired to only one amplifier for unbalanced applications.

The second problem with the Figure 8.24 circuit is that the output does not float. If any voltage difference (such as power-line hum) exists between the local ground and the ground at the device receiving the signal, it will appear as an addition to the signal. The only ground-noise rejection will be from the CMRR of the input stage at the receive end.

The preferred output stage is the electronically balanced and floating design shown in Figure 8.25. The circuit consists of two op-amps that are cross-coupled with positive and negative feedback. The output of each amplifier is dependent on the input signal

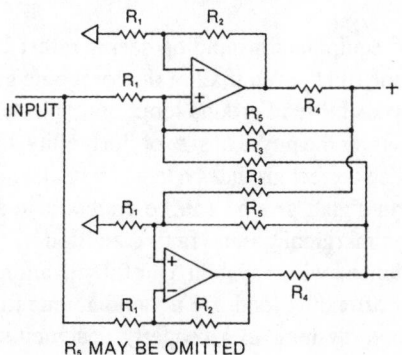

Figure 8.25 An electronically balanced and floating output circuit. A stage such as this will perform well even when driving unbalanced loads.

and the signal present at the output of the other amplifier. These designs may have gain or loss, depending on the selection of resistor values. The output impedance is set via appropriate selection of resistor values. Some resistance is needed from the output terminal to ground to keep the output voltage from floating to one of the power-supply rails. Care must be taken to properly compensate the devices. Otherwise, stability problems may result.

8.3.6 Cable Routing

Good engineering practice dictates that different signal levels be grouped and separated from each other. It is common practice to separate cables into the following groups:

- AC power
- Speaker lines
- Line-level audio
- Microphone-level audio
- Video lines
- Control and data lines

Always use two-conductor shielded cable for all audio signal cables. This includes both balanced and unbalanced circuits, and microphone-level and line-level cables. On any audio cable connecting two pieces of equipment, tie the shield at one end only. Connect at the receiving end of signal transmission. On video coaxial cables running to outlet jacks mounted on plates, isolate the connector from the plate. The shield should connect to ground only at the equipment input/output or patch panel. For data cables, carefully follow the recommendations of the equipment manufacturer. The preferred interconnection method for long data cables is fiber optics, which eliminates ground-loop problems.

8.3.7 Overcoming Ground-System Problems

Although the concept of equipment grounding seems rather basic, it can become a major headache if not done right. Even if all of the foregoing guidelines are followed to the letter, there is the possibility of ground loops and objectionable noise on audio, video, or data lines. The larger the physical size of the facility, the greater the potential for problems. An otherwise perfect ground system can be degraded by a single wiring error. An otherwise clean signal ground can be contaminated by a single piece of equipment experiencing a marginal ground fault condition.

If problems are experienced with a system, carefully examine all elements to track down the wiring error or offending load. Do not add ac line filters and/or signal line filters to correct a problem system. In a properly designed system, even one in a high-RF field, proper grounding and shielding techniques will permit reliable operation. Adding filters merely hides the problem. Instead, correct the problem at its source.

Do not deviate from the single-point grounding concept. Multiple ground paths will compound the problems being experienced, further hiding the real culprit. In a highly complex system such as a data processing facility, the necessity to interconnect a large number of systems may require the use of fiber-optic transmitters and receivers. It is far better, and less expensive in the long run, to use fiber-optic lines than to string multiple grounds in an effort to eliminate a noise problem. Multiple grounds will not eliminate the problem; they will just hide it.

8.4 BIBLIOGRAPHY

Benson, K. B., and J. Whitaker: *Television and Audio Handbook for Engineers and Technicians*, McGraw-Hill, New York, 1989.

Block, Roger: "The Grounds for Lightning and EMP Protection," PolyPhaser Corporation, Gardnerville, NV, 1987.

Davis, Gary, and Ralph Jones: *Sound Reinforcement Handbook*, Yamaha Music Corp., Hal Leonard Publishing, Milwaukee, 1987.

Fardo, S., and D. Patrick: *Electrical Power Systems Technology*, Prentice-Hall, Englewood Cliffs, NJ, 1985.

Federal Information Processing Standards Publication no. 94, *Guideline on Electrical Power for ADP Installations*, U.S. Department of Commerce, National Bureau of Standards, Washington, DC, 1983.

Lanphere, John: "Establishing a Clean Ground," *Sound & Video Contractor* magazine, Intertec Publishing, Overland Park, KS, August 1987.

Lawrie, Robert: *Electrical Systems for Computer Installations*, McGraw-Hill, New York, 1988.

Mullinack, Howard G.: "Grounding for Safety and Performance," *Broadcast Engineering* magazine, Intertec Publishing, Overland Park, KS, October 1986.

9

STANDBY POWER SYSTEMS

9.1 INTRODUCTION

When utility company power problems are discussed, most people immediately think of blackouts. The lights go out, and everything stops. With the facility down and in the dark, there is nothing to do but sit and wait until the utility company finds the problem and corrects it. This process generally takes only a few minutes. There are times, however, when it can take hours. In some remote locations, it can even take days.

Blackouts are, without a doubt, the most troublesome utility company problem that a facility will have to deal with. Statistics show that power failures are, generally speaking, a rare occurrence in most areas of the country. They are also short in duration. Studies have shown that 50 percent of blackouts last 6 s or less, and 35 percent are less than 11 min long. These failure rates usually are not cause for concern to commercial users, except where computer-based operations, transportation control systems, medical facilities, and communications sites are concerned.

When continuity of operation is critical, redundancy must be carried throughout the system. The site never should depend upon one critical path for ac power. For example, if the facility is fed by a single step-down transformer, a lightning flash or other catastrophic event could result in a transformer failure that would bring down the entire site. A replacement could take days or even weeks.

9.1.1 Blackout Effects

A facility that is down for even 5 min can suffer a significant loss of productivity or data that may take hours or days to rebuild. A blackout affecting a transportation or medical center could be life-threatening. Coupled with this threat is the possibility of extended power-service loss due to severe storm conditions. Many broadcast and communications relay sites are located in remote, rural areas or on mountaintops. Neither of these kinds of locations are well-known for their power reliability. It is not uncommon in mountainous areas for utility company service to be out for days after a major storm. Few operators are willing to take such risks with their business. Most choose to install standby power systems at appropriate points in the equipment chain.

The cost of standby power for a facility can be substantial, and an examination of the possible alternatives should be conducted before any decision on equipment is made. Management must clearly define the direct and indirect costs and weigh them appropriately. Include the following items in the cost-vs.-risk analysis:

- Standby power-system equipment purchase and installation cost.
- Exposure of the system to utility company power failure.
- Alternative operating methods available to the facility.
- Direct and indirect costs of lost uptime because of blackout conditions.

A distinction must be made between *emergency* and *standby* power sources. Strictly speaking, emergency systems supply circuits legally designated as being essential for safety to life and property. Standby power systems are used to protect a facility against the loss of productivity resulting from a utility company power outage.

9.2   STANDBY POWER OPTIONS

To ensure the continuity of ac power, many commercial/industrial facilities depend upon either two separate utility services or one utility service plus on-site generation. Because of the growing complexity of electrical systems, attention must be given to power-supply reliability.

The engine-generator shown in Figure 9.1 is the classic standby power system. An automatic transfer switch monitors the ac voltage coming from the utility company line for power failure conditions. Upon detection of an outage for a predetermined period of time (generally 1 to 10 s), the standby generator is started; once the generator is up to speed, the load is transferred from the utility to the local generator. Upon return of the utility feed, the load is switched back, and the generator is stopped. This basic type of system is used widely in industry and provides economical protection against prolonged power outages (5 min or more).

9.2.1   Dual Feeder System

In some areas, usually metropolitan centers, two utility company power drops can be brought into a facility as a means of providing a source of standby power. As shown in Figure 9.2, two separate utility service drops — from separate power-distribution systems — are brought into the plant, and an automatic transfer switch changes the load to the backup line in the event of a main-line failure. The dual feeder system provides an advantage over the auxiliary diesel arrangement in that power transfer from main to standby can be made in a fraction of a second if a static transfer switch

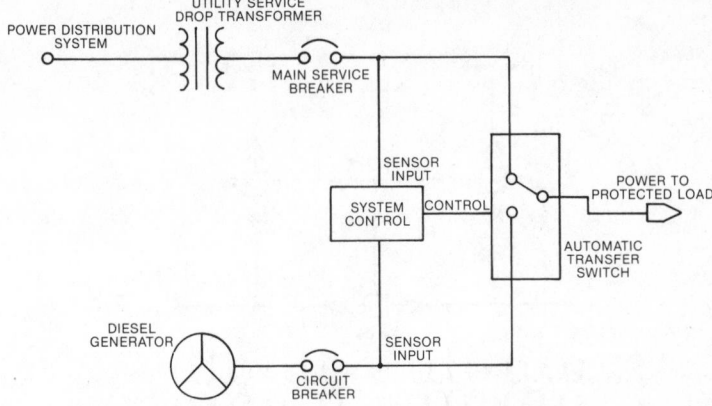

Figure 9.1   The classic standby power system using an engine-generator set. This system protects a facility from prolonged utility company power failures.

is used. Time delays are involved in the diesel generator system that limit its usefulness to power failures lasting more than several minutes.

The dual feeder system of protection is based on the assumption that each of the service drops brought into the facility is routed via different paths. This being the case, the likelihood of a failure on both power lines simultaneously is remote. The dual feeder system will not, however, protect against areawide power failures, which may occur from time to time.

The dual feeder system is limited primarily to urban areas. Rural or mountainous regions generally are not equipped for dual redundant utility company operation. Even in urban areas, the cost of bringing a second power line into a facility can be high, particularly if special lines must be installed for the feed. If two separate utility services

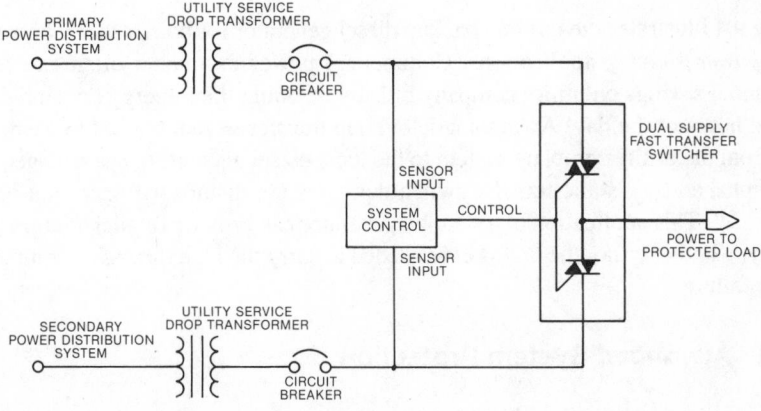

Figure 9.2   The dual utility feeder system of ac power loss protection. An automatic transfer switch changes the load from the main utility line to the standby line in the event of a power interruption.

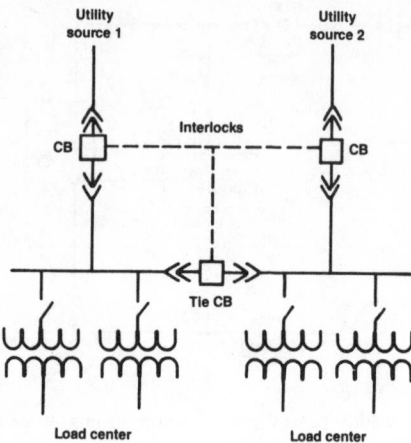

Figure 9.3   A dual utility feeder system with interlocked circuit breakers.

are available at or near the site, redundant feeds generally will be less expensive than engine-driven generators of equivalent capacity.

Figure 9.3 illustrates a dual feeder system that utilizes both utility inputs simultaneously at the facility. Notice that during normal operation, both ac lines feed loads, and the "tie" circuit breaker is open. In the event of a loss of either line, the circuit-breaker switches reconfigure the load to place the entire facility on the single remaining ac feed. Switching is performed automatically; manual control is provided in the event of a planned shutdown on one of the lines.

9.2.2   Peak Power Shaving

Figure 9.4 illustrates the use of a backup diesel generator for both standby power and *peak power shaving* applications. Commercial power customers often can realize substantial savings on utility company bills by reducing their energy demand during certain hours of the day. An automatic overlap transfer switch is used to change the load from the utility company system to the local diesel generator. The changeover is accomplished by a static transfer switch that does not disturb the operation of load equipment. This application of a standby generator can provide financial return to the facility, whether or not the unit is ever needed to carry the load through a commercial power failure.

9.2.3   Advanced System Protection

A more sophisticated power-control system is shown in Figure 9.5, where a dual feeder supply is coupled with a motor-generator set to provide clean, undisturbed ac power to the load. The m-g set will smooth over the transition from the main utility feed to the standby, often making a commercial power failure unnoticed by on-site personnel.

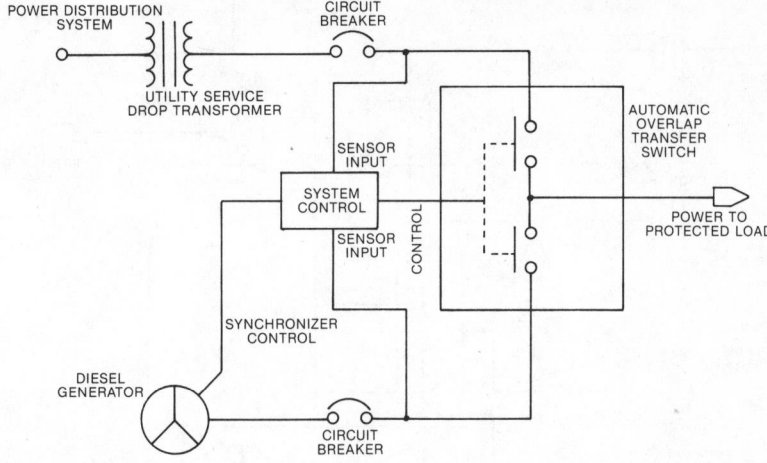

Figure 9.4   The use of a diesel generator for standby power and peak power shaving applications. The automatic overlap (static) transfer switch changes the load from the utility feed to the generator instantly so that no disruption of normal operation is encountered.

As discussed in Section 5.2, an m-g typically will give up to 1/2 s of power fail ride-through, more than enough to accomplish a transfer from one utility feed to the other. This standby power system is further refined in the application illustrated in Figure 9.6, where a diesel generator has been added to the system. With the automatic

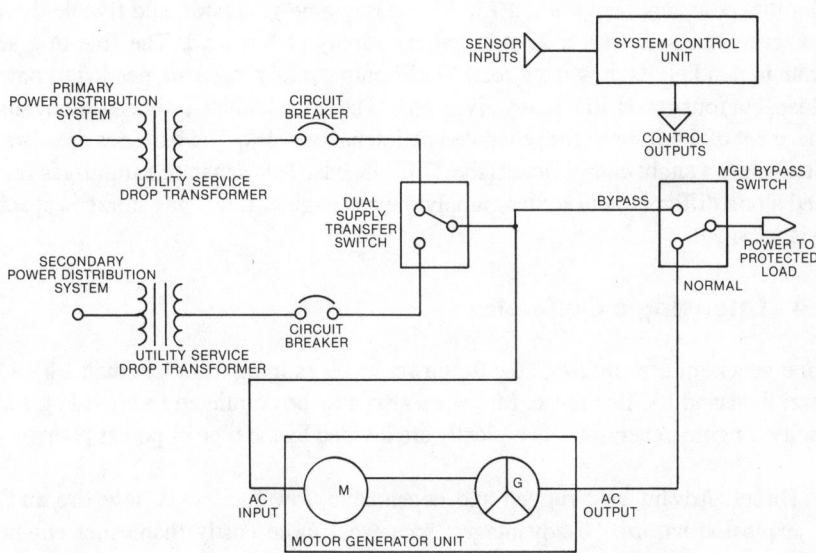

Figure 9.5   A dual feeder standby power system using a motor-generator set to provide power fail ride-through and transient-disturbance protection. Switching circuits allow the m-g set to be bypassed, if necessary.

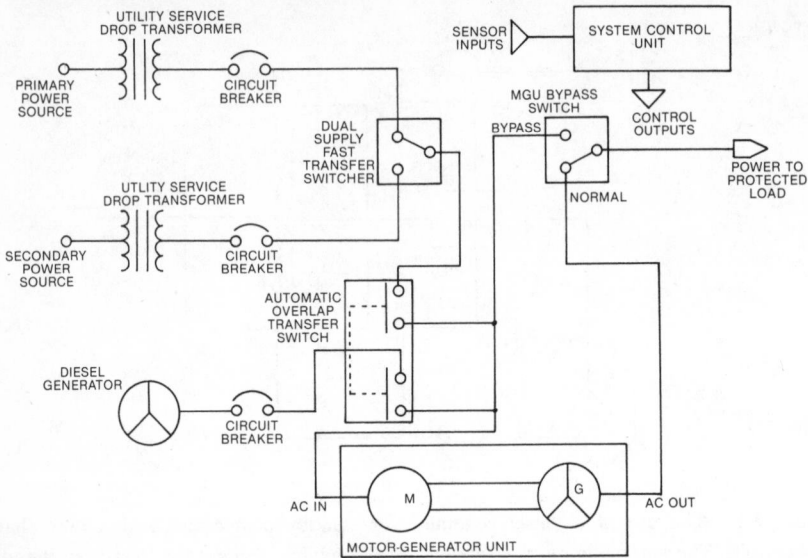

Figure 9.6 A premium power-supply backup and conditioning system using dual utility company feeds, a diesel generator, and a motor-generator set. An arrangement such as this would be used for critical loads that require a steady supply of clean ac.

overlap transfer switch shown at the generator output, this arrangement also can be used for peak demand power shaving.

Figure 9.7 shows a simplified schematic diagram of a 220 kW UPS system utilizing dual utility company feed lines, a 750 kVA gas engine-generator, and five dc-driven motor-generator sets with a 20-min battery supply at full load. The five m-g sets operate in parallel. Each is rated for 100 kW output. Only three are needed to power the load, but four are on-line at any given time. The fifth machine provides redundancy in the event of a failure or for scheduled maintenance work. The batteries are always on-line under a slight charge across the 270 V dc bus. Two separate natural-gas lines, buried along different land routes, supply the gas engine. Local gas storage capacity also is provided.

9.2.4 Choosing a Generator

Engine-generator sets are available for power levels ranging from less than 1 kVA to several thousand kVA or more. Machines also may be paralleled to provide greater capacity. Engine-generator sets typically are divided by the type of power plant used:

- Diesel. Advantages: rugged and dependable, low fuel costs, low fire and/or explosion hazard. Disadvantages: somewhat more costly than other engines, heavier in smaller sizes.
- Natural and liquefied-petroleum gas. Advantages: quick starting after long shutdown periods, long life, low maintenance. Disadvantage: availability of natural gas during areawide power failure subject to question.

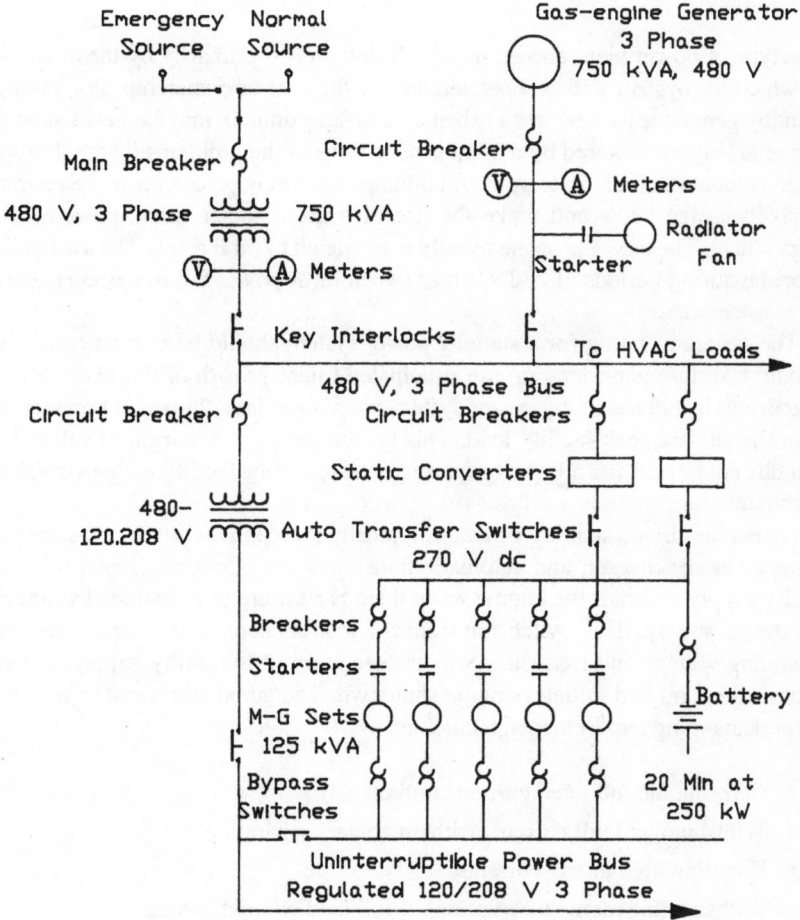

Figure 9.7 Simplified installation diagram of a high-reliability power system incorporating dual utility feeds, a standby gas-engine generator, and five battery-backed dc m-g sets. (Adapted from: Robert Lawrie, *Electrical Systems for Computer Installations*, McGraw-Hill, New York, 1988.)

- Gasoline. Advantages: rapid starting, low initial cost. Disadvantages: greater hazard associated with storing and handling gasoline, generally shorter mean time between overhaul.
- Gas turbine. Advantages: smaller and lighter than piston engines of comparable horsepower, rooftop installations practical, rapid response to load changes. Disadvantages: longer time required to start and reach operating speed, sensitive to high input air temperature.

The type of power plant chosen usually is determined primarily by the environment in which the system will be operated and by the cost of ownership. For example, a standby generator located in an urban area office complex may be best suited to the use of an engine powered by natural gas, because of the problems inherent in storing large amounts of fuel. State or local building codes may place expensive restrictions on fuel-storage tanks and make the use of a gasoline- or diesel-powered engine impractical. The use of propane usually is restricted to rural areas. The availability of propane during periods of bad weather (when most power failures occur) also must be considered.

The generator rating for a standby power system should be chosen carefully and should take into consideration the anticipated future growth of the plant. It is good practice to install a standby power system rated for at least 25 percent greater output than the current peak facility load. This headroom gives a margin of safety for the standby equipment and allows for future expansion of the facility without overloading the system.

An engine-driven standby generator typically incorporates automatic starting controls, a battery charger, and automatic transfer switch. Control circuits monitor the utility supply and start the engine when there is a failure or a sustained voltage drop on the ac supply. The switch transfers the load as soon as the generator reaches operating voltage and frequency. Upon restoration of the utility supply, the switch returns the load and initiates engine shutdown. The automatic transfer switch must meet demanding requirements, including:

- Carrying the full rated current continuously
- Withstanding fault currents without contact separation
- Handling high inrush currents
- Withstanding many interruptions at full load without damage

The nature of most power outages requires a sophisticated monitoring system for the engine-generator set. Most power failures occur during periods of bad weather. Most standby generators are unattended. More often than not, the standby system will start, run, and shut down without any human intervention or supervision. For reliable operation, the monitoring system must check the status of the machine continually to ensure that all parameters are within normal limits. Time-delay periods usually are provided by the controller that require an outage to last from 5 to 10 s before the generator is started and the load is transferred. This prevents false starts that needlessly exercise the system. A time delay of 5 to 30 min usually is allowed between the restoration of utility power and return of the load. This delay permits the utility ac lines to stabilize before the load is reapplied.

The transfer of motor loads may require special consideration, depending upon the size and type of motors used at a plant. If the residual voltage of the motor is out of phase with the power source to which the motor is being transferred, serious damage may result to the motor. Excessive current draw also may trip overcurrent protective devices. Motors above 50 hp with relatively high load inertia in relation to torque requirements, such as flywheels and fans, may require special controls. Restart time delays are a common solution.

Automatic starting and synchronizing controls are used for multiple-engine-generator installations. The output of two or three smaller units can be combined to feed the load. This capability offers additional protection for the facility in the event of a failure in any one machine. As the load at the facility increases, additional engine-generator systems can be installed on the standby power bus.

9.2.5 UPS Systems

An uninterruptible power system is an elegant solution to power outage concerns. The output of the UPS inverter may be a sine wave or pseudo sine wave. (See Section 5.3.2.) When shopping for a UPS system, consider the following:

- Power reserve capacity for future growth of the facility.
- Inverter current surge capability (if the system will be driving inductive loads, such as motors).
- Output voltage and frequency stability over time and with varying loads.
- Required battery supply voltage and current. Battery costs vary greatly, depending upon the type of units needed.
- Type of UPS system (forward-transfer type or reverse-transfer type) required by the particular application. Some sensitive loads may not tolerate even brief interruptions of the ac power source.
- Inverter efficiency at typical load levels. Some inverters have good efficiency ratings when loaded at 90 percent of capacity, but poor efficiency when lightly loaded.
- Size and environmental requirements of the UPS system. High-power UPS equipment requires a large amount of space for the inverter/control equipment and batteries. Battery banks often require special ventilation and ambient temperature control.

9.2.6 Standby Power-System Noise

Noise produced by backup power systems can be a serious problem if not addressed properly. Standby generators, motor-generator sets, and UPS systems produce noise that can disturb building occupants and irritate neighbors and/or landlords.

The noise associated with electrical generation usually is related to the drive mechanism, most commonly an internal combustion engine. The amplitude of the noise produced is directly related to the size of the engine-generator set. First consider whether noise reduction is a necessity. Many building owners have elected to tolerate the noise produced by a standby power generator because its use is limited to emergency situations. During a crisis, when the normal source of power is unavailable, most people will tolerate noise associated with a standby generator.

If the decision is made that building occupants can live with the noise of the generator, care must be taken in scheduling the required testing and exercising of the unit. Whether testing occurs monthly or weekly, it should be done on a regular schedule.

If it has been determined that the noise should be controlled, or at least minimized, the easiest way to achieve this objective is to physically separate the machine from occupied areas. This may be easier said than done. Because engine noise is predominantly low-frequency in character, walls and floor/ceiling construction used to contain the noise must be massive. Lightweight construction, even though it may involve several layers of resiliently mounted drywall, is ineffective in reducing low-frequency noise. Exhaust noise is a major component of engine noise but, fortunately, it is easier to control. When selecting an engine-generator set, select the highest-quality exhaust muffler available. Such units often are identified as "hospital-grade" mufflers.

Engine-generator sets also produce significant vibration. The machine should be mounted securely to a slab-on-grade or an isolated basement floor, or it should be installed on vibration isolation mounts. Such mounts usually are specified by the manufacturer.

Because a UPS system or motor-generator set is a source of continuous power, it must run continuously. Noise must be adequately controlled. Physical separation is the easiest and most effective method of shielding occupied areas from noise. Enclosure of UPS equipment usually is required, but noise control is significantly easier than for an engine-generator because of the lower noise levels involved. Nevertheless, the low-frequency 120 Hz fundamental of a UPS system is difficult to contain adequately; massive constructions may be necessary. Vibration control also is required for most UPS and m-g gear.

9.2.7 Critical System Bus

Many facilities do not require the operation of all equipment during a power outage. Rather than use one large standby power system, key pieces of equipment can be protected with small, dedicated, uninterruptible power systems. Small UPS units are available with built-in battery supplies for microcomputer systems and other hardware. If cost prohibits the installation of a systemwide standby power supply (using generator or solid-state UPS technology), consider establishing a *critical load bus* that is connected to a UPS system or generator via an automatic transfer switch. This separate power supply is used to provide ac to critical loads, thus keeping the protected systems up and running. The concept is illustrated in Figure 9.8. Unnecessary loads are dropped in the event of a power failure.

A standby system built on the critical load principle can be a cost-effective answer to the power-failure threat. The first step in implementing a critical load bus is to accurately determine the power requirements for the most important equipment. Typical power consumption figures can be found in most equipment instruction

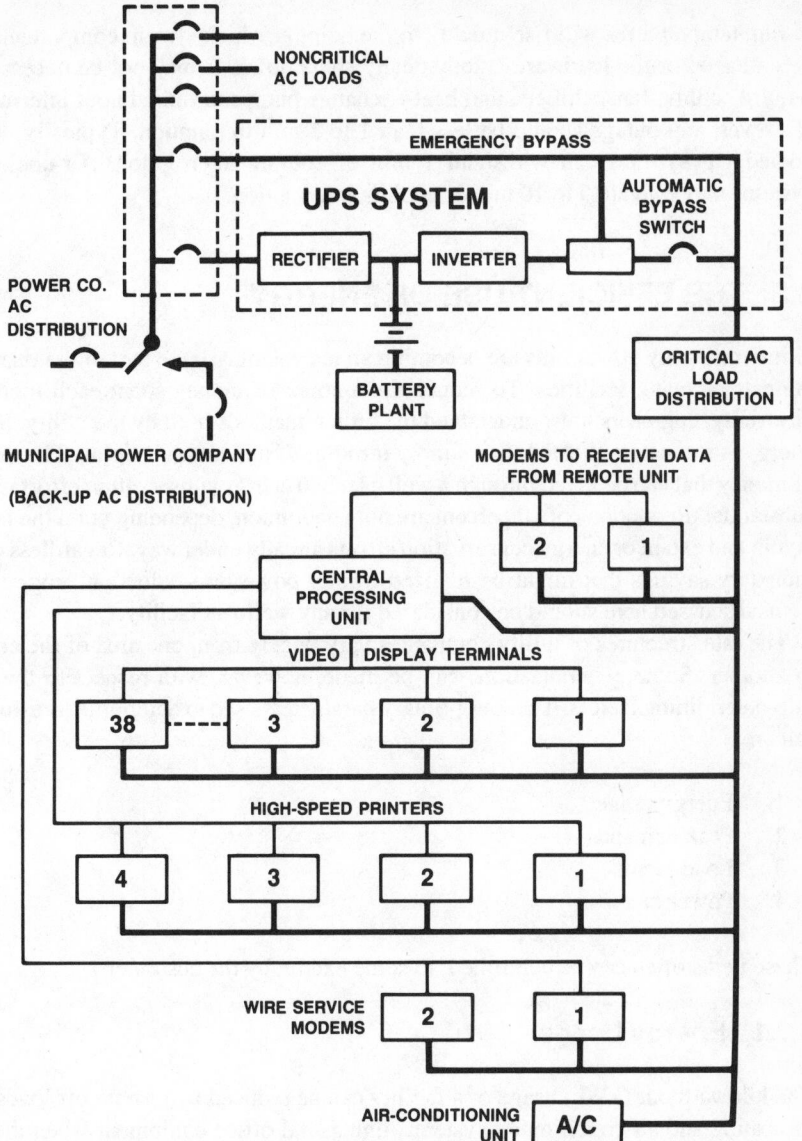

Figure 9.8   An application of the critical-load power bus concept. In the event of a power failure, all equipment necessary for continued operation is powered by the UPS equipment. Noncritical loads are dropped until commercial ac returns.

manuals. If the data is not listed or available from the manufacturer, it can be measured using a wattmeter.

When planning a critical load bus, be certain to identify accurately which loads are critical, and which can be dropped in the event of a commercial power failure. If air conditioning is interrupted but the computer equipment at a large DP center continues to run, temperatures will rise quickly to the point at which system components may be damaged or the hardware automatically shuts down. It may not be necessary to require cooling fans, chillers, and heat-exchange pumps to run without interruption. However, any outage should be less than 1 to 2 min in duration. Typically, liquid-cooled DP systems can withstand 1 min of cooling interruption. Air-cooled DP systems may tolerate 5 to 10 min of cooling interruption.

9.3 THE EFFICIENT USE OF ENERGY

Utility company power bills are becoming an increasingly large part of the operating budgets of many facilities. To reduce the amount of money spent each month on electricity, engineers must understand the billing methods used by the utility. Saving energy is more complicated than simply turning off unnecessary lights. The amount of money that can be saved through a well-planned energy conservation effort is often substantial. Reductions of 20 percent are not uncommon, depending upon the facility layout and extent of energy conservation efforts already under way. Regardless of any monetary savings that might be realized from a power-use-reduction program, the items discussed here should be considered for any well-run facility.

The rate structures of utility companies vary widely from one area of the country to another. Some generalizations can be made, however, with respect to the basic rate-determining factors. The four primary parameters used to determine a customer's bill are:

1. Energy usage
2. Peak demand
3. Load factor
4. Power factor

These items often can be controlled, to some extent, by the customer.

9.3.1 Energy Usage

The kilowatthour (kWh) usage of a facility can be reduced by turning off loads such as heating and air conditioning systems, lights, and office equipment when they are not needed. The installation of timers, photocells, or sophisticated computer-controlled energy-management systems can make substantial reductions in facility kWh demand each month. Common sense will dictate the conservation measures applicable to a particular situation. Obvious items include reducing the length of time high-power equipment is in operation, setting heating and cooling thermostats to reasonable levels, keeping office equipment turned off during the night, and avoiding excessive amounts of indoor or outdoor lighting.

Although energy conservation measures should be taken in every area of facility operation, the greatest savings generally can be found where the largest energy users are located. Transmitter plants, large machinery, and process drying equipment consume a large amount of power, so particular attention should be given to such hardware. Consider the following:

- Use the waste heat from equipment at the site for other purposes, if practical. In the case of high-power RF generators, room heating can be accomplished with a logic-controlled power amplifier exhaust-air recycling system.
- Have a knowledgeable consultant plan the air conditioning and heating system at the facility for efficient operation.
- Check thermostat settings on a regular basis, and consider installing time-controlled thermostats.
- Inspect outdoor-lighting photocells regularly for proper operation.
- Examine carefully the efficiency of high-power equipment used at the facility. New designs may offer substantial savings in energy costs.

The efficiency of large power loads, such as mainframe computers, transmitters, or industrial RF heaters, is an item of critical importance to energy conservation efforts. Most systems available today are significantly more efficient than their counterparts of just 10 years ago. Plant management often can find economic justification for updating or replacing an older system on the power savings alone. In virtually any facility, energy conservation can best be accomplished through careful selection of equipment, thoughtful system design, and conscientious maintenance practices.

9.3.2 Peak Demand

Conserving energy is a big part of the power bill reduction equation, but it is not the whole story. The *peak demand* of the customer load is an important criterion in the utility company's calculation of rate structures. The peak demand figure is a measure of the maximum load placed on the utility company system by a customer during a predetermined billing cycle. The measured quantities may be kilowatts, kilovolt-amperes, or both. Time intervals used for this measurement range from 15 to 60 min. Billing cycles may be annual or semiannual. Figure 9.9 shows an example of varying peak demand.

If a facility operated at basically the same power consumption level from one hour to the next and one day to the next, the utility company could predict accurately the demand of the load, then size its equipment (including the allocation of energy reserves) for only the amount of power actually needed. For the example shown in the figure, however, the utility company must size its equipment (including allocated energy reserves) for the peak demand. The area between the peak demand and the actual usage is the margin of inefficiency that the customer forces upon the utility. The peak demand factor is a method used by utility companies to assess penalties for such operation, thereby encouraging the customer to approach a more efficient state of operation (from the utility's viewpoint).

Load shedding is a term used to describe the practice of trimming peak power demand to reduce high-demand penalties. The goal of load shedding is to schedule the operation of nonessential equipment so as to provide a uniform power load from

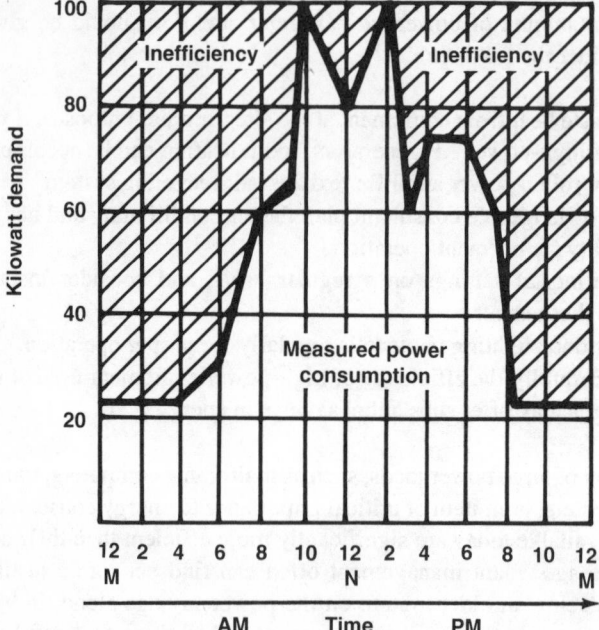

Figure 9.9   The charted power consumption of a facility not practicing energy-management techniques. Note the inefficiency that the utility company must absorb when faced with a load such as this.

the utility company, and thereby a better kWh rate. Nearly any operation has certain electric loads that can be rescheduled on a permanent basis or deferred as power demand increases during the day. Figure 9.10 illustrates the results of a load-shedding program. This more efficient operation has a lower overall peak demand and a higher average demand.

Peak demand reduction efforts can cover a wide range of possibilities. It would be unwise from an energy standpoint, for example, to test high-power standby equipment on a summer afternoon, when air conditioning units may be in full operation. Morning or evening hours would be a better choice, when the air conditioning is off and the demand of office equipment is reduced. Each operation is unique and requires an individual assessment of load-shedding options.

A computerized power-demand controller provides an effective method of managing peak demand. A controller can analyze the options available and switch loads as needed to maintain a relatively constant power demand from the utility company. Such systems are programmed to recognize which loads have priority and which loads are nonessential. Power demand then is automatically adjusted by the computer, based upon the rate schedule of the utility company. Many computerized demand control systems also provide the customer a printout of the demand profile of the plant, further helping managers analyze and reduce power costs.

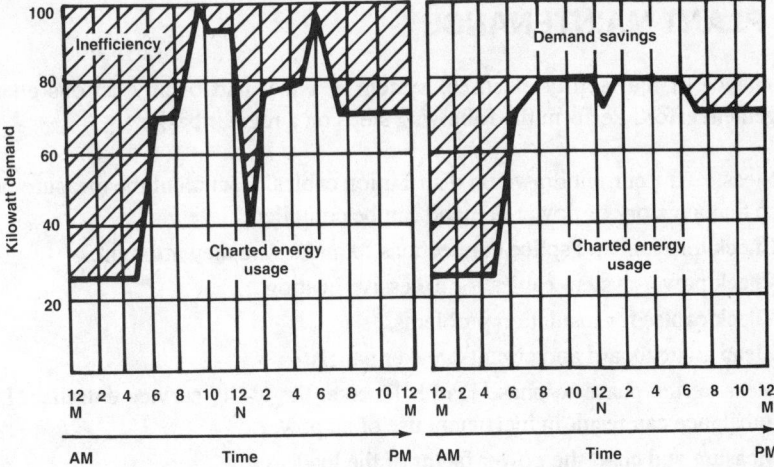

Figure 9.10   An example of the successful application of a load-shedding program. Energy usage has been spread more evenly throughout the day, resulting in reduced demand and, consequently, a better rate from the utility company.

9.3.3   Load Factor

The load factor on an electric utility company bill is a product of the peak demand and energy usage. It usually is calculated and applied to the customer's bill each month. Reducing either the peak demand or energy usage levels, or both, will decrease this added cost factor. Reducing power factor penalties also will help to reduce load factor charges.

9.3.4   Power Factor

Power factor (PF) charges are the result of heavy inductive loading of the utility company system. A poor PF will result in excessive losses along utility company feeder lines because more current is required to supply a particular load with a low PF than would be demanded if the load had a PF close to unity. (The technical aspects of power factor are discussed in Section 1.3.2.) The power factor charge is a penalty that customers pay for the extra current needed to magnetize motors and other inductive loads. This magnetizing current does not show up on the service drop wattmeter. It is, instead, measured separately or prorated as an additional charge to the customer. The power factor penalty sometimes can be reduced through the addition of on-site PF correction capacitors.

Power factor meters are available for measurement of a given load. It is usually less expensive in the long run, however, to hire a local electrical contractor to conduct a PF survey and recommend correction methods. Possible sources of PF problems include transmitters, blowers, air conditioners, heating equipment, and fluorescent and high-intensity discharge lighting-fixture ballasts.

9.4 PLANT MAINTENANCE

Maintenance of the facility electrical system is a key part of any serious energy-management effort. Perform the following steps on a regular basis:

- Measure the current drawn on distribution cables. Document the measurements so that a history of power demand can be compiled.
- Check terminal and splice connections to make sure they are tight.
- Check power-system cables for excessive heating.
- Check cables for insulation problems.
- Clean switchboard and circuit-breaker panels.
- Measure the phase-to-phase load balance at the utility service entrance. Load imbalance can result in inefficient use of ac power.
- Measure and chart the power factor of the load.

Develop and post a simplified one-line schematic of the entire power network as well as other building systems, including heating, air conditioning, security, and alarm functions. A *mimic board* is helpful in this process. Construct the mimic board control panel so that it depicts the entire ac power-distribution system. The board should have active indicators that show what loads or circuit breakers are turned on or off, what functions have been disabled, and key operating parameters, including input voltage, load current, and total kVA demand. Safety considerations require that machinery not be activated from the mimic board. Permit machinery to be energized only at the apparatus. As an alternative, remote control of machines can be provided, if a *remote/local* control switch is provided at the apparatus.

Environmental control systems should be monitored closely. Air conditioning, heating, and ventilation systems often represent a significant portion of the power load of a facility. Computer-based data-logging equipment with process control capability can be of considerable help in monitoring the condition of the equipment. The logger can be programmed to record all pertinent values periodically, and to report abnormal conditions.

9.4.1 Switchgear Maintenance

All too often, ac power switchgear is installed at a facility and forgotten — until a problem occurs. A careless approach to regular inspection and cleaning of switchgear has resulted in numerous failures, including destructive fires. The most serious fault in any switchgear assembly is arcing involving the main power bus. Protective devices often fail to open, or open only after a considerable delay. The arcing damage to busbars and enclosures can be significant. Fire often ensues, compounding the damage.

Moisture, combined with dust and dirt, is the greatest deteriorating factor insofar as insulation is concerned. Dust and/or moisture are thought to account for as much as half of switchgear failures. Initial leakage paths across the surface of bus supports result in flashover and sustained arcing. Contact overheating is another common cause of switchgear failure. Improper circuit-breaker installation or loose connections can result in localized overheating and arcing.

An arcing fault is destructive because of the high temperatures present (more than 6000°F). An arc is not a stationary event. Because of the ionization of gases and the presence of vaporized metal, an arc can travel along bare busbars, spreading the damage and sometimes bypassing open circuit breakers. It has been observed that most faults in three-phase systems involve all phases. The initial fault that triggers the event may involve only one phase, but because of the traveling nature of an arc, damage quickly spreads to the other lines.

Preventing switchgear failure is a complicated discipline, but consider the following general guidelines:

- Install insulated busbars for both medium-voltage and low-voltage switchgear. Each phase of the bus and all connections should be enclosed completely by insulation with electrical, mechanical, thermal, and flame-retardant characteristics suitable for the application.
- Establish a comprehensive preventive maintenance program for the facility. Keep all switchboard hardware clean from dust and dirt. Periodically check connection points for physical integrity.
- Maintain control over environmental conditions. Switchgear exposed to contaminants, corrosive gases, moist air, or high ambient temperatures may be subject to catastrophic failure. Conditions favorable to moisture condensation are particularly perilous, especially when dust and dirt are present.
- Accurately select overcurrent trip settings, and check them on a regular basis. Adjust the trip points of protection devices to be as low as possible, consistent with reliable operation.
- Divide switchgear into compartments that isolate different circuit elements. Consider adding vertical barriers to bus compartments to prevent the spread of arcing and fire.
- Install ground-fault protection devices at appropriate points in the power-distribution system. (See Section 10.2.2.)
- Adhere to all applicable building codes.

9.4.2   Ground-System Maintenance

Out of sight, out of mind does not — or, at least, *should* not — apply to a facility ground system. Grounding is a crucial element in achieving reliable operation of electronic equipment. If a ground system has been buried for 10 years or more, it is due for a complete inspection. If the system has been in place longer than 15 years, it is probably due for replacement. Soil conditions vary widely across the country, but few areas have soil that permits a ground system to last more than 15 years.

The method of construction and bonding of the ground network also can play a significant role in the ultimate life expectancy of the system. For example, ground conductors secured only by mechanical means (screws and bolts, crimping, and rivets) can quickly break down when exposed to even mild soil conditions. Unless silver-

Figure 9.11   Ground system inspection: (a) Even though a buried copper strap may appear undamaged, give it a pull to be sure. This strap came apart with little effort. (b) Acidic soil conditions created holes in this ground screen.

[c]

Figure 9.11   Ground system inspection (*continued*): (c) Small pieces of copper strap were used in this ground system to attach radials to the ground screen around the base of a tower. Proper installation procedures would have incorporated a solid piece of strap around the perimeter of the screen for such connections.

soldered or bonded using a Cadwelding method, such connections soon will be useless for all practical purposes.

The inspection process involves uncovering portions of the ground system to check for evidence of failure. Pay particular attention to interconnection points, where the greatest potential for problems exists. In some cases, a good metal detector will help identify portions of the ground system. It will not, however, identify breaks in the system. Portions of the ground system still will need to be uncovered to complete the inspection. Accurate documentation of the placement of ground-system components will aid the inspection effort greatly.

Check any buried mechanical connections carefully. Bolts that have been buried for many years may be severely deteriorated. Carefully remove several bolts, and inspect their condition. If a bolt is severely oxidized, it may twist off as it is removed. After uncovering representative portions of the ground system, document the condition of the ground through notes and photographs. These will serve as a reference point for future observation. The photos in Figure 9.11 illustrate some of the problems that can occur with an aging ground system. Note that many of the problems experienced with the system shown in the photographs resulted from improper installation of the components.

9.5 BIBLIOGRAPHY

Angevine, Eric: "Controlling Generator and UPS Noise," *Broadcast Engineering* magazine, Intertec Publishing, Overland Park, KS, March 1989.

Baietto, Ron: "How to Calculate the Proper Size of UPS Devices," *Microservice Management* magazine, Intertec Publishing, Overland Park, KS, March 1989.

Federal Information Processing Standards Publication no. 94, *Guideline on Electrical Power for ADP Installations*, U.S. Department of Commerce, National Bureau of Standards, Washington, DC, 1983.

Highnote, Ronnie L.: *The IFM Handbook of Practical Energy Management*, Institute for Management, Old Saybrook, CT, 1979.

Lawrie, Robert: *Electrical Systems for Computer Installations*, McGraw-Hill, New York, 1988.

Smith, Morgan: "Planning for Standby AC Power," *Broadcast Engineering* magazine, Intertec Publishing, Overland Park, KS, March 1989.

Stuart, Bud: "Maintaining an Antenna Ground System," *Broadcast Engineering* magazine, Intertec Publishing, Overland Park, KS, October 1986.

10

SAFETY AND PROTECTION SYSTEMS

Portions of this chapter were contributed by Brad Dick, editor, *Broadcast Engineering* magazine, Overland Park, KS.

10.1 INTRODUCTION

Safety is critically important to engineering personnel who work around powered hardware, especially if they work under considerable time pressures. Safety is not something to be taken lightly. *Life safety* systems are those designed to protect life and property. Such systems include emergency lighting, fire alarms, smoke exhaust and ventilating fans, and site security.

10.1.1 Facility Safety Equipment

Personnel safety is the responsibility of the facility manager. Proper life safety procedures and equipment must be installed. Safety-related hardware includes the following:

- *Emergency power off (EPO) button.* EPO push buttons are required by safety code for data processing centers. One must be located at each principal exit from the DP room. Other EPO buttons may be located near operator workstations. The EPO system, intended only for emergencies, disconnects all power to the room, except for lighting.
- *Smoke detector.* Two basic types of smoke detectors commonly are available. The first compares the transmission of light through air in the room with light through a sealed optical path into which smoke cannot penetrate. Smoke causes

a differential or *backscattering* effect that, when detected, triggers an alarm after a preset threshold has been exceeded. The second type of smoke detector senses the ionization of combustion products rather than visible smoke. A mildly radioactive source, usually nickel, ionizes the air passing through a screened chamber. A charged probe captures ions and detects the small current that is proportional to the rate of capture. When combustion products or material other than air molecules enter the probe area, the rate of ion production changes abruptly, generating a signal that triggers the alarm.

- *Flame detector.* The flame sensor responds not to heated surfaces or objects, but to infrared when it flickers with the unique characteristics of a fire. Such detectors, for example, will respond to a lighted match, but not to a cigarette. The ultraviolet light from a flame also is used to distinguish between hot, glowing objects and open flame.
- *Halon.* The Halon fire-extinguishing agent is a low-toxicity, compressed gas that is contained in pressurized vessels. Discharge nozzles in DP rooms and other types of equipment rooms are arranged to dispense the entire contents of a central container or of multiple smaller containers of Halon when actuated by a command from the fire control system. The discharge is sufficient to extinguish flame and stop combustion of most flammable substances. Halon is one of the more common fire-extinguishing agents used for DP applications. Halon systems usually are not practical, however, in large, open-space computer centers.
- *Water sprinkler.* Although water is an effective agent against a fire, activation of a sprinkler system will cause damage to the equipment it is meant to protect. Interlock systems must drop all power (except for emergency lighting) before the water system is discharged. Most water systems use a two-stage alarm. Two or more fire sensors, often of different design, must signal an alarm condition before water is discharged into the protected area. Where sprinklers are used, floor drains and EPO controls must be provided.
- *Fire damper.* Dampers are used to block ventilating passages in strategic parts of the system when a fire is detected. This prevents fire from spreading through the passages and keeps fresh air from fanning the flames. A fire damper system, combined with the shutdown of cooling and ventilating air, enables Halon to be retained in the protected space until the fire is extinguished.

Many life safety system functions can be automated. The decision of what to automate and what to operate manually requires considerable thought. If the life safety control panels are accessible to a large number of site employees, most functions should be automatic. Alarm-silencing controls should be maintained under lock and key. A mimic board can be used to readily identify problem areas. Figure 10.1 illustrates a well-organized life safety control system. Note that fire, HVAC (heating, ventilation, and air conditioning), security, and EPO controls all are readily accessible. Note also that operating instructions are posted for life safety equipment, and an evacuation route is shown. Important telephone numbers are posted, and a direct-line telephone (not via the building switchboard) are provided. All equipment is located adjacent to a lighted emergency exit door.

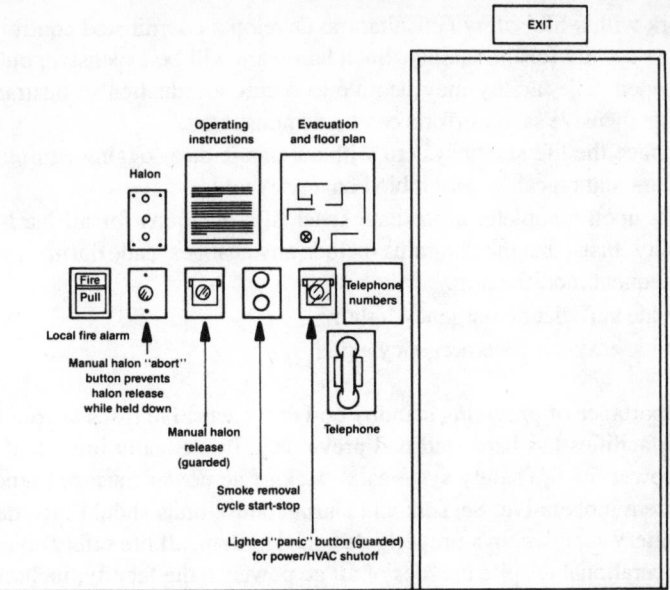

Figure 10.1 A well-organized life safety control station. (Adapted from: Federal Information Processing Standards Publication no. 94, *Guideline on Electrical Power for ADP Installations*, U.S. Department of Commerce, National Bureau of Standards, Washington, DC, 1983.)

Life safety equipment must be maintained just as diligently as the computer system that it protects. Conduct regular tests and drills. It is, obviously, not necessary or advisable to discharge Halon or water during a drill.

Configure the life safety control system to monitor not only the premises for dangerous conditions, but also the equipment designed to protect the facility. Important monitoring points include HVAC machine parameters, water and/or Halon pressure, emergency battery-supply status, and other elements of the system that could compromise the ability of life safety equipment to carry out its functions. Basic guidelines for life safety systems include the following:

- Carefully analyze the primary threats to life and property within the facility. Develop contingency plans to meet each threat.
- Prepare a life safety manual, and distribute it to all employees at the facility. Require them to read it.
- Conduct drills for employees at random times without notice. Require acceptable performance from employees.
- Prepare simple, step-by-step instructions on what to do in an emergency. Post the instructions in a conspicuous place.
- Assign after-hours responsibility for emergency situations. Prepare a list of supervisors that operators should contact if problems arise. Post the list with phone numbers. Keep the list accurate and up-to-date. Always provide the names of three individuals who may be contacted in an emergency.

- Work with a life safety consultant to develop a coordinated control and monitoring system for the facility. Such hardware will be expensive, but it must be provided. The facility may be able to secure a reduction in insurance rates if comprehensive safety efforts can be demonstrated.
- Interface the life safety system with automatic data-logging equipment so that documentation can be assembled on any event.
- Insist upon complete, up-to-date schematic diagrams for all hardware at the facility. Insist that the diagrams include any changes made during installation or subsequent modification.
- Provide sufficient emergency lighting.
- Provide easy-access emergency exits.

The importance of providing standby power for sensitive loads at commercial and industrial facilities has been outlined previously. It is equally important to provide standby power for life safety systems. A lack of ac power must not render the life safety system inoperative. Sensors and alarm control units should include their own backup battery supplies. In a properly designed system, all life safety equipment will be fully operational despite the loss of all ac power to the facility, including backup power for sensitive loads.

Place cables linking the life safety control system with remote sensors and actuators in a separate conduit containing only life safety conductors. Study the National Electrical Code and all applicable local and federal codes relating to safety. Follow them to the letter.

10.2 ELECTRIC SHOCK

It takes surprisingly little current to injure a person. Studies at Underwriters' Laboratories (UL) show that the electrical resistance of the human body varies with the amount of moisture on the skin, the muscular structure of the body, and the applied voltage. The typical hand-to-hand resistance ranges from 500 Ω to 600 kΩ, depending on the conditions. Higher voltages have the capability to break down the outer layers of the skin, which can reduce the overall resistance value. UL uses the lower value, 500 Ω, as the standard resistance between major extremities, as from the hand to the foot. This value generally is considered the minimum that would be encountered. In fact, it may not be unusual because wet conditions or a cut or other break in the skin significantly reduce human body resistance.

10.2.1 Effects on the Human Body

Table 10.1 lists some effects that typically result when a person is connected across a current source with a hand-to-hand resistance of 2.4 kΩ. The table shows that a current of 50 mA will flow between the hands, if one hand is in contact with a 120 V ac source and the other hand is grounded. The table indicates that even the relatively small current of 50 mA can produce *ventricular fibrillation* of the heart, and maybe even cause death. Medical literature describes ventricular fibrillation as very rapid,

Table 10.1   The Effects of Current on the Human Body

1 mA or less	No sensation, not felt
More than 3 mA	Painful shock
More than 10 mA	Local muscle contractions, sufficient to cause "freezing" to the circuit for 2.5 percent of the population
More than 15 mA	Local muscle contractions, sufficient to cause "freezing" to the circuit for 50 percent of the population
More than 30 mA	Breathing is difficult, can cause unconsciousness
50 mA to 100 mA	Possible ventricular fibrillation of the heart
100 mA to 200 mA	Certain ventricular fibrillation of the heart
More than 200 mA	Severe burns and muscular contractions; heart more apt to stop than to go into fibrillation
More than a few amperes	Irreparable damage to body tissues

uncoordinated contractions of the ventricles of the heart resulting in loss of synchronization between heartbeat and pulse beat. The electrocardiograms shown in Figure 10.2 compare a healthy heart rhythm with one in ventricular fibrillation. Unfortunately, once ventricular fibrillation occurs, it will continue. Barring resuscitation techniques, death will ensue within a few minutes.

The route taken by the current through the body greatly affects the degree of injury. Even a small current, passing from one extremity through the heart to another extremity, is dangerous and capable of causing severe injury or electrocution. There are cases in which a person has contacted extremely high current levels and lived to

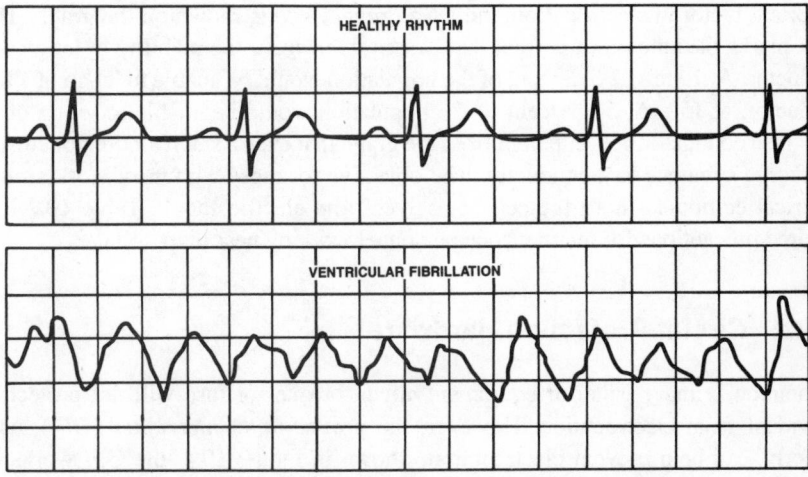

Figure 10.2   Electrocardiograms showing the healthy rhythm of a heart (*top*), and ventricular fibrillation of the heart (*bottom*).

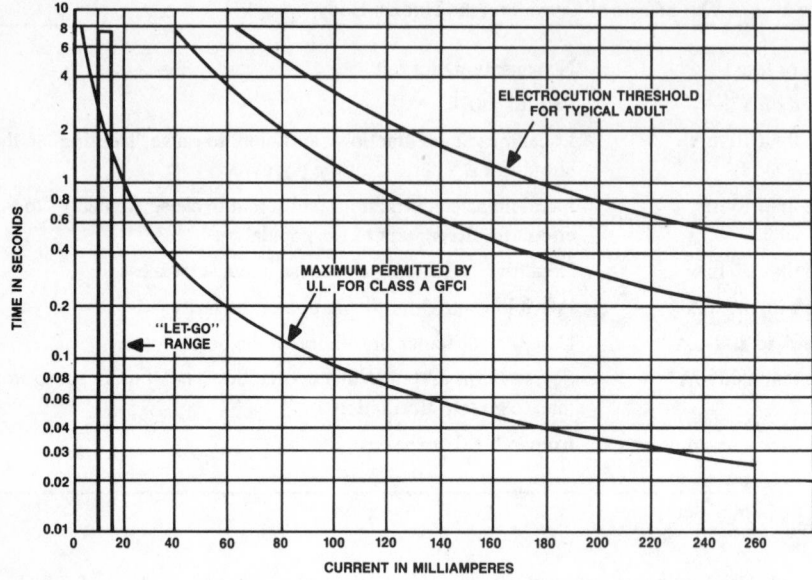

Figure 10.3   Effects of electrical current and time on the human body. Note the "let-go" range.

tell about it. However, when this happens, it is usually because the current passes only through a single limb and not through the entire body. In these instances, the limb is often lost but the person survives.

Current is not the only factor in electrocution. Figure 10.3 summarizes the relationship between current and time on the human body. The graph shows that 100 mA flowing through an adult human body for 2 s will cause death by electrocution. An important factor in electrocution, the *let-go range*, also is shown on the graph. This point marks the amount of current that causes *freezing*, or the inability to let go of a conductor. At 10 mA, 2.5 percent of the population would be unable to let go of a live conductor; at 15 mA, 50 percent of the population would be unable to let go of an energized conductor. It is apparent from the graph that even a small amount of current can freeze someone to a conductor. The objective for those who must work around electrical equipment is to protect themselves from electric shock. Table 10.2 lists required precautions for maintenance personnel working near high voltages.

10.2.2   Circuit-Protection Hardware

A common primary panel or equipment circuit breaker or fuse will not protect an individual from electrocution. However, the *ground-fault interrupter* (GFI), used properly, can help prevent electrocution. Shown in Figure 10.4, the GFI works by monitoring the current being applied to the load. It uses a differential transformer that senses an imbalance in load current. If a current (typically 5 mA, + 1 mA on a low-current 120 V ac line) begins flowing between the neutral and ground or between

Table 10.2   Required Safety Practices for Engineers Working Around High-Voltage Equipment

- Remove all ac power from the equipment. Do not rely on internal contactors or SCRs to remove dangerous ac.
- Trip the appropriate power-distribution circuit breakers at the main breaker panel.
- Place signs as needed to indicate that the circuit is being serviced.
- Switch the equipment being serviced to the *local control* mode as provided.
- Discharge all capacitors using the discharge stick provided by the manufacturer.
- Do not remove, short-circuit, or tamper with interlock switches on access covers, doors, enclosures, gates, panels, or shields.
- Keep away from live circuits.
- Allow any component to cool completely before attempting to replace it.
- If a leak or bulge is found on the case of an oil-filled or electrolytic capacitor, do not attempt to service the part until it has cooled completely.
- Know which parts in the system contain PCBs. Handle them appropriately.
- Minimize exposure to RF radiation.
- Avoid contact with hot surfaces within the system.
- Do not take chances.

the hot and ground leads, the differential transformer detects the leakage and opens the primary circuit (typically within 2.5 ms).

OSHA (Occupational Safety and Health Administration) rules specify that temporary receptacles (those not permanently wired) and receptacles used on construction

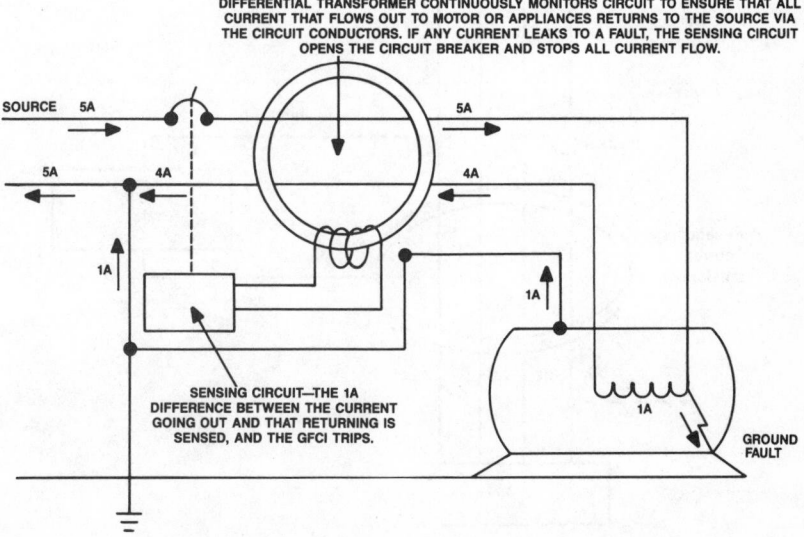

Figure 10.4   Basic design of a ground-fault interrupter (GFI).

sites be equipped with GFI protection. Receptacles on two-wire, single-phase portable and vehicle-mounted generators of not more than 5 kW, where the generator circuit conductors are insulated from the generator frame and all other grounded surfaces, need not be equipped with GFI outlets.

GFIs will not protect a person from every type of electrocution. If you become connected to both the neutral and the hot wire, the GFI will treat you as if you are merely a part of the load and will not open the primary circuit.

For large, three-phase loads, detecting ground currents and interrupting the circuit before injury or damage can occur is a more complicated preposition. The classic method of protection involves the use of a zero-sequence current transformer (CT). Such devices are basically an extension of the single-phase GFI circuit shown in Figure 10.4. Three-phase CTs have been developed to fit over bus ducts, switchboard buses, and circuit-breaker studs. Rectangular core-balanced CTs are able to detect leakage currents as small as several milliamperes when the system carries as much as 4 kA. "Doughnut-type" toroidal zero-sequence CTs also are available in varying diameters.

The zero-sequence current transformer is designed to detect the magnetic field surrounding a group of conductors. As shown in Figure 10.5, in a properly operating three-phase system, the current flowing through the conductors of the system, including the neutral, goes out and returns along those same conductors. The net magnetic flux detected by the CT is zero. No signal is generated in the transformer winding, regardless of current magnitudes — symmetrical or asymmetrical. If one phase conductor is faulted to ground, however, the current balance will be upset. The ground-fault-detection circuit then will trip the breaker and open the line.

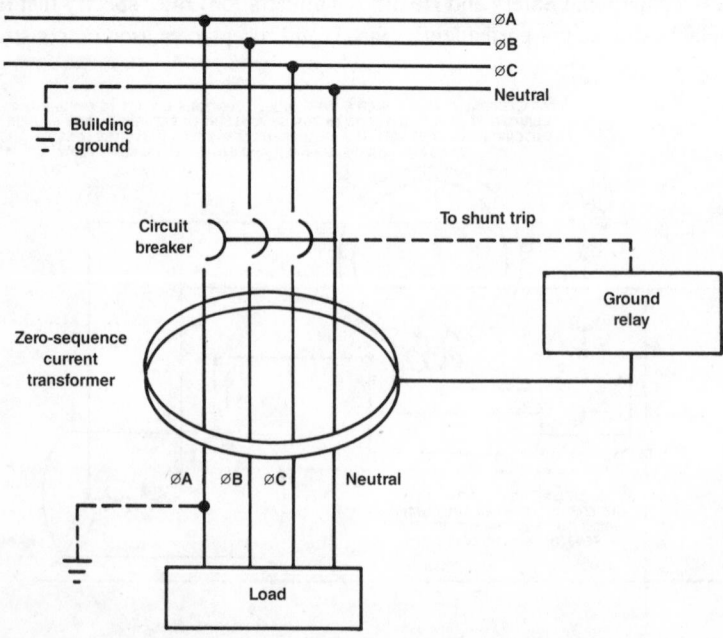

Figure 10.5   Ground-fault detection in a three-phase ac system.

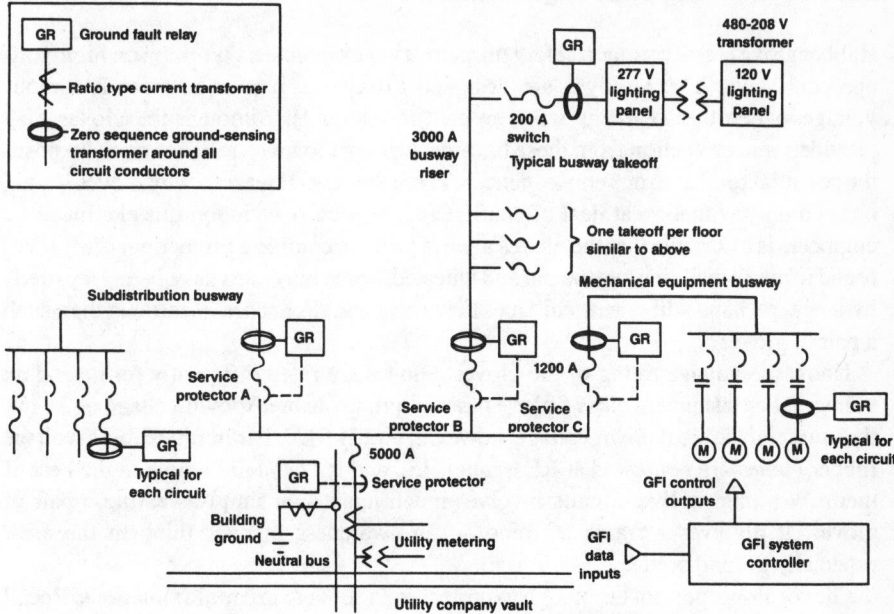

Figure 10.6   Ground-fault protection system for a large, multistory building.

For optimum protection in a large facility, GFI units are placed at natural branch points of the ac power system. It is, obviously, preferable to lose only a small portion of a facility in the event of a ground fault than it is to have the entire plant dropped. Figure 10.6 illustrates such a distributed system. Sensors are placed at major branch points to isolate any ground fault from the remainder of the distribution network. In this way, the individual GFI units can be set for higher sensitivity and shorter time delays than would be practical with a large, distributed load. The technology of GFI devices has improved significantly within the past few years. New integrated circuit devices and improved CT designs have provided improved protection components at a lower cost.

Sophisticated GFI monitoring *systems* are available that analyze ground-fault currents and isolate the faulty branch circuit. This feature prevents needless tripping of GFI units up the line toward the utility service entrance. For example, if a ground fault is sensed in a fourth-level branch circuit, the GFI system controller automatically locks out first-, second-, and third-level devices from operating to clear the fault. The problem, therefore, is safely confined to the fourth-level branch. The GFI control system is designed to operate in a fail-safe mode. In the event of a control-system shutdown, the individual GFI trip relays would operate independently to clear whatever fault currents may exist.

Any facility manager would be well-advised to hire an experienced electrical contractor to conduct a full ground-fault protection study. Direct the contractor to identify possible failure points, and to recommend corrective actions.

10.2.3 Working with High Voltage

Rubber gloves are a common safety measure used by engineers working on high-voltage equipment. These gloves are designed to provide protection from hazardous voltages when the wearer is working on "hot" circuits. Although the gloves may provide some protection from these hazards, placing too much reliance on them poses the potential for disastrous consequences. There are several reasons why gloves should be used only with a great deal of caution and respect. A common mistake made by engineers is to assume that the gloves always provide complete protection. The gloves found in many facilities may be old and untested. Some may even have been "repaired" by users, perhaps with electrical tape. Few tools could be more hazardous than such a pair of gloves.

Know the voltage rating of the gloves. Gloves are rated differently for ac and dc voltages. For instance, a *class 0* glove has a minimum dc breakdown voltage of 35 kV; the minimum ac breakdown voltage, however, is only 6 kV. Furthermore, high-voltage rubber gloves are not tested at RF frequencies, and RF can burn a hole in the best of them. Working on live circuits involves much more than simply wearing a pair of gloves. It involves a frame of mind — an awareness of everything in the area, especially ground points.

Gloves alone may not be enough to protect an individual in certain situations. Recall the axiom of keeping one hand in your pocket while working on a device with current flowing? The axiom actually is based on simple electricity. It is not the hot connection that causes the problem; it is the ground connection that permits current flow. A study in California showed that more than 90 percent of electrical equipment fatalities occurred when the grounded person contacted a live conductor. Line-to-line electrocution accounted for less than 10 percent of the deaths.

When working around high voltages, always look for grounded surfaces — and keep away from them. Even concrete can act as a ground if the voltage is high enough. If work must be conducted in live cabinets, consider using, in addition to rubber gloves, a rubber floor mat, rubber vest, and rubber sleeves. Although this may seem to be a lot of trouble, consider the consequences of making a mistake. Of course, the best troubleshooting methodology is never to work on any circuit unless you are sure no hazardous voltages are present. In addition, any circuits or contactors that normally contain hazardous voltages should be grounded firmly before work begins.

Another important safety rule is to never work alone. Even if a trained assistant is not available when maintenance is performed, someone should accompany you and be available to help in an emergency.

10.2.4 First Aid Procedures

Be familiar with first aid treatment for electrical shock and burns. Always keep a first aid kit on hand at the facility. Figure 10.7 illustrates the basic treatment for electric shock victims. Copy the information, and post it in a prominent location. Better yet, obtain more detailed information from your local heart association or Red Cross chapter. Personalized instruction on first aid usually is available locally. Table 10.3 lists basic first aid procedures for burns.

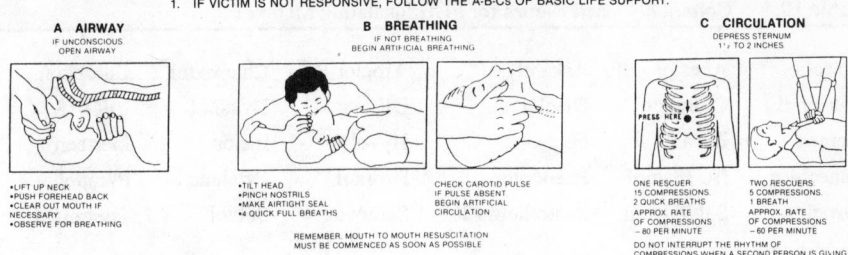

TREATMENT OF ELECTRICAL SHOCK

1.   IF VICTIM IS NOT RESPONSIVE, FOLLOW THE A-B-Cs OF BASIC LIFE SUPPORT.

A   AIRWAY
IF UNCONSCIOUS
OPEN AIRWAY

B   BREATHING
IF NOT BREATHING
BEGIN ARTIFICIAL BREATHING

C   CIRCULATION
DEPRESS STERNUM
1½ TO 2 INCHES

PRESS HERE

•LIFT UP NECK
•PUSH FOREHEAD BACK
•CLEAR OUT MOUTH IF NECESSARY
•OBSERVE FOR BREATHING

•TILT HEAD
•PINCH NOSTRILS
•MAKE AIRTIGHT SEAL
•4 QUICK FULL BREATHS

CHECK CAROTID PULSE
IF PULSE ABSENT
BEGIN ARTIFICIAL CIRCULATION

REMEMBER, MOUTH TO MOUTH RESUSCITATION
MUST BE COMMENCED AS SOON AS POSSIBLE

ONE RESCUER
15 COMPRESSIONS
2 QUICK BREATHS
APPROX. RATE
OF COMPRESSIONS
– 80 PER MINUTE

TWO RESCUERS
5 COMPRESSIONS
1 BREATH
APPROX. RATE
OF COMPRESSIONS
– 60 PER MINUTE

DO NOT INTERRUPT THE RHYTHM OF
COMPRESSIONS WHEN A SECOND PERSON IS GIVING
BREATH

2.   IF VICTIM IS RESPONSIVE, KEEP HIM WARM AND QUIET, LOOSEN CLOTHING AND PLACE IN RECLINING POSITION.
PLACE VICTIM FLAT ON HIS BACK ON A HARD SURFACE
CALL FOR MEDICAL ASSISTANCE AS SOON AS POSSIBLE

Figure 10.7   Basic first aid treatment for electric shock.

10.3 POLYCHLORINATED BIPHENYLS

Polychlorinated biphenyls (PCBs) belong to a family of organic compounds known as *chlorinated hydrocarbons*. Virtually all PCBs in existence today have been synthetically manufactured. PCBs are of a heavy, oil-like consistency and have a high boiling point, a high degree of chemical stability, low flammability, and low electrical

Table 10.3   Basic First Aid Procedures

For extensively burned and broken skin:

- Cover affected area with a clean sheet or cloth.
- Do not break blisters, remove tissue, remove adhered particles of clothing, or apply any salve or ointment.
- Treat victim for shock as required.
- Arrange for transportation to a hospital as quickly as possible.
- If victim's arms or legs are affected, keep them elevated.
- If medical help will not be available within an hour and the victim is conscious and not vomiting, prepare a weak solution of salt and soda. Mix 1 teaspoon of salt and 1/2-teaspoon of baking soda to each quart of tepid water. Allow the victim to sip slowly about 4 oz (half a glass) over a period of 15 min. Discontinue fluid intake if vomiting occurs. (Do not allow alcohol consumption.)

For less severe burns (first- and second-degree):

- Apply cool (not ice-cold) compresses using the cleanest available cloth article.
- Do not break blisters, remove tissue, remove adhered particles of clothing, or apply salve or ointment.
- Apply clean, dry dressing if necessary.
- Treat victim for shock as required.
- Arrange for transportation to a hospital as quickly as possible.
- If victim's arms or legs are affected, keep them elevated.

Table 10.4   Commonly Used Names for PCB Insulating Material

Apirolio	Abestol	Askarel	Aroclor B	Chlorextol	Chlophen
Chlorinol	Clorphon	Diaclor	DK	Dykanol	EEC-18
Elemex	Eucarel	Fenclor	Hyvol	Inclor	Inerteen
Kanechlor	No-Flamol	Phenodlor	Pydraul	Pyralene	Pyranol
Pyroclor	Sal-T-Kuhl	Santothern FR	Santovac	Solvol	Therminal

conductivity. These characteristics led to the past widespread use of PCBs in high-voltage capacitors and transformers. Commercial products containing PCBs were distributed widely from 1957 to 1977 under several trade names, including:

- Aroclor
- Pyroclor
- Sanotherm
- Pyranol
- Askarel

Askarel also is a generic name used for nonflammable dielectric fluids containing PCBs. Table 10.4 lists some common trade names for Askarel. These trade names typically are listed on the nameplate of a PCB transformer or capacitor.

10.3.1   Health Risk

PCBs are harmful because, once they are released into the environment, they tend not to break apart into other substances. Instead, PCBs persist, taking several decades to slowly decompose. By remaining in the environment, they can be taken up and stored in the fatty tissues of all organisms, from which they are released slowly into the bloodstream. Therefore, because of the storage in fat, the concentration of PCBs in body tissues can increase with time, even though PCB exposure levels may be quite low. This process is called *bioaccumulation*. Furthermore, as PCBs accumulate in the tissues of simple organisms, which are consumed by progressively higher organisms, the concentration increases. This process is called *biomagnification*. These two factors are especially significant because PCBs are harmful even at low levels. Specifically, PCBs have been shown to cause chronic (long-term) toxic effects in some species of animals and aquatic life. Well-documented tests on laboratory animals show that various levels of PCBs can cause reproductive effects, gastric disorders, skin lesions, and cancerous tumors.

PCBs may enter the body through the lungs, the gastrointestinal tract, and the skin. After absorption, PCBs are circulated in the blood throughout the body and stored in fatty tissues and skin, as well as in a variety of organs, including the liver, kidneys, lungs, adrenal glands, brain, and heart.

The health risk lies not only in the PCB itself, but also in the chemicals developed when PCBs are heated. Laboratory studies have confirmed that PCB by-products, including *polychlorinated dibenzofurans* (PCDFs) and *polychlorinated dibenzo-p-dioxins* (PCDDs), are formed when PCBs or chlorobenzenes are heated to temperatures ranging from approximately 900°F to 1300°F. Unfortunately, these products are more toxic than PCBs themselves.

The problem for the owner of PCB equipment is that the liability from a PCB spill or fire contamination can be tremendous. A fire involving a PCB large transformer in Binghampton, NY, resulted in $20 million in cleanup expenses. The consequences of being responsible for a fire-related incident with a PCB transformer may be monumental.

10.3.2   Governmental Action

Congress took action to control PCBs in October 1975, by passing the Toxic Substances Control Act (TSCA). A section of this law specifically directed the EPA to regulate PCBs. Three years later, the EPA issued regulations to implement a congressional ban on the manufacture, processing, distribution, and disposal of PCBs. Since, several revisions and updates have been issued by the EPA. One of these revisions, issued in 1982, specifically addressed the type of equipment used in industrial plants. Failure to properly follow the rules regarding the use and disposal of PCBs has resulted in high fines and some jail sentences.

Although PCBs no longer are being produced for most electrical products in the United States, the EPA estimates that more than 107,000 PCB transformers and 350 million PCB small capacitors were in use or in storage in 1984. Approximately 77,600 of these transformers were used in or near commercial buildings. Approximately 3.3 million PCB large capacitors were in use as late as 1981. The threat of widespread contamination from PCB fire-related incidents is one reason behind the EPA's efforts to reduce the number of PCB products in the environment. The users of high-power equipment are affected by the regulations primarily because of the widespread use of PCB transformers and capacitors. These components usually are located in older (pre-1979) systems, so this is the first place to look for them. However, some facilities also maintain their own primary power transformers. Unless these transformers are of recent vintage, it is quite likely that they too contain a PCB dielectric. Table 10.5 lists the primary classifications of PCB devices.

10.3.3   PCB Components

The two most common PCB components are transformers and capacitors. A PCB transformer is one containing at least 500 ppm (parts per million) PCBs in the dielectric fluid. An Askarel transformer generally has 600,000 ppm or more. A PCB transformer may be converted to a *PCB-contaminated device* (50 to 500 ppm) or a *non-PCB device* (less than 50 ppm) by being drained, refilled, and tested. The testing must not take place until the transformer has been in service for a minimum of 90 days. Note that this is *not* something that a maintenance technician can do. It is the exclusive domain of specialized remanufacturing companies.

Table 10.5   Definition of PCB Terms as Identified by the EPA

Term	Definition	Examples
PCB	Any chemical substance that is limited to the biphenyl molecule that has been chlorinated to varying degrees, or any combination of substances that contain such substances.	PCB dielectric fluids, PCB heat-transfer fluids, PCB hydraulic fluids, 2,2',4-trichlorobiphenyl
PCB article	Any manufactured article, other than a PCB container, that contains PCBs, and whose surface has been in direct contact with PCBs.	Capacitors, transformers, electric motors, pumps, pipes
PCB container	A device used to contain PCBs or PCB articles and whose surface has been in direct contact with PCBs.	Packages, cans, bottles, bags, barrels, drums, tanks
PCB article container	A device used to contain PCB articles or equipment, and whose surface has not been in direct contact with PCBs.	Packages, cans, bottles, bags, barrels, drums, tanks
PCB equipment	Any manufactured item, other than a PCB container or PCB article container, which contains a PCB article or other PCB equipment.	Microwave ovens, fluorescent light ballasts, electronic equipment
PCB item	Any PCB article, PCB article container, PCB container, or PCB equipment that deliberately or unintentionally contains, or has as a part of it, any PCBs.	See *PCB article*, *PCB article container*, *PCB container*, and *PCB equipment*.
PCB transformer	Any transformer that contains PCBs in concentrations of 500 ppm or greater.	High-power transformers
PCB contaminated	Any electrical equipment that contains more than 50, but less than 500 ppm, of PCBs. (Oil-filled electrical equipment, other than circuit breakers, reclosers, and cable, whose PCB concentration is unknown, must be assumed to be PCB-contaminated electrical equipment.)	Transformers, capacitors, circuit breakers, reclosers, voltage regulators, switches, cable, electromagnets

PCB transformers must be inspected quarterly for leaks. However, if an impervious dike (sufficient to contain all the liquid material) is built around the transformer, the inspections can be conducted yearly. Similarly, if the transformer is tested and found to contain less than 60,000 ppm, a yearly inspection is sufficient. Failed PCB transformers cannot be repaired; they must be disposed of properly.

If a leak develops, it must be contained and daily inspections must begin. A cleanup must be initiated as soon as possible, but no later than 48 hours after the leak is discovered. Adequate records must be kept of all inspections, leaks, and actions taken for 3 years after disposal of the component. Combustible materials must be kept a minimum of 5 m from a PCB transformer and its enclosure.

As of October 1, 1990, the use of PCB transformers (500 ppm or greater) is prohibited in or near commercial buildings when the secondary voltages are 480 V ac or higher. The use of radial PCB transformers is allowed if certain electrical protection is provided.

The EPA regulations also require that the operator notify others of the possible dangers. All PCB transformers (including those in storage for reuse) must be registered with the local fire department. Supply the following information:

- The location of the PCB transformer(s).
- Address(es) of the building(s). For outdoor PCB transformers, provide the outdoor location.
- Principal constituent of the dielectric fluid in the transformer(s).
- Name and telephone number of the contact person in the event of a fire involving the equipment.

Any PCB transformers used in a commercial building must be registered with the building owner. All owners of buildings within 30 m of such PCB transformers also must be notified. In the event of a fire-related incident involving the release of PCBs, immediately notify the Coast Guard National Spill Response Center at 1-800-424-8802. Also take appropriate measures to contain and control any possible PCB release into water.

Capacitors are divided into two size classes: *large* and *small*. The following are guidelines for classification:

- A PCB small capacitor contains less than 1.36 kg (3 lb) of dielectric fluid. A capacitor having less than 100 in^3 also is considered to contain less than the 3 lb of dielectric fluid.
- A PCB large capacitor has a volume of more than 200 in^3 and is considered to contain more than 3 lb of dielectric fluid. Any capacitor having a volume from 100 to 200 in^3 is considered to contain 3 lb of dielectric, provided the total weight is less than 9 lb.
- A PCB *large low-voltage capacitor* contains 3 lb or more of dielectric fluid and operates below 2 kV.
- A PCB *large high-voltage capacitor* contains 3 lb or more of dielectric fluid and operates at 2 kV or greater voltages.

The use, servicing, and disposal of PCB small capacitors is not restricted by the EPA unless there is a leak. In that event, the leak must be repaired or the capacitor disposed of. Disposal may be performed by an approved incineration facility, or the component may be placed in a specified container and buried in an approved chemical waste landfill. Currently, chemical waste landfills are only for disposal of liquids containing 50 to 500 ppm PCBs and for solid PCB debris. Items such as capacitors that are leaking oil containing greater than 500 ppm PCBs should be taken to an EPA-approved PCB disposal facility.

10.3.4   Identifying PCB Components

The first task for the facility manager is to identify any PCB items on the premises. Equipment built after 1979 probably does not contain any PCB-filled devices. Even so, inspect all capacitors, transformers, and power switches to be sure. A call to the manufacturer also may help. Older equipment (pre-1979) is more likely to contain PCB transformers and capacitors. A liquid-filled transformer usually has cooling fins, and the nameplate may provide useful information about its contents. If the transformer is unlabeled or the fluid is not identified, it must be treated as a PCB transformer. Untested (not analyzed) mineral-oil-filled transformers are assumed to contain at least 50 ppm, but less than 500 ppm PCBs. This places them in the category of PCB-contaminated electrical equipment, which has different requirements than PCB transformers. Older high-voltage systems are likely to include both large and small PCB capacitors. Equipment rectifier panels, exciter/modulators, and power-amplifier cabinets may contain a significant number of small capacitors. In older equipment, these capacitors often are Askarel-filled. Unless leaking, these devices pose no particular hazard. If a leak does develop, follow proper disposal techniques. Also, liquid-cooled rectifiers may contain Askarel. Even though their use is not regulated, treat them as a PCB article, as if they contain at least 50 ppm PCBs. Never make assumptions about PCB contamination; check with the manufacturer to be sure.

Any PCB article or container being stored for disposal must be date-tagged when removed, and inspected for leaks every 30 days. It must be removed from storage and disposed of within 1 year from the date it was placed in storage. Items being stored for disposal must be kept in a storage facility meeting the requirements of 40 CFR (Code of Federal Regulations), Part 761.65(b)(1), unless they fall under alternative regulation provisions. There is a difference between PCB items stored for disposal and those stored for reuse. Once an item has been removed from service and tagged for disposal, it cannot be returned to service.

10.3.5   Labeling PCB Components

After identifying PCB devices, proper labeling is the second step that must be taken by the facility manager. PCB article containers, PCB transformers, and large high-voltage capacitors must be marked with a standard 6-in x 6-in large marking label (ML) as shown in Figure 10.8. Equipment containing these transformers or capacitors also should be marked. PCB large low-voltage (less than 2 kV) capacitors need not be labeled until removed from service. If the capacitor or transformer is too small to

Figure 10.8 Marking label (ML) used to identify PCB transformers and PCB large capacitors.

hold the large label, a smaller 1-in • 2-in label is approved for use. Labeling each PCB small capacitor is not required. However, any equipment containing PCB small capacitors should be labeled on the outside of the cabinet or on access panels. Properly label any spare capacitors and transformers that fall under the regulations. Identify with the large label any doors, cabinet panels, or other means of access to PCB transformers. The label must be placed so that it can be read easily by firefighters. All areas used to store PCBs and PCB items for disposal must be marked with the large (6-in • 6-in) PCB label.

10.3.6 Record Keeping

Inspections are a critical component in the management of PCBs. EPA regulations specify a number of steps that must be taken and the information that must be recorded. Table 10.6, which summarizes the schedule requirements, can be used as a checklist for each transformer inspection. This record must be retained for 3 years. In addition to the inspection records, some facilities may need to maintain an annual report. This report details the number of PCB capacitors, transformers, and other PCB items on the premises. The report must contain the dates when the items were removed from service, their disposition, and detailed information regarding their characteristics. Such a report must be prepared if the facility uses or stores at least one PCB transformer containing greater than 500 ppm PCBs, 50 or more PCB large capacitors, or at least

Table 10.6   The Inspection Schedule Required for PCB Transformers and Other Contaminated Devices

PCB Transformers

- Standard PCB transformer — Quarterly
- If full-capacity impervious dike added — Yearly
- If retrofitted to < 60,000 ppm PCB — Yearly
- If leak discovered, cleanup ASAP — Daily
  (retain those records for 3 years)

PCB article or container stored for disposal — 30 days
(remove and dispose of within 1 year)

Retain all records for 3 years after disposing of transformer.

45 kg of PCBs in PCB containers. Retain the report for 5 years after the facility ceases using or storing PCBs and PCB items in the prescribed quantities. Table 10.7 lists the information required in the annual PCB report.

10.3.7 Disposal

Disposing of PCBs is not a minor consideration. Before contracting with a company for PCB disposal, verify its license with the area EPA office. That office also can supply background information on the company's compliance and enforcement history.

The fines levied for improper disposal are not mandated by federal regulations. Rather, the local EPA administrator, usually in consultation with local authorities,

Table 10.7   Inspection Checklist for PCB Components

Transformer location: _____

Date of visual inspection: _____

Leak discovered? _____ (Yes/No)

   If yes, date discovered (if different from inspection date): _____

Location of leak: _____

Person performing inspection: _____

Estimate of the amount of dielectric fluid released from leak: _____

Date of cleanup, containment, repair, or replacement: _____

Description of cleanup, containment, or repair performed: _____

Results of any containment and daily inspection required for uncorrected active leaks:

determines the cleanup procedures and costs. Civil penalties for administrative complaints issued for violations of the PCB regulations are determined according to a matrix provided in the PCB penalty policy. This policy, published in the Federal Register, considers the amount of PCBs involved and the potential for harm posed by the violation.

10.3.8 Proper Management

Properly managing the PCB risk is not difficult. The keys are to understand the regulations and to follow them carefully. A PCB management program should include the following steps:

- Locate and identify all PCB devices. Check all stored or spare devices.
- Properly label PCB transformers and capacitors according to EPA requirements.
- Perform the required inspections, and maintain an accurate log of PCB items, their location, inspection results, and actions taken. These records must be maintained for 3 years after disposal of the PCB component.
- Complete the annual report of PCBs and PCB items by July 1 of each year. This report must be retained for 5 years. (See Table 10.8.)
- Arrange for any necessary disposal through a company licensed to handle PCBs. If there are any doubts about the company's license, contact the EPA.
- Report the location of all PCB transformers to the local fire department and owners of any nearby buildings.

The importance of following the EPA regulations cannot be overstated.

10.4 OSHA SAFETY REQUIREMENTS

The federal government has taken a number of steps to help improve safety within the workplace. OSHA, for example, helps industries to monitor and correct safety practices. The agency's records show that electrical standards are among the most frequently violated of all safety standards. Table 10.9 lists 16 of the most common electrical violations, which include these areas:

- Protective covers
- Identification and marking
- Extension cords
- Grounding

10.4.1 Protective Covers

Exposure of live conductors is a common safety violation. All potentially dangerous electric conductors should be covered with protective panels. The danger is that someone may come into contact with the exposed, current-carrying conductors. It also

Table 10.8   Required Information for PCB Annual Report

I. PCB device background information:
   a. Dates when PCBs and PCB items are removed from service.
   b. Dates when PCBs and PCB items are placed into storage for disposal, and are placed into transport for disposal.
   c. The quantities of the items removed from service, stored, and placed into transport are to be indicated using the following breakdown:
      1) Total weight, in kilograms, of any PCB and PCB items in PCB containers, including identification of container contents (such as liquids and capacitors).
      2) Total number of PCB transformers and total weight, in kilograms, of any PCBs contained in the transformers.
      3) Total number of PCB large high- or low-voltage capacitors.
II. The location of the initial disposal or storage facility for PCBs and PCB items removed from service, and the name of the facility owner or operator.
III. Total quantities of PCBs and PCB items remaining in service at the end of calendar year per the following breakdown:
   a. Total weight, in kilograms, of any PCB and PCB items in PCB containers, including the identification of container contents (such as liquids and capacitors).
   b. Total number of PCB transformers and total weight, in kilograms, of any PCBs contained in the transformers.
   c. Total number of PCB large high- or low-voltage capacitors.

is possible for metallic objects such as ladders, cable, or tools to contact a hazardous voltage, creating a life-threatening condition. Open panels also present a fire hazard.

10.4.2   Identification and Marking

Properly identify and label all circuit breakers and switch panels. The labels for breakers and equipment switches may be years old, and may no longer describe the equipment that is actually in use. This confusion poses a safety hazard. Improper labeling of the circuit panel may lead to unnecessary damage — or worse, casualties — if the only person who understands the system is unavailable in an emergency. If there are a number of devices connected to a single disconnect switch or breaker, provide a diagram or drawing for clarification. Label with brief phrases, and use clear, permanent, and legible markings.

Equipment marking is a closely related area of concern. This is not the same thing as equipment identification. Marking equipment means labeling the equipment breaker panels and ac disconnect switches according to device rating. Breaker boxes should contain a nameplate showing the manufacturer name, rating, and other pertinent electrical factors. The intent of this rule is to prevent devices from being subjected to excessive loads or voltages.

Table 10.9   Sixteen Common OSHA Violations (*National Electrical Code, NFPA no. 70*)

Fact Sheet Number	Subject	NEC Reference
1	Guarding of live parts	110-17
2	Identification	110-22
3	Uses allowed for flexible cord	400-7
4	Prohibited uses of flexible cord	400-8
5	Pull at joints and terminals must be prevented	400-10
6-1	Effective grounding, Part 1	250-51
6-2	Effective grounding, Part 2	250-51
7	Grounding of fixed equipment, general	250-42
8	Grounding of fixed equipment, specific	250-43
9	Grounding of equipment connected by cord and plug	250-45
10	Methods of grounding, cord and plug-connected equipment	250-59
11	AC circuits and systems to be grounded	250-5
12	Location of overcurrent devices	240-24
13	Splices in flexible cords	400-9
14	Electrical connections	110-14
15	Marking equipment	110-21
16	Working clearances about electrical equipment	110-16

10.4.3   Extension Cords

Extension (flexible) cords often are misused. Although it may be easy to connect a new piece of equipment with a flexible cord, be careful. The National Electrical Code lists only eight approved uses for flexible cords.

The use of a flexible cord where the cable passes through a hole in the wall, ceiling, or floor is an often-violated rule. Running the cord through doorways, windows, or similar openings also is prohibited. A flexible cord should not be attached to building surfaces or concealed behind building walls or ceilings. These common violations are illustrated in Figure 10.9.

Along with improper use of flexible cords, failure to provide adequate strain relief on connectors is a common problem. Whenever possible, use manufactured cable connections.

10.4.4   Grounding

OSHA regulations describe two types of grounding: *system grounding* and *equipment grounding*. System grounding actually connects one of the current-carrying conductors (such as the terminals of a supply transformer) to ground. See Figure 10.10. Equipment grounding connects all the noncurrent-carrying metal surfaces together and to ground. From a grounding standpoint, the only difference between a grounded

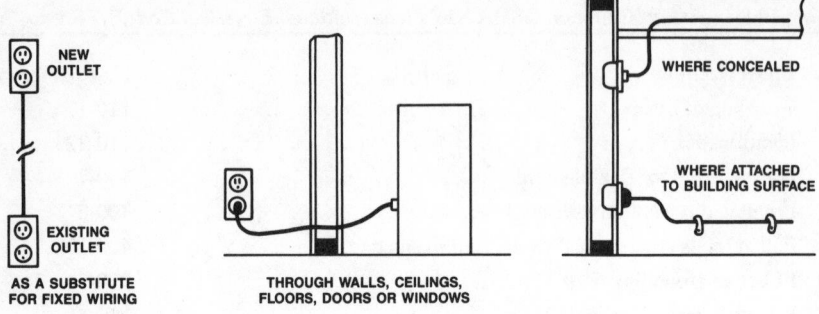

Figure 10.9   Flexible cord uses prohibited under NEC rules.

electrical system and an ungrounded electrical system is that the *main-bonding jumper* from the service equipment ground to a current-carrying conductor is omitted in the ungrounded system.

The system ground performs two tasks:

- It provides the final connection from equipment-grounding conductors to the grounded circuit conductor, thus completing the ground-fault loop.
- It solidly ties the electrical system and its enclosures to their surroundings (usually earth, structural steel, and plumbing). This prevents voltages at any source from rising to harmfully high voltage-to-ground levels.

It should be noted that equipment grounding — bonding all electric equipment to ground — is required whether or not the system is grounded. System grounding should be handled by the electrical contractor installing the power feeds.

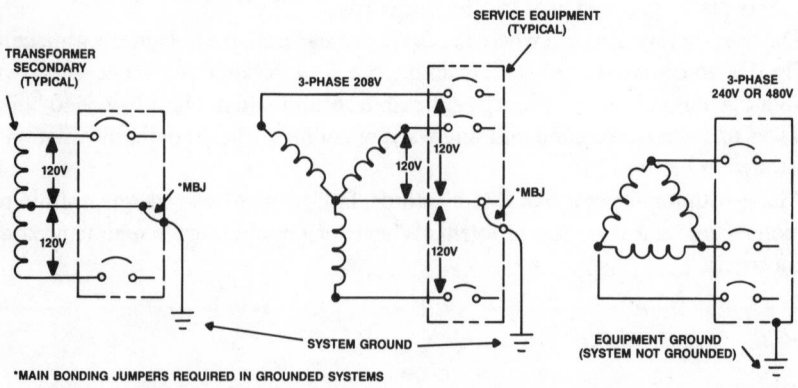

Figure 10.10   Even though regulations have been in place for many years, OSHA inspections still uncover violations in the grounding of primary electrical service systems.

Equipment grounding serves two important functions:

- It bonds all surfaces together so that there can be no voltage differences among them.
- It provides a ground-fault current path from a fault location back to the electrical source, so that if a fault current develops, it will rise to a level high enough to operate the breaker or fuse.

The National Electrical Code is complex, and it contains numerous requirements concerning electrical safety. If the facility electric wiring system has gone through many changes over the years, have the system inspected by a qualified consultant. The fact sheets listed in Table 10.9 provide a good starting point for a self-evaluation. The fact sheets are available from any local OSHA office.

10.4.5 Management Responsibility

The key to operating a safe facility is diligent management. A carefully thought-out plan ensures a coordinated approach to protecting staff members from injury and the facility from potential litigation. Facilities that have effective accident-prevention programs follow seven basic guidelines. Although the details and overall organization may vary from workplace to workplace, these practices are summarized in Table 10.10.

If managers are concerned about safety, it is likely that employees also will be. Display safety pamphlets, and recruit employee help in identifying hazards. Reward workers for good safety performance. Often, an incentive program will help to encourage safe work practices. Eliminate any hazards identified, and obtain OSHA forms and any first aid supplies that would be needed in an emergency. The OSHA *Handbook for Small Business* outlines the legal requirements imposed by the Occupational Safety and Health Act of 1970. The handbook, which is available from OSHA, also suggests ways in which a company can develop an effective safety program.

Table 10.10   Major Points to Consider When Developing a Facility Safety Program

- Management assumes the leadership role regarding safety policies.
- Responsibility for safety- and health-related activities is clearly assigned.
- Hazards are identified, and steps are taken to eliminate them.
- Employees at all levels are trained in proper safety procedures.
- Thorough accident/injury records are maintained.
- Medical attention and first aid is readily available.
- Employee awareness and participation is fostered through incentives and an ongoing, high-profile approach to workplace safety.

Table 10.11   Sample Checklist of Important Safety Items

Refer regularly to this checklist to maintain a safe facility. For each category shown, be sure that:

ELECTRICAL SAFETY

- Fuses of the proper size have been installed.
- All ac switches are mounted in clean, tightly closed metal boxes.
- Each electrical switch is marked to show its purpose.
- Motors are clean and free of excessive grease and oil.
- Motors are maintained properly and provided with adequate overcurrent protection.
- Bearings are in good condition.
- Portable lights are equipped with proper guards.
- All portable equipment is double-insulated or properly grounded.
- The facility electrical system is checked periodically by a contractor competent in the NEC.
- The equipment-grounding conductor or separate ground wire has been carried all the way back to the supply conductor.
- All extension cords are in good condition, and the grounding pin is not missing or bent.
- Ground-fault interrupters are installed as required.

EXITS AND ACCESS

- All exits are visible and unobstructed.
- All exits are marked with a readily visible, properly illuminated sign.
- There are sufficient exits to ensure prompt escape in the event of an emergency.

FIRE PROTECTION

- Portable fire extinguishers of the appropriate type are provided in adequate numbers.
- All remote vehicles have proper fire extinguishers.
- Fire extinguishers are inspected monthly for general condition and operability, which is noted on the inspection tag.
- Fire extinguishers are mounted in readily accessible locations.
- The fire alarm system is tested annually.

Free on-site consultations also are available from OSHA. A consultant will tour the facility and offer practical advice about safety. These consultants do not issue citations, propose penalties, or routinely provide information about workplace conditions to the federal inspection staff. Contact the nearest OSHA office for additional information. Table 10.11 provides a basic checklist of safety points for consideration.

10.5 BIBLIOGRAPHY

Code of Federal Regulations, 40, Part 761.

"Current Intelligence Bulletin no. 45," National Institute for Occupational Safety and Health Division of Standards Development and Technology Transfer, February 24, 1986.

"Electrical Standards Reference Manual," U.S. Department of Labor, Washington, DC.

Federal Information Processing Standards Publication no. 94, *Guideline on Electrical Power for ADP Installations*, U.S. Department of Commerce, National Bureau of Standards, Washington, DC, 1983.

Hammar, Willie: *Occupational Safety Management and Engineering*, Prentice-Hall, New York.

Hazardous Waste Consultant, January/February 1984.

Lawrie, Robert: *Electrical Systems for Computer Installations*, McGraw-Hill, New York, 1988.

"Occupational Injuries and Illnesses in the United States by Industry," OSHA Bulletin 2278, U.S. Department of Labor, Washington, DC, 1985.

OSHA "Electrical Hazard Fact Sheets," U.S. Department of Labor, Washington, DC, January 1987.

OSHA "Handbook for Small Business," U.S. Department of Labor, Washington, DC.

Pfrimmer, Jack: "Identifying and Managing PCBs in Broadcast Facilities," *NAB Engineering Conference Proceedings*, National Association of Broadcasters, Washington, D.C., 1987.

"Toxic Information Series," Office of Toxic Substances, July 1983.

ACKNOWLEDGMENT

Appreciation is expressed to Charles J. Fuhrman, Fuhrman Investigations, Inc., and William S. Watkins, P.E., for their assistance in the preparation of portions of this chapter.

INDEX

A

Ac line disturbances
  assessment of
    A/D conversion, 68–69
    capturing transient waveforms, 70
    digital measurement instruments,
      66–68
    digital monitor features, 69–70
  carrier storage, 58–59
  classification of disturbances, 61–65
  coupling transient energy, 50–52
  electromagnetic pulse radiation,
    49–50
  electrostatic discharge, 47–49
  lightning, 37–47
  switch contact arcing, 54–56
  telephone system transients, 56–58
  transient-generated noise, 59–61
  utility system faults, 54
Ac-power systems
  elements of
    capacitors, 17–20
    control and switching systems,
      21–22

    power generators, 14–17
    power transformers, 5, 7–13
    transmission circuits, 20–21
  power relationships, 4–5
  terminology related to, 1–5
  utility ac power systems, 22–34
  vector representations, 4
Actual current, 26
A/D conversion, ac line disturbances,
  68–69
A/D converter, 67
  types of, 68
Alpha multiplication, 80
Alternator, defined, 1
Apparent power, 5, 25
Arcing time, 128
Arcs, characteristics of, 183
Area grid, 23
Area loads, 23
Armortisseur winding, 146–147
Atmospheric energy, sources of, 38–41
Avalanche breakdown, 80, 81
Avalanche rectifiers, 109–111

B

Balun, 180
Battery supply
   types of batteries, 165–166
   uninterruptable power systems,
       164–166
Bioaccumulation, 304
Biomagnification, 304
Blackouts, 273–274
Boost rectifier circuits, 115
Building codes, 249–250
Bulkhead grounding, 241–242
   bulkhead panel, in facility grounding
       system, 237–242
   wye-to-wye arrangement,
       237–241

C

Cable grounding. *See* Signal-carrying
      cable grounding
Cadwelding, 222–223
Capacitive reactance, 18–20
Capacitor failure
   and capacitor life span, 124–125
   electrolyte failures, 123–124
   electrolytic capacitors, 121
   mechanical factors, 121–122
   tantalum capacitors, 125
   temperature cycling, 122
Capacitors, 17–20
   equivalent series resistance, 20
   unit of capacitance, 17
Capacity factor, 17
Carrier storage
   ac line disturbances, 58–59
   carrier storage effect, 108
Charge-coupled device, 68
Chemical ground rods, 215–219
Chip protection, semiconductor failure,
      99–101
Circuit breakers, 127–128
Circuit protection hardware, 298–301
Circular mil, defined, 1
Common-mode noise, 169
   defined, 1

Common-mode rejection ratio, 266–269
Common-mode voltage, 263–264
Common stator machines, 148–150
Compensating winding, 167
Cone of protection, defined, 2
Constant current mode, 165
Constant voltage mode, 165
Control circuits, thyristor servo
      systems, 118
Conversion resolution, 69
Cosmic rays, 38
   defined, 2
Coulomb, defined, 2
Counter-electromotive force, 9–12
   defined, 2
Counterpoise, defined, 2
Coupling transient energy, 50–52
Critical electric field, 83
Critical system bus, standby power,
      282–284
Crowbar devices, 182–183
Current sharing, 109, 115

D

Dedicated protection systems
   ferroresonant transformers,
      166–169
   isolation transformer, 169–170
   line conditioner, 175–176
   tap-changing regulator, 170–174
Delay-trip considerations, and fault
      protectors, 130–131
Delta configuration, 14
Delta magnetic inverter, 158–159
Delta-to-delta arrangement, 32
Delta-wye arrangement, 32
Demand meter, defined, 2
Derating, types of, 79
Dielectric, defined, 2
Dielectric absorption, 123
Difference signal, 262
Differential gain, 266
Differential input circuit, 262
Digital measurement instruments,
      ac line disturbances, 66–68

Digital monitors, 69–70
  benefits of memory storage, 69–69
  features of, 69
Digital storage oscilloscope, 67–68
Discrete transient-suppression
      hardware, 179–208
  crowbar devices, 182–183
  factors in selection of, 186
  filter devices, 179–181
  performance testing, 187–190
  voltage-clamping devices, 183–186
Dissipation factor, 20
Dissipation region, 76
Dual feeder system, standby power,
      274–276

E

Earth current, 42
Eddy currents, 7
  defined, 2
Efficiency, defined, 2
Electric shock, 296–302
  circuit protection hardware, 298–301
  effects on body, 296–298
  first aid for, 302
  guidelines for working with high
      voltage, 302
Electrolyte failures, capacitor failure,
      123–124
Electrolytic capacitors, 121
Electromagnetic generation, 38–39
Electromagnetic induction, 5
Electromagnetic pulse radiation, 49–50
Electrostatic discharge, 47–49
  failure mechanisms, 92–95
  failure modes, 91–92
  latent failures, 93–94
  triboelectric effects, 47–49
Emergency power off button, 293
Energy use
  determination of energy use, 284–285
    load factor, 287
    peak demand, 285–286
    power factor, 287
Equipment racks, grounding, 257–260

F

Facility grounding
  in bare rock, rock-based radial
      elements, 230–231
  bonding elements
    cadwelding, 222–223
    facility ground interconnection,
      227–230
    grounding tower elements, 225
    ground-system inductance, 225
    ground wire dressing, 226–227
  design of
    bulkhead grounding, 241–242
    bulkhead panel, 237–241
    checklist of proper grounding,
      244–246
    grounding arrangement
      installation, 243–244
    lightning arresters, 242–243
  earth ground
    chemical ground rods, 215–219
    grounding interface, 212–215
    Ufer ground system, 219–222
  purposes of, 211
  transmission-system grounding
    cable considerations, 233
    satellite antenna grounding,
      233–234
    transmission line, 231–232
Facility protection
  circuit-level
    device application cautions,
      207–208
    for high-voltage supplies, 203–204
    for inductive load switching, 207
    for logic circuits, 205
    for low-voltage supplies, 202
    for telco lines, 205–207
  facility wiring, 191–193
  power-system protection
    design cautions, 198–199
    lightning arresters, 195–197
    single-phasing, 199–201
    staging, 197–198
    utility service entrance, 193–195

Farad, 17
Fault protectors
    application considerations, 128–131
    circuit breakers, 127–128
    and delay-trip considerations,
            130–131
    fuses, 126–127
    semiconductor fuses, 128
    and transient currents, 129–130
Ferroresonant inverter, 157
Ferroresonant transformers, 166–169
    magnetic-coupling-controlled
            voltage regulator, 168–169
    windings, 166–167
Filter devices, 179–181
Fire damper, 294
Flame detector, 294
Flash converter, 68
Float voltage level, 165
Flyback, 84
Forward bias safe operating area, 77
Forward-current rating, types of, 107
Forward-transfer mode, 153
Full load percent impedance, 13
Fuses, 126–127
Fusing, thyristor servo systems,
            117–118

G

Gas-gaps, 182
Gas tube, 206
Gate trigger sensitivity, 89
Generators
    automatic transfer switch, 280–281
    defined, 2
    for standby power, 278–281
    and type of power plant, 278, 280
Geothermal systems, 16
Glitch capture, 68
Gloves, rubber, 302
Ground-fault interrupters, 22, 298–301
Grounding a facility. *See* Facility
        grounding
Grounding conductor size, 254–255

Grounding practices
    building codes, 249–250
    grounding equipment racks, 257–260
    isolation transformers, 256–257
    OSHA regulations, 313–315
    power-center grounding, 255–256
    signal-carrying cable grounding,
            260–271
    single-point ground, 250–260
Ground loop, defined, 2
Ground-system maintenance, 289–291

H

Halon, 294
Henry, 10–11
Horsepower, defined, 2
HVAC, defined, 2
Hybrid ferroresonant transformer, 175
Hybrid suppression circuits, 184–186
Hybrid transient suppressor, 176
Hysterisis loss, defined, 2

I

Impedance, defined, 2
Induced voltage, defined, 2
Inductive loads, thyristor servo
            systems, 113–114
Inductive reactance, 11–12
Insulation breakdown, 101–102
Interdigited geometry technique, 78
Inverter-fed L/C tank, 159–160
Isokeranuic map, 135
Isolated redundant mode, 163–164
Isolation transformers, 169–170,
            256–257
Isulation resistance, 20

J

Joule, defined, 2

K

Keranuic number, 135

L

Lagging power factor, 25
Lead-antimony batteries, 165-166
Lead calcium batteries, 165
Leading power factor, 25
Leakage current, 20
Life safety system, defined, 3
Lightning, 37-47
    assessment of hazard, 135-136
    categories of, 41
    characteristics of, 41-43
    cloud-to-cloud activity, 42-43
    lightning arresters, 21-22, 195-197,
        242-243
    positive and negative strikes, 40
    protection against, 44-47
    protection area, 46-47
Lightning flash, defined, 3
Linear amplifier correction system, 175
Line conditioner, 175-176
    types of, 175-176
Load factor, 17
    energy use, 287
Load shedding, energy use, 285-286
Local ground point, 251
Logic circuits, protection of, 205

M

Magnetic-coupling-controlled voltage
        regulator, 168-169
Main ground point, 250, 258
Melting time, 128
Metallization failure, 82-83
Metal-oxide varistor, defined, 3
Mimic board, 288
Momentary power interruption, 61, 62
MOSFET devices
    failure modes, 86-88
    safe operating area, 83-86
    specifications for, 83
Motor generator set
    attributes of, 141-143
    maintenance considerations, 149

motor-design configuration, 146-149
motor-generator UPS, 149-151
single-shaft systems, 148-149
system configuration, 144-146
MOV, 188-190

N

Negative resistance characteristic, 88
Negative resistance effect, 183
Network system, 24
Neutralizing winding, 167
Nickel-cadmium batteries, 166
Noise
    analyzing noise currents, 261-262
    common mode noise, 169
    and contact arcing, 60
    ESD noise, 59-60
    normal-mode noise, 169
    and SCR switching, 60-61
    standby power, 281-282
    transient-generated noise, 59-61
    types of, 263-264
Normal-mode noise, 169
    defined, 3
Normal-mode voltage, 263-264

O

Occupational Safety and Health Act
        requirements, 311-316
    extension cords, 313
    grounding, 313-315
    identification and marking, 312
    management responsibility, 315-316
    protective covers, 311-312
Open-delta arrangement, 32, 33
Output transfer switch, uninterruptable
        power systems, 164

P

Parallel redundant mode, 162-163
Patch-bay grounding, 265-266
Peak-accumulation mode, 68
Peak demand, energy use, 285-286

Peak detection, 68
Peak power sharing, 276
Perimeter ground, 226–227
Permeability, defined, 3
Phase angle, 10
Phase modulation inverter, 162
Phase-to-phase balance, 33–34
Phasor diagram, 4
Plant exposure factor, 136
Plant maintenance
 ground-system maintenance,
  289–291
 guidelines for, 288
 switchgear maintenance, 288–289
Point discharge theory, 44–46
Polarity reversal, 83
Polychlorinated biphenyls, 303–311
 common names for, 304
 components containing, 305,
  307–308
 disposal of, 310–311
 governmental action, 305
 health risk, 304–305
 identification of PCB components,
  308
 labeling PCB components, 308–309
 management program, 311
 record keeping related to, 309–310
 terminology, 306
Pony motors, 145
Power-center grounding, 255–256
Power factor, 5
 defined, 3
 energy use, 287
Power-follow problems, 182
Power generators, 14–17
Power rectifiers
 operating rectifiers in parallel, 109
 operating rectifiers in series, 107–108
 reverse-voltage ratings, 107
 silicon avalanche rectifiers, 109–111
Power-supply components, 105–106
Power transformers, 5, 7–13
 counter-electromotive force, 9–12
 full load percent impedance, 13

Protection methods
 assessment of, 139–140
 dedicated protection systems,
  166–177
 discrete device approach, 136–137
 discrete transient-suppression
  hardware, 179–190
 facility protection, 190–210
 key tolerance envelope, 134–135
 and system specification, 137–138
 transient protection alternatives,
  136–138, 141
 uninterruptible power system,
  153–164
Protective covers, 311–312
Punch-through, 80

Q

Quanitzation process, 67
Quasi-square wave inverter, 160–161

R

Radial system, 25
Radioactive decay, 38
Radio frequency interference, defined, 3
Rated voltage, 124
Reactance, defined, 3
Reactive current, 27
Reactive power, defined, 3
Reciprocal double exponential
  waveform, 39
Reclosers, 22
Recovery time, 89
Rectifiers. See Power rectifiers
Redundant operation, uninterruptable
  power systems, 162–164
Remote/local control switch, 288
Reverse bias safe operating area, 78
Reverse-transfer mode, 153
Reverse voltage, 125
Reverse-voltage ratings, types of,
  107
Ring system, 25
Ripple current/voltage, 20, 125

Rock, grounding on. *See* Facility
    grounding

S

Safe operating area, 76
    defined, 3
    MOSFET devices, 83–86
Safety
    basic guidelines, 295–296
    electric shock, 296–302
    Occupational Safety Health Act
        requirements, 311–316
    polychlorinated biphenyls, 303–311
    safety equipment, types of, 293–296
Salting, and soil resistivity, 215–216
Sampling interval, 67
Second breakdown region, 76, 77
Self inductance, defined, 3
Semiconductor devices
    failure modes
        avalanche-related failure, 81
        and device ruggedness, 76
        and forward bias safe operating
            area, 77
        metallization failure, 82–83
        polarity reversal, 83
        and power handling capacity,
            78–79
        and reverse bias safe operating
            area, 78
        and semiconductor derating, 79
        thermal runaway, 81–82
        thermal second breakdown, 82
        types of, 76, 79–80
Semiconductor failure
    chip protection, 99–101
    overvoltage sources, 96
Semiconductor fuses, 128
Signal-carrying cable grounding,
        260–271
    analyzing noise currents, 261–262
    cable routing, 270
    inputs/output circuits, 266–270
    overcoming ground system
        problems, 270–271

patch-bay grounding, 265–266
    skin effect, 265
    types of noise, 263–264
Single-phasing, 199–201
    defined, 3
Single-point ground, 250–260
    facility ground system, 250–255
    grounding conductor size, 254–255
    star grounding system, 250–253
Single-shaft systems, motor generator
        set, 148–149
Skin effect, 225, 265
Smoke detector, 293–294
Soil resistivity, 214
    and salting, 215–216
Solar wind, 38
    defined, 3
Space charge region, 80
Staging, 197–198
Standby power
    advanced system protection, 276–278
    choosing generator for, 278–281
    critical system bus, 282–284
    dual feeder system, 274–276
    peak power sharing, 276
    system noise, 281–282
    UPS systems, 281
Star grounding system, 250–253
Static capacitors, 28–29
Static electricity, 38
Step wave inverter, 161–162
Storage time, 89
Strokes, 37
Successive approximations converter,
        68
Surge voltage, 20
Switch-arcing effects, insulation
        breakdown, 101–102
Switch contact arcing, ac line
        disturbances, 54–56
Switch failure, reasons for, 101
Switchgear maintenance, 288–289
Switching safe operating area, 84
Synchronous capacitors, 29

T

Tantalum capacitors, 125
Tap-changing high-isolation
  transformer, 175
Tap-changing regulator, 170–174
  variable ratio regulator, 173–174
Telco lines, protection of, 205–207
Telephone system transients, ac line
  disturbances, 56–58
Temperature
  and capacitor failure, 122
  and transformer failure, 119
Thermal runaway, 81–82
Thermal second breakdown, 82
Thermoplastic, 191–192
Thermosetting, 191–192
Thousand circular mils, 21
Thyristors
  application considerations, 90–91
  device parameters, 88–89
  failure modes, 89–90
  nature of, 88
Thyristor servo systems
  control circuits, 118
  fusing, 117–118
  inductive loads, 113–114
  triggering circuits, 115–117
Total harmonic distortion, 141
Transfer switches, configuration
  options, 164
Transformer failure
  mechanical factors, 120–121
  thermal considerations, 119
  voltage considerations, 119–120
Transient currents, and fault protectors,
  129–130
Transient disturbance, 61, 62
  defined, 3
Transient-generated noise, 59–61
Transmission circuits, 20–21
Transmission system grounding.
  See Facility grounding
Transverse-mode voltage, 263

Trapping centers, 81
Triboelectric effects, electrostatic
  discharge, 47–49
Triggering circuits, thyristor servo
  systems, 115–117
Trip point, 203–204
True power, 5, 25

U

Ufer ground system, 219–222
Unbalanced systems, 261
Uninterruptible power systems
  basic designs for, 153
  battery supply, 164–166
  defined, 3
  output transfer switch, 164
  power-conversion methods,
    156–162
    delta magnetic inverter, 158–159
    ferroresonant inverter, 157
    inverter-fed L/C tank, 159–160
    phase modulation inverter, 162
    quasi-square wave inverter,
      160–161
    step wave inverter, 161–162
  redundant operation, 162–164
  standby power, 281
  UPS configuration, 154–155
Utility ac power systems, 22–34
  on-site power factor correction,
    30–31
  phase-to-phase balance, 33–34
  power distribution, 23–24
  power factor, 25–31
  reliability considerations, 34
  utility company interfacing, 32–33
  voltage configurations, 22
Utility service entrance, facility
  protection, 193–195
Utility system faults, ac line
  disturbances, 54

V

Variable ratio regulator, 173–174
Vector representations, 4
Voltage, and transformer failure,
      119–120
Voltage-clamping devices, 183–186
   hybrid suppression circuits, 184–186
Voltage regulation, defined, 3
Voltage sag, 61, 62
Voltage surge, 61, 62

W

Water sprinkler, 294
Windings, ferroresonant transformer,
      166–167
Windstorm effect, 43
Wind systems, 16
Wiring
   facility wiring, 191–193
   insulation for, 191–193
Wye configuration, 14
Wye-to-wye arrangement, 32, 237–241

ABOUT THE AUTHOR

Jerry Whitaker is a technical writer and editorial consultant for *Broadcast Engineering* magazine and *Video Systems* magazine, published by Intertec Publishing Corp. of Overland Park, KS. He is a Fellow of the Society of Broadcast Engineers, and an SBE-certified senior AM-FM engineer. He is also a member of the SMPTE, AES, ITVA, and IEEE (Broadcast Society, Power Electronics Society, and Reliability and Maintainability Society). He has written and lectured extensively on the topic of electronic systems installation and maintenance. Mr. Whitaker is a former radio station chief engineer and television news producer. He is the author of the CRC Press publication, *Maintaining Electronic Systems*. Mr. Whitaker is also author of the McGraw-Hill publication, *Radio Frequency Transmission Systems: Design and Operation*, and co-author of McGraw-Hill's *Television and Audio Handbook for Technicians and Engineers*. Mr. Whitaker is a contributor to the McGraw-Hill *Audio Engineering Handbook* and to the National Association of Broadcasters' *NAB Engineering Handbook, 7th and 8th Editions*. He has twice received a Jesse H. Neal Award Certificate of Merit from the Association of Business Publishers for editorial excellence.

CONTRIBUTORS

Bradley Dick, *Broadcast Engineering* magazine, Intertec Publishing, Overland Park, KS.

Roger Block, PolyPhaser Corporation, Gardnerville, NV.